GeoPlanet: Earth and Planetary Sciences

More information about this series at https://link.springer.com/bookseries/8821

C. P. Rajendran · Kusala Rajendran

Earthquakes of the Indian Subcontinent

Seismotectonic Perspectives

Springer

C. P. Rajendran
National Institute of Advanced Studies
Bengaluru, Karnataka, India

Kusala Rajendran
Indian Institute of Science
Bengaluru, Karnataka, India

The GeoPlanet: Earth and Planetary Sciences Book Series is in part a continuation of Monographic Volumes of Publications of the Institute of Geophysics, Polish Academy of Sciences, the journal published since 1962 (http://pub.igf.edu.pl/index.php).

ISSN 2190-5193 ISSN 2190-5207 (electronic)
GeoPlanet: Earth and Planetary Sciences
ISBN 978-981-16-4747-5 ISBN 978-981-16-4748-2 (eBook)
https://doi.org/10.1007/978-981-16-4748-2

This Springer imprint is published by the registered company Springer Nature Singapore Pte Ltd.
The registered company address is: 152 Beach Road, #21-01/04 Gateway East, Singapore 189721, Singapore

Preface

This may not be the first time a book is published on earthquakes of the Indian subcontinent. Why another book on this topic? There is none that incorporates a comprehensive current understanding of the earthquakes that occurred in and around India. This book is intended to fill that gap. The earthquake scene within and around India is ever-changing, as data are being added and new insights are gained from recent earthquakes. During the last two decades some notable earthquakes have occurred here in the interplate and intraplate tectonic settings, and each of them makes good case studies. Any book on the Indian earthquakes like the present one will have to do justice to the accumulation of a newer database. Our primary goal for this book is to highlight the fresh insights into the significant earthquakes that occurred during historical and modern times. The other equally important aim is to present a highly accessible and concise discussion of these earthquakes so that the book can be used as supplementary teaching material for courses in seismology. This book is also written for a wider readership consisting of geoscientists, engineers, and policymakers or even non-academics who are keen to know more about the Indian earthquakes.

Some of the historical earthquakes in the Indian subcontinent, particularly the 1819 Kachchh and the 1897 Shillong events, are considered to have led to the initial awakenings of modern earthquake science. Back in 1830s, Charles Lyell in his classic "Principles of Geology" discussed the 1819 Kachchh earthquake, treating it as an example to demonstrate coseismic surface faulting. R. D. Oldham, an officer in the Geological Survey of India had identified for the first time, the distinct arrival of P, S and surface waves from the seismogram of the 1897 (M_w 8.2) Shillong earthquake, based on their differing travel times. In a scientific paper published in 1906, Oldham linked these observations to develop a theory of how waves propagate within the Earth and inferred the presence of a core at the center of the Earth. Indian earthquakes have always been baffling as well as educative, forming new benchmarks in broadening the range of our knowledge of earthquake generation. Recent in the series of great earthquakes, the 2004 Andaman-Sumatra mega-event with an attendant transoceanic tsunami, though not a part of this book, continues to teach us newer lessons on the mechanisms of subduction zone earthquakes. The 2015 earthquake in Nepal can be considered as a modern primer on the earthquake processes in the Himalayan tectonic environments. This book on the earthquakes in India and its neighborhood intends to bring out some of the salient features of those awesome and, at same time, dangerous events that had caused untold personal and societal miseries and what we have learned of their mechanisms and seismotectonic contexts.

Any book would reflect the biases and predilections of the authors to some extent, particularly when the authors themselves have been involved in the research on Indian earthquakes for a few decades. This book is no exception. However, we have tried to ensure that this book is a fair balance of the multitude of views, giving the reader the choice of looking at the broad pictures that finally emerge at the end of each chapter.

Bengaluru, India

C. P. Rajendran
Kusala Rajendran

Acknowledgements

The science of earthquakes has been our passion during much of our professional life. This book has resulted from our research on earthquakes for over three decades. Our studies of earthquakes would not have happened without someone footing the bill or providing the much-needed resources. Earthquake studies that we conducted after we returned from the United States in 1993 were mostly supported by the Department of Science and Technology (DST) under the Government of India until 2007, and we thank the late G. D. Gupta, then director and coordinator of the seismology division of the DST, for his enthusiastic support. Since 2007, we have been supported by the Ministry of Earth Sciences, Government of India and we thank Shailesh Nayak (former Secretary, Ministry of Earth Sciences, Govt. of India, currently the director of the Institute of Advanced Studies in Bangalore, for his continued support.

Another happy task is to thank our guardian angels who were in the forefront, providing us with a helping hand when we faltered for want of resources and inspiration. We shall merely name them below in temporal order with respect to our career trajectories: Professors the late C. Karunakaran (Founder Director, Centre for Earth Science Studies), Pradeep Talwani (Professor, University of South Carolina), the late S. Ramaseshan, P. Balaram (both former Directors, Indian Institute of Science), C. N. R. Rao (Founder President, Jawaharlal Nehru Centre for Advanced Scientific Research), and the late K. S. Valdiya (Professor, Jawaharlal Nehru Centre for Advanced Scientific Research). Many students, post-doctoral researchers, and associates have been part of our long and arduous journey. Among them, we want to specially mention John Paul, Malay Mukul, Jaishri Sanwal, Anil Earnest, Biju John, Revathy Parameswaran, Rishav Mallik, Thulasiraman Natarajan, Ananya Divyadarshini, Mahesh Thakkar, Tejpal Singh and Ananda Sabari. Some of them have read parts of this book and offered their comments on the initial drafts. We are especially thankful to Swapnil Mache, our young student associate at Indian Institute of Science (IISc), for helping us with the figures, and reading through the entire text, working most diligently under the shadow of a deadly pandemic, often through online exchanges.

We owe a special debt to many of our international colleagues for interactions during the past several years. Among them, Arch Johnston (CERI, Memphis), Brian Atwater (USGS), Jeff Freymueller (Univ. of Alaska; Univ. of Michigan), Mike Sandiford (Univ. of Melbourne) and Steve Wesnousky (Univ. of Nevada) need special mention. Roger Bilham (Univ. of Colorado), who has conducted pioneering work on Indian earthquakes, has been a great inspiration for us. We have worked parallelly with him on some of the Indian earthquakes, sometimes ending up with divergent conclusions, but his inspiring presence in meetings and discussions that followed have always led towards newer ways of thinking. We also thank him for reviewing the chapters on 1819 Kachchh and 1897 Shillong earthquakes and providing his critical feedback. Lastly, we would like to thank Debasish Roy, Professor, Civil Engineering Department, IISc, for hosting us in the Centre of Excellence in Advanced Mechanics of Materials, supported by the Indian Space Applications Research Organization. We are also thankful to our colleagues in the National Centre for Earth Science Studies, Indian Institute of Science, Jawaharlal Nehru Centre for Advanced Research, and National Institute of Advanced Studies, with which both of us have been variously associated either together or independently during different phases of our professional lives. We are grateful to Claudio Vita-Finzi, who was a professor at the University College London and Prof. Wojciech Dębski from the Institute of Geophysics, Polish Academy of Sciences for reviewing the manuscript and offering their comments and suggestions.

The idea of this book was seeded during an Asia-Oceania Geosciences Society (AOGS) meeting in Singapore back in 2015 when we met Loyola D'Silva, Publishing Editor, Springer, and the project took five long years because of our professional preoccupations. The Springer have been extremely patient, putting up with our deadline extensions. We also want to acknowledge the services rendered by Anna Dziembowska, Managing Editor of the GeoPlanet Book Series. We thank them for their interest in this book.

C. P. Rajendran, currently an adjunct professor at the National Institute of Advanced Studies, Bangalore and **Kusala Rajendran**, formerly a professor at the Indian Institute of Science, Bangalore are husband and wife seismologist team. They have over thirty-five years of experience in research and teaching in geosciences and have written over hundred scientific papers on seismotectonic aspects of major historical and recent earthquakes in India and contiguous areas.

Contents

Chapter 1
Introduction

The Indian subcontinent as a singular geographic entity includes Bangladesh, India, Nepal, and Pakistan, separated from the rest of Asia by the Himalaya mountain system. The geodynamical positioning of India is unique in that it borders the most active continent-continent and ocean-continent convergent margins that have resulted from the India-Eurasia collision. The Himalaya mountain range, raised by the subduction-cum-collisional processes (40–45 Ma), runs from west-northwest to east-southeast in a ~2500-km-long arc and forms a spectacular geographical marker that defines the India-Eurasia plate boundary. The western anchor 'Nanga Parbat' lies to the south of the northernmost bend of the Indus River and the eastern anchor 'Namcha Barwa' is located to the west of the bend of the Tsangpo-Brahmaputra River. Bordered on the northwest by the Karakoram and the Hindu Kush ranges, the mountain chain on the north is separated from the Tibetan Plateau by the Indus-Tsangpo Suture, a 50–60 km wide tectonic valley. To the west of the arcuate continental collision boundary lies the Makran Subduction Zone (MSZ), where the Arabian plate subducts beneath the Eurasian plate. To the east, the collision boundary takes a southerly swing along the Arakan-Yoma Range (AYR) and joins the Andaman-Sumatra Subduction Zone (ASSZ), where the Indian ocean plate slides beneath the Eurasia plate. Together, these segments of the collision zone between the India and Eurasia are among the most seismically vulnerable regions in the world. Some of the greatest and the most damaging earthquakes in the history of the earth have occurred along this plate boundary (Fig. 1.1).

From a seismotectonic perspective, the Indian subcontinent falls broadly under three domains. One, the ~2500 km long Himalayan arc in the north, stretching from the Hindu-Kush mountains in the northwest to the Arakan-Yoma range in the east, extending further down and joining with the Andaman-Sumatra subduction zone. Two, the Indo-Gangetic Plains, located between the foothills of the Himalaya and the Indian peninsula. Three, the Indian peninsula that comprises the shield region (e.g., Dasgupta et al., 2000). All these tectonic domains, whether associated with

C. P. Rajendran and K. Rajendran, *Earthquakes of the Indian Subcontinent*,
GeoPlanet: Earth and Planetary Sciences,
https://doi.org/10.1007/978-981-16-4748-2_1

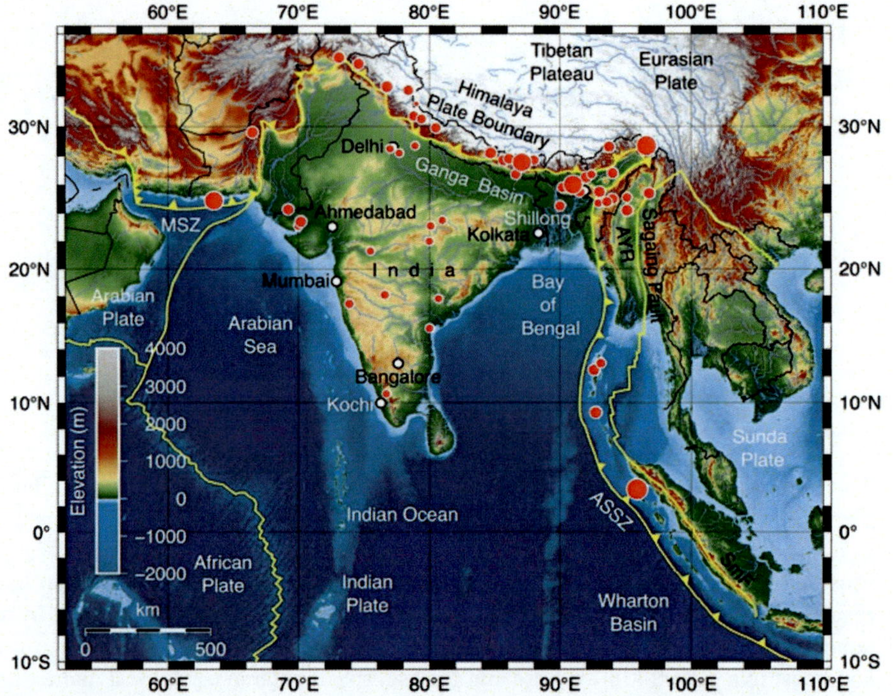

Fig. 1.1 Distribution of earthquakes (M 5.5 and above) from 1800 to 2021, in India and adjoining areas. MSZ: Makran Subduction Zone; AYR: Arakan Yoma Range; ASSZ: Andaman-Sumatra Subduction Zone

the plate boundaries, paleo-rift systems, or non-rifted regions within the peninsular region, have generated moderate to large/great earthquakes (Fig. 1.1; Table 1.1).

The Himalayan arc has witnessed several large and great earthquakes in the last 200 years or more; the most recent being the pair of M_w 7.8 and 7.3 events that nucleated in the Nepal Himalaya during April–May 2015. Significant among the historically documented earthquakes along the Himalayan arc and its vicinity are the 1803 Uttarkashi (M_w ~ 7.5), 1905 Kangra (M_w 7.8), 1934 Nepal-Bihar (M_w 8.2), 1897 Shillong (M_w 8.2), 1950 Assam (M_w 8.6) and 2005 Kashmir (M_w 7.8) events (Fig. 1.2). The 2004 earthquake (M_w 9.1) that originated on the Andaman-Sumatra plate boundary is significant as a rare tsunamigenic megathrust event to have occurred in this part of the globe and the giant transoceanic tsunami that followed affected all the Indian Ocean rim countries, including India. Two other large earthquakes, the 1881 (M_w 7.9), off Car Nicobar and the 1941 (M_w ~7.6), off South/Middle Andaman, partly occupy northern segments ruptured in the 2004 earthquake (see Table 1.1 and Fig. 1.1). However, none of these events had generated a transoceanic tsunami, comparable to that of the 2004 event. The 1945 (M_w 8.2) Makran earthquake is another great tsunamigenic event that has affected some parts of the Indian coastal areas of Gujarat and Maharashtra, but with a greater intensity on the coasts of Pakistan

Table. 1.1 Major earthquakes (M 5.5 and above) 1800–2021, in India and contiguous areas

Year	Mag.	Location/source area °N: °E	Fatalities/damage intensity	References
1803, September 1	~7.5	Srinagar 31.50:79.00	~500	Ambraseys and Douglas (2004)
1819, June 10	~7.5	Kachchh 24.25:69.25	>1500	Rajendran and Rajendran (2001)
1833, August 26	~7.7	Kathmandu 27.70:85.70	500	Ambraseys and Douglas (2004)
1869, January 10	~7.5	Cachar 25.5:93.0	Heavy damage, not many casualties	Ambraseys and Douglas (2004)
1881, December 31	~7.9	Car Nicobar 9.25:92.70	No reported casualties	Bilham et al. (2005)
1885, May 30	7.5	Kashmir 34.1:74.6	3000	Ambraseys and Douglas (2004)
1885, July 14	6.9	Bengal 24.5:90.0	47	Ambraseys and Douglas (2004)
1897, June 12	~8.2	Shillong, 26.0, 91.0[1] 25.7:91.1[2]	>1500	[1]Macroseismic [2]Ambraseys and Bilham (2003)
1900, February 7	~5.5	Near Coimbatore 10.7:76.7	Not known	Basu (1963)
1905, April 05	7.8	Kangra, 32.63:76.78	20,000	Szeliga and Bilham (2017)
1916, August 28	<7.0	Dharchula 29.9:80.5	Not known	Ambraseys and Douglas (2004)
1918, July 8	>7.6	Srimangal 26.50:92.00	Heavy damage, not many casualties	Ambraseys and Douglas (2004)
1923, September 9	7.1	Dabigiri 25.50:91.50	Not known	Ambraseys and Douglas (2004)
1927, June 2	~5.5	Umeria, Son Valley 23.5:81.0	Damage and casualties not known	ASC
1930, July 2	>7.0	Dhubri 25.8:90.2	Heavy damage, not many casualties	Ambraseys and Douglas (2004)
1931, January 27	7.6	Assam 25.4:96.8	Destruction of property	IMD
1934, January 15	8.2	Bihar-Nepal 27.55:87.09	~10,700	Ambraseys and Douglas (2004)
1935, May 31	7.7	Quetta 29.60:66.50	~30,000	Ambraseys and Douglas (2004)
1938, March 14	~5.5	Satpura 21.3:75.5	No reports available	Mukherjee (1942)

(continued)

Table. 1.1 (continued)

Year	Mag.	Location/source area °N: °E	Fatalities/damage intensity	References
1941, June 26	~7.6	Middle Andaman 12.5:92.57	~7000	Jhingran (1953)
1945, November 28	8.1	Makran 24.5:63.0	~4000	ISC
1943, October 23	7.2	Hojai (Kopili Valley) 26.80:94.00	Not known	Ambraseys and Douglas, (2004)
1947, July 29	7.3	Ziro, Arunachal 28.63:93.73	Slight damage in some parts of the Upper Assam	Chen and Molnar (1977, 1983)
1950, August 15	8.6	Upper Assam 28.70:96.60	~1526	Ambraseys and Douglas (2004)
1954, March 21	7.7	Manipur 24.2:95.1	Not known	Ambraseys and Douglas (2004)
1956, July 21	6.0	Anjar 23.1:70.0	156	IMD
1956, October 10	~5.5	Kurja 28.2:77.7	No reports of fatalities	ASC
1957, July 1	7.0	Indo-Burma 25.0:94.0	Not known	IMD
1957, August 25	~5.5	Balaghat 22.0:80.0	25	ASC
1960, August 27	5.3–6.0	Near Delhi 28.5:77.0	No reports of fatalities	Nath et al. (1967)
1967, December 10	6.3	Koyna, 17.41, 73.86	~200	ISC
1966, August 15	5.5	Moradabad 28.7:78.9	14	ISC
1967, March 27	5.8	Ongole 15.6:80.0	No reports of fatalities	USGS
1969, April 13	5.7	Bhadrachalam, 17.8:80.7	No reports of fatalities	ASC
1975, January 19	6.8	Kinnaur, 32.4,78.4	47	Khatri et al. (1978)
1980, November 19	6.0	Gangtok, 27.4, 88.8	No reports	Drukpa et al. (2006)
1984, December 30	5.6	Cachar 24.7:92.9	20	ISC
1988, August 20	6.8	Udaipur 26.71:86.62	700	ISC

(continued)

Table. 1.1 (continued)

Year	Mag.	Location/source area °N: °E	Fatalities/damage intensity	References
1988, August 6	6.6	Near Manipur 25.1:95.1	2	ISC
1991, October 20	6.8	Uttarkashi 30.7:78.8	2000	ISC
1993, September 29	6.3	Killari 18.1:76.6	9748	ISC
1997, May 22	5.8	Jabalpur, 23.1:80.1	39	ISC
1999, March 29	6.6	Chamoli, 30.5:79.4	103	ISC
2001, January 26	7.7	Bhuj 23.4:70.2	20,085	ISC
2002, August 13	6.5	Diglipur 13.0:93.1	No reports of fatalities	ISC
2005, October 8	7.7	Kashmir 34.5:73.7	86,000	ISC
2011, September 18	6.9	Sikkim 27.7:88.0	111	ISC
2015, April 25	7.8	Nepal-India border 28.2:84.7	9000	ISC
2015, May 12	7.3	Nepal-India border 27.8:86.1	169	ISC
2016, January 4	6.7	Imphal (Manipur) 24.8:93.56	11	ISC
2021, April 28	6.0	Sontipur (Assam) 26.7:92.4	No fatalities reported	IMD

IMD India Meteorological Department, *ISC* International Seismological Centre; *ASC* Amateur Seismic Centre

and Iran. We have not included these subduction zone events in this book because the unique characteristics of these ocean floor earthquakes demand an independent treatment, which is beyond the scope of its current format.

It is now widely accepted that the Stable Continental Region (SCR) earthquakes occur by the reactivation of ancient pre-existing faults within rifted or non-rifted crust (e.g., Johnston & Kanter, 1990; Crone et al., 1992). Due to the low deformation rate and the long inter-seismic intervals (tens of thousands of years, as against 100s of years for the plate boundary), these earthquakes appear to be isolated in time and space. During the last three decades, the Indian peninsula has been unusually active, generating significant earthquakes within the Narmada, Godavari and Kachchh rift systems (Fig. 1.3). Interestingly, the 2001 Bhuj earthquake (M_w 7.7), the largest

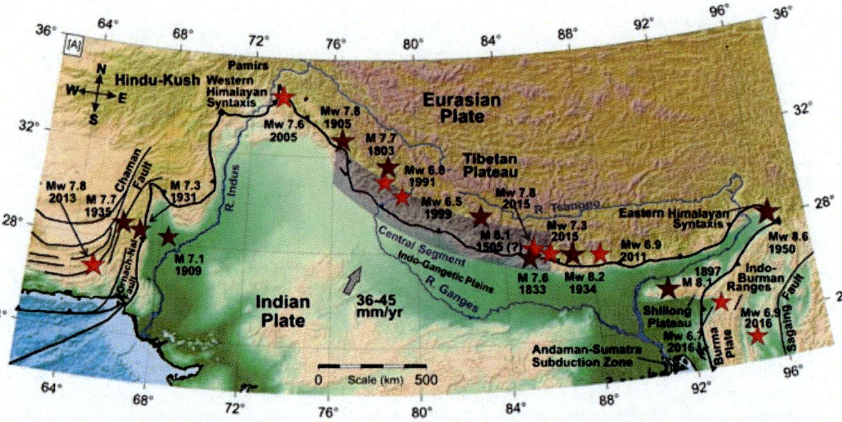

Fig. 1.2 Significant earthquakes along the India-Eurasia plate boundary. Red stars: Modern-day events ($M_w \geq 6.5$). Brown stars: Historic earthquakes ($M \geq 7$). The thick black barbed line shows the convergent plate boundary, and the black lines show major fault systems (modified after Rajendran et al., 2017)

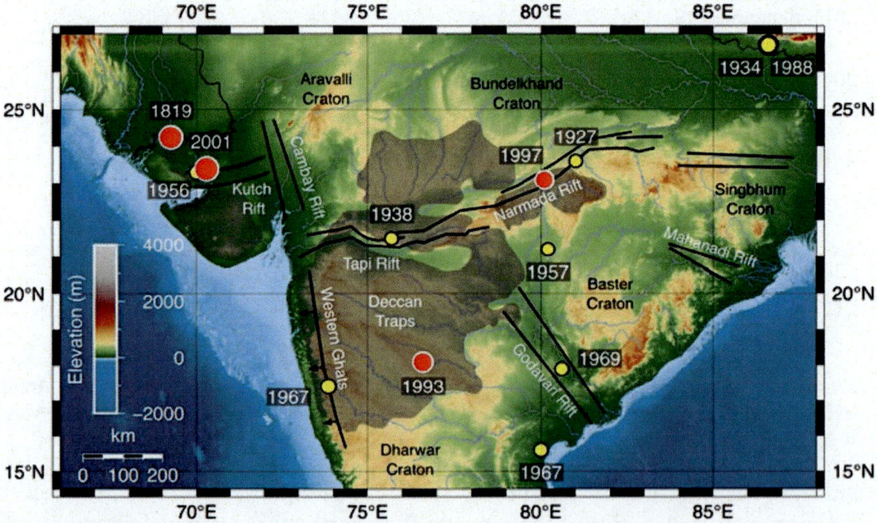

Fig. 1.3 Distribution of significant earthquakes and major geological structures (rifts) in the peninsular part of India. Earthquakes shown by solid red circles are discussed in Chaps. 3–6

modern-day paleo-rift-related earthquake globally, was preceded by another comparable size event in 1819. The September 30, 1993, Killari (Latur) earthquake (M_w 6.3), the most damaging among any globally documented continental interior region events, is believed to have occurred on a NW-SE oriented fault system, in a region, not known for any damaging earthquakes in the recent or historic times. In addition to these zones of recent activity, there are a few other hotspots in southern peninsular India. Coimbatore, in Tamil Nadu, is one such site, where a moderate (~M 5.5) event had occurred in 1900 CE, although the region has no such prior history. Occurrences of a few other moderate earthquakes (M $\geq$ 5) associated with pre-existing fault systems in central and south India indicate the potential of this region to generate occasional earthquakes. The earthquake doublet (ML 5.0; and M_w 4.7) of December 2000, in the central midland Kerala are noteworthy in this context (Bhattacharya & Dattatrayam, 2002; Rajendran et al., 2009).

The Indian subcontinent is also known for earthquakes caused by the influence of large artificial reservoirs. In fact, the most compelling global example of reservoir-triggered seismicity has been documented at the Shivaji Sagar Lake, impounded by the Koyna Dam, in western India. The case of seismicity at Koyna stands out globally, as one of the most remarkable, among the four cases of earthquakes of M > 6 near artificial reservoirs. A site where seismicity shows consistent spatial and temporal correlation with the loading and unloading history of the reservoir, it remains a unique example of sustained seismic activity from 1967 onwards (Gupta, 2002). However, we have excluded this case because a discussion on the mechanism of reservoir-triggered-seismicity and the extensive work carried out at Koyna would require an independent treatment.

The understanding of the nature of earthquake sources, their style of deformation, faulting and response to shaking effects are important for developing strategies for mitigating future hazards. With this idea in view, we will be discussing some of the significant recent and historical earthquakes that took place in the Indian subcontinent. Newly gained understanding, especially from the studies of modern-day events, provides new insights and benchmarks in the extrapolation of their predecessors. With multiple research groups exploring the source zones of older events, diverse views and interpretations have emerged. In this book, we have tried to represent dissimilar views and their respective supporting arguments, taking care to avoid any biases.

Following the above discussion, we have selected the large and great earthquakes along the Himalaya, the northeast India and from the peninsular India. Due to the varied nature of their seismic sources, we discuss these earthquakes as two categories, namely, intracontinental and interplate. We provide a broad overview of their respective seismogenic regions and discuss significant modern-day examples from each. Finally, in the epilogue, we recapitulate the earthquake hazard scenario in India. Lessons learned from each of these earthquakes are crucial for developing a set of measures to be taken at the individual, organizational and societal levels to minimize impact of any future earthquakes. We believe that this journey through some of the well-known earthquakes and source zones in India, in the backdrop of their respective tectonic environments and seismic histories should serve as guidelines for hazard mitigation efforts by planners, policymakers, administrators and the public. A preliminary assessment on the status of the various seismogenic zones in India can obtained from the Seismic Zonation Map of India. This map divides the country as four seismic zones on the basis of seismotectonics, frequency of earthquakes, and intensity experienced since historic times. Thus Zone V is considered to be the most vulnerable, while zone II falls in the zone of lowest level of activity (Fig. 1.4).

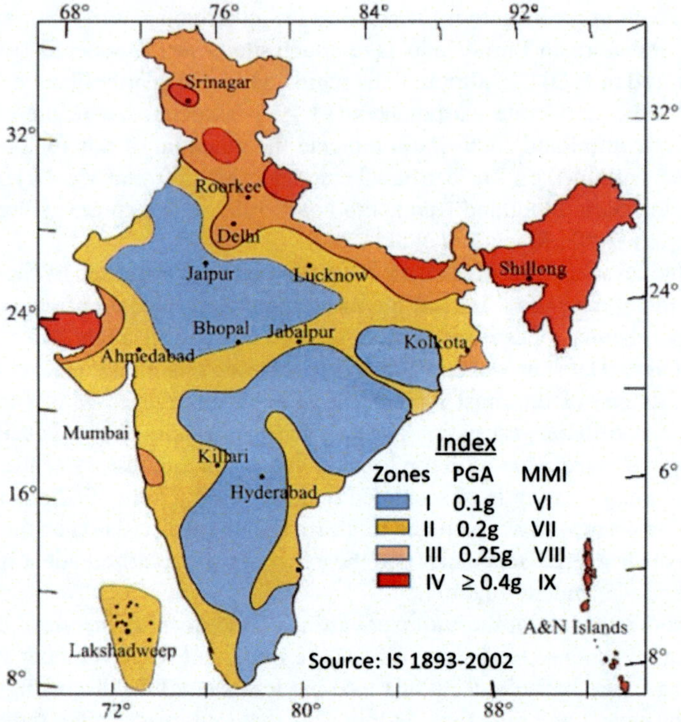

Fig. 1.4 The seismic zonation map of India (Bureau of Indian Standards BIS, 2002)

References

Ambraseys, N., & Bilham, R. (2003). Re-evaluated intensities for the great Assam earthquake of 12 June 1897, Shillong India. *Bulletin of the Seismological Society of America, 93*(2), 655–673.

Ambraseys, N., & Douglas, J. J. (2004). Magnitude calibration of north Indian earthquakes. *Geophysical Journal International, 159*, 165–206.

Basu, K. L. (1963). A note on Coimbatore earthquake of February, 1900. *Indian Journal of Meteorological Geophysics, 15*, 281–286.

Bhattacharya, S. N., & Dattatrayam, R. S. (2002). Earthquake sequences in Kerala during December 2000 and January 2001. *Current Science, 82L*, 1275–1278.

Bilham, R., Engdahl, R., Feldl, N., & Satyabala, S. P. (2005). Partial and complete rupture of the Indo-Andaman plate boundary 1847–2004. *Seismological Research Letters, 76*(3), 299–311. https://doi.org/10.1785/gssrl.76.3.299

Chen, W. P., & Molnar, P. (1977). Seismic moments of major earthquakes and the average rate of slip in central Asia. *Journal of Geophysical Research, 82*, 2945–2969.

Chen, W.-P., & Molnar, P. (1983). Focal depths of intracontinental and intraplate earthquakes and their implications for the thermal and mechanical properties of the lithosphere. *Journal of Geophysical Research, 88*, 4183–4214.

Crone, A. J., Machette, M. N., & Bowman, J. R. (1992). Geologic investigations of the 1988 Tennant Creek, Australia earthquakes—Implications for paleoseismicity in stable continental regions. *U.S. Geological Survey Bulletin 2032-A*, 51 pp.

Dasgupta, S. et al. (2000). Seismotectonic atlas of India and its environs. *Geological Survey of India Special Publication*, (59), 87p.

Drukpa, D., Velasco, A. A., & Doser, D. I. (2006). Seismicity in the Kingdom of Bhutan (1937–2003): Evidence for crustal transcurrent deformation. *Journal of Geophysics Research, 111*, B06301. https://doi.org/10.1029/2004JB00308

Gupta, H. K. (2002). A review of recent studies of triggered earthquakes by artificial water reservoirs with special emphasis on earthquakes in Koyna, India. *Earth- Science Reviews, 58*, 279–310.

Jhingran, A. G. (1953). A note on an earthquake in the Andaman Islands (26th June 1941). *Records of the Geological Survey of India, 82*, 300–307.

Johnston, A. C., & Kanter, L. R. (1990). Earthquakes in stable continental crust. *Scientific American, 262*(3), 68–75.

Khattri, K. N., Rai, K., Jain, A. K., Sinvhal, H., Gaur, V. K., & Mithal, R. S. (1978). The Kinnaur earthquake, Himachal Pradesh, India, of 19 January, 1975. *Tectonophysics, 49*, 1–21.

Mukherjee, S. M. (1942). Seismological features of the Satpura earthquakes of the 14th March 1938. *Proceedings of the Indian Academy of Sciences, 16*, 167–175.

Nath, M., Narain, K., & Srivastava, J. P. (1967). The Delhi earthquake of 27th August 1960. *Records of the Geological Survey of India, 95*(Pt. 2), 367–382.

Rajendran, C. P., & Rajendran, K. (2001). Characteristics of deformation and past seismicity associated with the 1819 Kutch earthquake, Northwestern India. *Bulletin of the Seismological Society of America, 91*(3), 407–426.

Rajendran, C. P., John, B., Sreekumari, K., & Rajendran, K. (2009). Reassessing the earthquake hazard in Kerala based on the historical and current seismicity. *Journal of Geological Society of India, 73*, 785–802.

Rajendran, K., Parameswaran, R., & Rajendran, C. P. (2017). Seismotectonic perspectives on the Himalayan arc and contiguous areas: Inferences from past and recent earthquakes. *Earth-Science Reviews, 173*, 1–30.

Szeliga, W., & Bilham, R. (2017). New constraints on the mechanism and rupture area for the 1905 Mw 7.8 kangra earthquake, northwest himalaya: New constraints on the mechanism and rupture area for the 1905 Mw 7.8 kangra earthquake. *Bulletin of the Seismological Society of America, 107*, 2467–2479.

Part I
Intra-Continental Earthquakes

Chapter 2
Earthquakes in the Continental Interiors: An Overview

2.1 Introduction

The broad framework of plate tectonics explains how the elastic strain accumulated within the continental interiors is released by moderate to large earthquakes (Richardson et al., 1976). It is now widely known that over time, the tectonic processes build up stresses on faults because of steady plate motions, and that eventually, when the frictional strength of faults is exceeded, they are released in the form of earthquakes (Scholtz, 1989). Geodetic data along plate boundaries show that steady plate motions are accommodated by localized elastic deformation of the crust, leading to earthquakes (e.g., Stevens & Avouac, 2016). Talwani (2017) describes continental intraplate regions as those located away from active plate boundaries and therefore characterized by uniform stresses, for distances of over thousands of kilometres. Accordingly, earthquakes occur on pre-existing structures when the local stresses become comparable with their regional counterparts (i.e., hundreds of megapascals) and they are treated as stress perturbations.

Given this broad tectonic framework, it is also important to make a distinction between regions that are proximal to, and away from plate boundaries, because ultimately it is the forces generated by plate motions that drive their respective seismic productivities. To describe the different tectonic and seismogenic regions within the plate tectonic framework, one may follow the description by Stein (2007), based on plate velocities. Thus, the regions that move faster with respect to the stable interior of a plate (i.e., more than ~2–3 mm/year), may belong to the diffused plate boundary zone. By this definition, regions which move slower than 2–3 mm/year are treated as part of the plate interior.

Although plate velocities can provide a useful criterion, global examples suggest that description of plate interiors based on their velocities alone is rather inadequate. It has been noted that along with the long-wavelength tectonic stresses arising from plate boundary forces, sub-lithospheric density and lateral strength contrasts may also contribute to the reactivation of the pre-existing faults within plate interiors (Ghosh

C. P. Rajendran and K. Rajendran, *Earthquakes of the Indian Subcontinent*, GeoPlanet: Earth and Planetary Sciences,
https://doi.org/10.1007/978-981-16-4748-2_2

et al., 2019). For example, India's collision with Tibet and the resultant transmission of stresses are regarded as underlying factors for India's intraplate seismicity (Gowd et al., 1992). Bilham et al. (2003) introduced a model in which the forces applied at the plate boundary generates flexural deformation in the central part of the Peninsular India. The flexural bulge has a wavelength of 670 km and is manifested in the free-air gravity anomaly and the geoid. It is suggested that the observed pattern of distribution of earthquakes is consistent with the inferred flexural stress field. Vita-Finzi (2004) developed this thesis further with the help of neotectonics data and related the major local earthquakes with elastic buckling in Peninsular India. However, these models need to be tested with more examples of earthquakes, their mechanisms and the proposed geometry of the flexures.

Although local perturbations may facilitate earthquakes within the continental interiors, they are generally considered as stable. However, regions adjoining subduction zones that undergo compressive deformation during the interseismic intervals are not so stable and they generate intermediate and deep focus earthquakes. Such events, originating within the "old, dense and cold" subducting slab, on the landward part of the trench, are referred to as "intraslab" or as "intraplate" earthquakes (e.g., southwest Japan; Seno, 1979). In this context it may be noted that the term "intraplate earthquake" is used in the literature to refer to all types of non-plate boundary earthquakes. Thus, the common usage of the term "intraplate" makes no distinction between regions that are proximal to active plate boundaries and those located thousands of kilometers away. A distinction is therefore necessary, as the continental interiors are much more stable, and the earthquakes occur much less frequently, as discussed in the rest of this chapter.

2.2 Terminologies and Characteristics of Continental Interior Earthquakes

Johnston (1989) treated cratonic regions as Stable Continental Interiors (SCRs), and he had defined them as "*areas where the continental crust is largely unaffected by currently active plate boundary processes.*" According to Johnston (1992), SCRs are clearly different from other intraplate tectonic provinces, a distinction based mainly on the age of the crust (2.5–0.5 Byr). Another criterion is the low deformation rate in such regions, which is an order of magnitude less than that of active intraplate and plate boundary regions. Johnston (1989, 1996) identified nine continental-scale SCRs in Africa, Antarctica, Asia, Australia, China, India and North and South Americas, associated with Precambrian cratons, Paleozoic fold belts, passive continental margins and failed intra-cratonic rifts. A more recent review suggests that large earthquakes are a rarity in SCRs, with just about two dozen events of magnitude ≥ 6 reported in the historical record worldwide (Fig. 2.1; Calais et al., 2016).

SCRs generally show low strain rates, as evident from the weak expressions of topography, an observation reiterated from space geodetic measurements (Calais

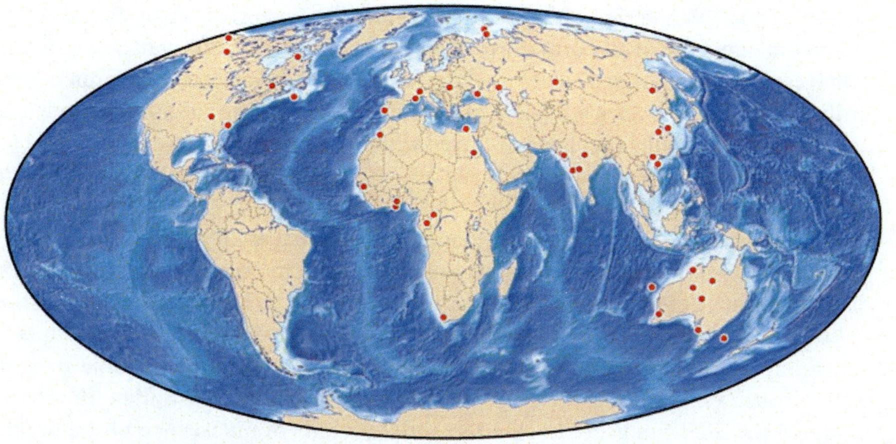

Fig. 2.1 Global distribution of continental intraplate earthquakes (M > 6) shown by red circles (figure modified from Calais et al., 2016)

et al., 2006). Studies on the surface expressions of SCR earthquakes became a topic of interest in the 1990s motivated by a few examples in Australia. These are: 1968 Meckering (M_w 6.5; Langston, 1987), 1986 Marryat Creek (Ms 5.8; Machette et al., 1993), 1988 Tennant Creek (three earthquakes Ms 6.3–6.7; Bowman, 1992), and in northern Canada (1989 Ungava; Ms 6.3; Adams et al., 1991). A few of these events had generated perceptible surface scarps attributed to the reactivation of pre-existing faults within the Precambrian crust. Some other events had occurred in landscapes that did not show any apparent or latent geomorphological features indicative of past surface-rupturing earthquakes, at least for hundreds of thousands of years (Calais et al., 2016 and references therein). Thus, poor surface expressions of causative structures and long elapsed time (time since the last similar earthquake) were considered as distinctive characteristics, at least for the Australian SCR (Crone et al., 1997).

Based on the global reviews, it can be surmised that the SCR earthquakes are generally associated with regions that show varied geologic and tectonic characteristics. These include: (a) passive rifts (e.g., New Madrid, Missouri; Johnston & Kanter, 1990); (b) edges of cratons (e.g., Central and Eastern United States; Van Lanen & Mooney, 2007); (c) hot spot tracks (Chu et al., 2013) and (d) regions with unusual physical properties such as higher porosity, within the fault zones (Costain et al., 1987). There are many earthquakes that are associated with non-rifted crust and the Australian SCR provides some typical examples (Crone et al., 1997). At high latitudes and in some passive margins affected by the Pleistocene glaciation, crustal strain fields originating from deglaciation or glacial unloading, are known to promote earthquakes (Muir-Wood, 2000). Some of the passive margins, such as those in North America, might have undergone glacial isostatic adjustment, resulting in localized margin stresses and/or dynamic topography (Stein et al., 1989).

Schulte and Mooney (2005) have defined SCR earthquakes under five categories based on their locations: (a) interior rifts, (b) rifted continental margins, (c) non-rifted crust, (d) possible interior rifts, and (e) possible rifted margins. Their data show that approximately 27% of all events have occurred in the interior rifts and 25% of them in rifted continental margins. Nearly 36% of earthquakes have occurred within the non-rifted crust, and 12% does not fall in any distinct group. These authors concluded that although many earthquakes have occurred within the rifted crust, non-rifted crust also holds significant potential for moderate activity, as exemplified by the 1993 Killari (Latur) earthquake (Rajendran et al., 1996).

Tesauro et al. (2015) suggested that SCR earthquakes in North America occur along the edges of cratons where tectonic stress tend to accumulate, but this correlation does not hold for other active locations (Schulte & Mooney, 2005). Although failed/passive rifts have generated some of the largest continental earthquakes, damaging events have also occurred in regions without any perceptible evidence of previous faulting, let alone rifting, as exemplified in the Australian hinterlands. Some of the SCR faults are known to exhibit periods of quiescence of the order of 10,000–100,000 years or more (Crone et al., 1997), which make the earthquakes appear as totally unexpected. Some notable examples of seismic activity in SCRs appear to be episodic (Crone et al., 2003; Liu & Stein, 2016). These episodes of clustered earthquakes separated by long intervals of inactive periods running into 10,000–100,000 years became apparent on the two historically quiescent faults- the approximately 30-km-long Roopena fault (South Australia) and the Hyden fault (Western Australia.) The paleoseismicity data from the Hyden fault scarp (32 km long; 2.5 m high) - a historically aseismic fault-located ~350 km east of Perth in Western Australia- is particularly interesting as it provides evidence for episodic activity during the Quaternary (Crone et al., 2003). Such long inter-event intervals also give a false semblance of aseismicity, leading to low levels of preparedness. The 1993 Killari (Latur) earthquake is often described as an example of an 'out-of-the-blue' event that shook a region that was the least prepared for a moderate earthquake. Understandably, the ground shaking led to total collapse of the poorly built and non-engineered rural abodes.

Various mechanisms have been proposed to explain the mechanism of SCR earthquakes. A widely followed model is based on stress concentrations at intersecting faults (Talwani, 1999), or around buried intrusions in the crust (Campbell, 1978; Pollitz et al., 2001). Stress concentrations at the tip of a low-velocity upper mantle (Zhan et al., 2016); local weakening of the lower crust either thermally (Kenner & Segall, 2000) or geochemically (Chen et al., 2016), are other viable mechanisms. Tesauro et al. (2015) proposed bulk weakening as a possible mechanism in regions where the mechanically strong mantle lithosphere is absent. Although these mechanisms may locally perturb the long-term tectonic stress field of the continental interior regions, they do not explain the episodicity of earthquake occurrences, and their long and variable inter-event intervals. Intervals of clustered activity separated by long quiescent periods reported from Australia are explained as being due to pore-pressure fluctuations within fault zones that are favourably oriented for failure in the regional stress field. It has been suggested that during quiescent periods, volumes of SCR crust would be strained in a gradual manner, due to the regional far-field stresses. As the strain builds up leading to deformation of rocks, critically stressed faults in the rock volume would also experience the build-up of pore pressure and eventually they would fail, marking the beginning of an active phase in the faults' history. After a few earthquakes that might occur within geologically short intervals (in the order of tens of thousands of years), the surrounding volume of the crust would be sufficiently relaxed, to be succeeded by another long interval of quiescence (Crone et al., 2003).

Some recent studies argue that SCR earthquakes occur as a result of the release of elastic energy from a pre-stressed lithosphere through transient perturbations of local stress (Calais et al., 2016). Following this new paradigm, events can occur in regions with no previous history of earthquakes and with no evidence for cycles of strain accumulation and release. By that argument, SCR earthquakes need not necessarily repeat, as the tectonic loading rate is close to zero. It follows that in a stress regime of zero tectonic loading (as within the plate interiors), the idea of "recurrence" does not apply, and thus the notions of "seismic cycle" or "slip rate" have no relevance. As these terms generally relate to quantifying the regional seismic hazard, their application to SCR seismicity can be quite challenging. The most difficult quantity to be estimated is the time elapsed after the last earthquake. It is known that their infrequent occurrences in regions of low strain rates leave no distinctive topographic expressions. With erosion stripping off any residual signatures of fault scarps during the long intervening quiet periods, the resultant featureless topography can be quite deceptive, and with no information on the time elapsed, the hazard estimation becomes very challenging.

Perceptions of long recurrence intervals notwithstanding, studies point to predecessors of historically known events at relatively close intervals, for some SCRs. The 1811–1812 (New Madrid; M_w 7.0–7.6), the 1819 (Kachchh; $M_w \geq 7.5$), the 1886 (Charleston; M_w 7.3) earthquakes are such examples of SCRs where earthquakes are known to have repeated within 500–1000 years. The New Madrid Seismic zone in the Midwestern United States, within a failed rift, ~1000 km away from the nearest plate boundary, is one of the best-studied SCRs globally. Two large earthquakes occurred here in a sequence (1811 and 1812) and an average recurrence time of ~500 years has been suggested (Tuttle et al., 2002). The Kachchh seismic zone in western India is another example of a failed rift in a passive continental margin where two large earthquakes occurred in a span of 182 years (1819 and 2001). This region is believed to have hosted another large earthquake prior to 1819 although its location is not well-established (Bilham & Lodi, 2010; Rajendran & Rajendran, 2001). The source region of the 1886 Charleston (eastern United States) is reported to have produced similar events and an average recurrence interval of 500–600 years has been suggested for $M > 7$ earthquakes (Obermeier et al., 1985; Rajendran & Talwani, 1993). It must be noted that the recurrence estimates in all these three regions, which are highly vulnerable to seismically induced liquefaction, are based primarily on palaeoliquefaction features.

Mid and lower crustal earthquakes reported from the ancient continental rift systems are considered as a special class of SCR earthquakes. The source zones of these earthquakes are marked by the presence of mantle pillows, and remnants of the active rifting phase. Thus, stress loading through ductile creeping of the weaker lower crust is considered a potential mechanism in the Japan back-arc basin (Kato et al., 2009). Zoback and Richardson (1996) suggested rift pillow model for the Manaus (45 km) and Amazon basin (23 km), central Brazil earthquakes. In their model, excess mass of the rift pillow frozen into the lower crust induces deviatoric stresses and cause earthquakes. Stuart et al. (1997) followed a similar argument to suggest that stress concentration above the pillow is responsible for the nucleation of earthquakes in the New Madrid rift. Kenner and Seagall (2000) propose that rift pillows can relax the weak zones due to thermal/fluid pressure/transient perturbations in the regional stress. The 1997 Jabalpur earthquake (M_w 5.8), at hypocentral depth of >35 km, is also considered as related to a rift-pillow. The 1989 Ayers Rock (m_b 5.8; 31 km), Australia and the 1983 Solberg, Sweden (m_b 4.1; 40 km) are examples of lower crustal continental earthquakes in non-rifted crust (Chen, 1988; Johnston, 1996).

From the perspective of the genesis of rift-related earthquakes, an interesting question is whether there is any commonality between the mechanisms of Kachchh- and New Madrid-type of earthquakes. The fact that both these earthquakes have originated in a failed rift tectonic environment under inverted stress regimes make them somewhat analogous. However, a major difference is that the New Madrid earthquakes were sourced in an area located >1000 km away from the nearest plate boundary, whereas the 1819 Kachchh and 2001 Bhuj earthquake sources are just 400 km away from the plate boundary. With its proximity to the Himalaya plate boundary, the Kachchh region has been described as a diffused plate boundary (Stein et al., 2002). If the components of stress transferred from the plate boundary, are overprinted local stresses, it is a challenge to understand which of these is dominant in refuelling the faults in such terrains. These are the pertinent questions concerning SCRs that are relatively proximal to plate boundaries.

This brief overview provides the background for discussing the examples from the Peninsular India (Fig. 1.3). Two earthquakes discussed in the succeeding chapters (1819 $M_w \geq 7.5$, 2001 M_w 7.7) occurred within the Mesozoic intracontinental Kachchh rift in western India. The 1819 earthquake is the first continental earthquake to produce a notable surface scarp, as discussed in the earliest modern geology books (e.g., Lyell, 1857). The 2001 earthquake is the first modern-day example of a large event in a continental paleorift setting, and one that has been studied extensively based on a large database and modern tools of analysis. The subsequent chapters discuss the 1993 (M_w 6.2), Killari (Latur) earthquake and the 1997 Jabalpur earthquake (M_w 5.8). While the studies on the 1819 earthquake are based mostly on historical, archaeological and geologic evidence, the 1993, 1997 and 2001 events are among the best instrumentally documented modern-day SCR earthquakes globally. Together, these examples provide an insight into the diverse characteristics of SCR earthquakes.

References

Adams, J., Wetmiller, R. J., Hasegava, H. S., & Drysdale, J. (1991). The first surface faulting from historical intraplate earthquakes in North America. *Nature, 352*, 617–619.

Bilham, R., & Lodi, S. (2010). The door knockers of Mansurah: Strong shaking in a region of low perceived seismic risk, Sindh, Pakistan. In M. Sintubin, I.S. Stewart, T.M. Niemi, & E. Altunel (Eds.), *Ancient Earthquakes: Geological Society of America Special Paper* (Vol. 471, pp. 29–37).

Bilham, R., Bendick, R., & Wallace, K. (2003). Flexure of the Indian plate and intraplate earthquakes. *Proceedings of Indian Academy of Sciences (Earth Planet Science) 112*, 315–329.

Bowman, J.R. (1992). The 1988 Tennant Creek, Northern Territory, earthquakes: A synthesis. *Australian Journal of Earth Science, 39*, 651–669. https://doi.org/10.1080/08120099208728056.

Calais, E., Dong, L., Wang, M., Shen, Z., & Vergnolle, M. (2006). Continental deformation in Asia from a combined GPS solution. *Geophysical Research Letters, 33*, L24319. https://doi.org/10.1029/2006GL028433

Calais, E., Camelbeeck, T., Stein, S., Liu, M, & Craig, T. J. (2016). A new paradigm for large earthquakes in stable continental plate interiors. *Geophysical Research Letters*. https://doi.org/10.1002/2016GL070815.

Campbell, D. (1978). Investigation of the stress–concentration mechanism for intraplate earthquakes. *Geophysical Research Letters, 5*, 477–479.

Chen, W. P. (1988). A brief update on the focal depths of intracontinental earthquakes and their correlations with heatflow and tectonic age. *Bulletin of the Seismological Society of America, 59*, 263–272.

Chen, C., Gilbert, H., Andronicos, C., Hamburger, M., Larson, T., Marshak, S., Pavlis, G., & Yang, X. (2016). Shear velocity structure beneath the central United States: Implications for the origin of the Illinois Basin and intraplate seismicity. *Geochemistry, Geophysics, Geosystems, 17*, 1020–1041.

Chu, R., Leng, W., Helmberger, D., et al. (2013). Hidden hotspot track beneath the eastern United States. *Nature Geoscience, 6*, 963–966.

Costain, J. K., Bellinger, G. A., & Speer, J. A. (1987). Hydroseismicity—A hypothesis for the role of water in the generation of intraplate seismicity. *Geology, 15*(7), 618–621.

Crone, A. J., Machette, M. N., & Bowman, J. R. (1997). Episodic nature of earthquake activity in stable continental regions revealed by palaeoseismicity studies of Australian and North American quaternary faults. *Australian Journal of Earth Sciences, 44*(2), 203–214.

Crone, A. J., De Martini, P. M., Machette, M. N., et al. (2003). Paleoseismicity of two historically quiescent faults in Australia: Implications for fault behavior in Stable Continental Regions. *Bulletin of Seismological Society of America, 93*, 1913–1934.

Ghosh, A., Holt, W. E., & Bahadori, A. (2019). Role of large-scale tectonic forces in intraplate earthquakes of Central and Eastern North America. *Geochemistry, Geophysics, Geosystems, 20*, 2134–2156.

Gowd, T. N., Sriramaro, S. V., & Gaur, V. K. (1992). Tectonic stress field in the Indian subcontinent. *Journal of Geophysical Research, 97*, 11789–11888.

Johnston, A. C. (1989). Seismicity of stable continental interiors. In S. Gregersen & P. W. Basham (Eds.), *Earthquakes at North-Atlantic passive margins: Neotectonics and postglacial rebound* (pp. 299–327). Kluwer Academic Publishers.

Johnston, A. C. (1992). Intraplate not always stable. *Nature, 355*, 213–214.

Johnston, A. C. (1996). Seismic moment assessment of earthquakes in stable continental regions I. *Geophysical Journal International, 124*, 381–414.

Johnston, A. C., & Kanter, L. R. (1990). Earthquakes in stable continental crust. *Scientific American, 262*(3), 68–75.

Kato, A., Kurashimo, E., Igarashi, T., Sakai, S., Iidaka, T., Shinohara, M., Kanazawa, T., Yamada, T., Hirata, N., & Iwasaki, T. (2009). Reactivation of ancient rift systems triggersdevastating intraplate earthquakes. *Geophysical Research Letters, 36*, L05301.

Kenner, S. J., & Segall, P. (2000). A mechanical model for intraplate earthquakes; application to the New Madrid seismic zone. *Science, 289*, 2329–2332.

Langston, C. A. (1987). Depth of faulting during the 1968 Meckering, Australia, Earthquake sequence determined from waveform analysis of local seismograms. *Journal of Geophysical Research, 92*, 11561–11574.

Liu, M., & Stein, S. (2016). Mid-continental earthquakes: Spatiotemporal occurrences, causes, and hazards. *Earth-Science Reviews, 162*(2016), 364–386.

Lyell, C. (1857). *Principles of geology* (Eleventh, p. 834). Appleton & Co.

Machette, M. N., Crone, A. J., & Bowman, J. R. (1993). Geologic investigations of the 1986 Marryat Creek, Australia earthquake—Implications for paleoseismicity in stable continental regions. *U.S. Geological Survey Bulletin 2032-B*, 29 p.

Muir-Wood. (2000). Deglaciation seismotectonics: A principle influence on intraplate seismogenesis at high latitudes. *Quaternary Science Reviews, 19*, 1399–1411.

Obermeier, S. F., Gohn, G. S., Weems, R. E., Gelinas, R. L., & Rubin, M. (1985). Geological evidence for recurrent moderate to large earthquakes near Charleston. *South Carolina, Science, 227*, 408–411.

Pollitz, F. F., Kel logg, L., & Bürgmann, R. (2001). Sinking mafic body in a reactivated lower crust: A mechanism for stress concentration at the New Madrid Seismic Zone. *Bulletin of the Seismological Society of America, 91*(6), 1882–1897.

Rajendran, C. P., & Rajendran, K. (2001). Characteristics of deformation and past seismicity associated with the 1819 Kutch earthquake, northwestern India. *Bulletin of the Seismological Society of America, 91*(3), 407–426.

Rajendran, C. P., & Talwani, P. (1993). Paleoseismic indicators near Bluffton, South Carolina: An appraisal of their tectonic implications. *Geology, 21*(11), 987–990.

Rajendran, C. P., Rajendran, K., & John, B. (1996). The 1993 Killari (Latur), Central India earthquake: An example of fault reactivation in the Precambrian crust. *Geology, 24*, 651–654.

Richardson, R. M., Solomon, S. C., & Sleep, N. H. (1976). Intraplate stresses as indicator of Plate Tectonic Driving Forces. *Journal of Geophysics Research, 81*(11), 1847–1856.

Scholtz, C. H. (1989). Mechanics of faulting. *Annual Review Earth and Planet Science, 17*, 309–334.

Schulte, S. M., & Mooney, W. D. (2005). 2005, An updated global earthquake catalogue for stable continental regions: Reassessing the correlation with ancient rifts. *Geophysical Journal International, 161*, 707–721.

Seno, T. (1979). Patterns of intraplate seismicity in southwest Japan before and after great earthquakes. *Tectonophysics, 57*, 267–283.

Stein, S., Sella, G. F., & Okal, E. A. (2002). The January 26, 2001, Bhuj earthquake and the diffuse western boundary of the Indian plate. In S. Stein & J. Freymueller (Eds.), *Plate Boundary Zones: Washington* (pp. 243–254). American Geophysical Union.

Stein, S., Cloetingh, S., Sleep, N., & Wortel, R. (1989). Passive margin earthquakes, stresses, and rheology. In S. Gregerson & P. Basham (Eds.), *Earthquakes at North Atlantic Passive Margins: Neotectonics and Postglacial Rebound* (pp. 231–260). Dordrecht: Kluwer.

Stein, S. (2007). Approaches to continental intraplate earthquake issues. *The Geological Society of America Special Paper 425*.

Stevens, V. L., & Avouac, J.-P. (2016). Millenary Mw > 9.0 earthquakes required by geodetic strain in the Himalaya. *Geophysical Research Letters, 43*, 1118–1123.

Stuart, D. W., Hildenbrand, T. G., & Simpson, R. W. (1997). Stressing of the New Madrid seismic zone by a lower crust detachment fault. *Journal of Geophysical Research, 102*, 27623–27633.

Talwani, P. (1999). Fault geometry and earthquakes in continental interiors. *Tectonophysics, 305*, 371–379.

Talwani, P. (2017). On the nature of intraplate earthquakes. *Journal of Seismology, 21*(1). https://doi.org/10.1007/s10950-016-9582-8.

Tesauro, M., Kaban, M. K., & Mooney, W. D. (2015). Variations of the lithospheric strength and elastic thickness in North America. *Geochemistry, Geophysics, Geosystems, 16*, 2197–2220.

Tuttle, M. P., Hengesh, J., Tucker, K. B., Lettis, W., Deaton, S. L. & Frost, J. D. (2002). Observations and comparisons of liquefaction features and related effects induced by the Bhuj earthquake. *Earthquake Spectra, 18*(suppl. A), 79–100.

Van Lanen, X., & Mooney, W. D. (2007). Integrated geologic and geophysical studies of North American continental intraplate seismicity. In S. M. S. Stein (Ed.), *The Geological Society of America* (pp. 101–112). Boulder CO.

Vita-Finzi, C. (2004). Buckle-controlled seismogenic faulting in peninsular India. Quat Sci Reviews 23: 2405–2412.

Zhan, Y., Hou, G., Kusky, T., & Gregg, P. (2016). Stress development in heterogenetic lithosphere: Insights into earthquake processes in the New Madrid Seismic Zone. *Tectonophysics, 671*, 56–62.

Zoback, M. L., & Richardson, R. M. (1996). Stress perturbation associated with the Amazonas and other ancient continental rifts. *Journal of Geophysical Research, 101*, 5459–5475.

Chapter 3
Kachchh 1819

3.1 Introduction

The 1819 earthquake (M $\geq$ 7.5), located in the northern part of the Kachchh (Kutch) rift basin in the state of Gujarat, is the largest and one of the most well-documented continental earthquakes prior to seismic instrumentation. Although located in a region that was sparsely populated, it caused severe impact around its source as well as in distant locations like Hyderabad (Sindh Province in Pakistan) and Ahmedabad in Gujarat (Fig. 3.1). The earthquake triggered soil liquefaction and ground failures in the vast swathes of the Rann of Kachchh. As one of the earliest documented continental earthquakes, it was widely quoted for its manifest surface deformation (Baker, 1846; Burnes, 1835; Lyell, 1857; Oldham, 1898, 1926). Particularly interesting was the coseismically formed surface scarp, known as the "*Allah Bund*" (the Mound of God). This ridge acted as a natural barrier that blocked the flow of the Nara (Puran) River, an east-flowing distributary of the Indus River, disrupting an active riverine trade route connecting the hinterland. In his treatise, "The Principles of Geology," Lyell, (1857) used the surface deformation from this earthquake to explain how repeated faulting events could act as agents of landform changes.

Much of our early understanding of about the impact of the 1819 earthquake (also referred to as Kachchh earthquake) is derived from the accounts published by Mac Murdo (1824), who was the resident commissioner stationed in Bhuj. It is reported that shaking from the earthquake lasted for two to three minutes and was accompanied by a 'heavy and appalling' noise, followed by violent undulatory motion of the ground, causing the buildings to crash. The ground surveys conducted subsequently by several investigators documented the ground level changes and other coseismically formed features (e.g., Baker, 1846; Burnes, 1835; Wynne, 1872). The findings recorded in these early reports were summarized later by Oldham (1926). A comprehensive report describing the effects of the earthquake, reproduced from the original transcripts, is provided by Dasgupta and Mukhopadhyay (2014).

C. P. Rajendran and K. Rajendran, *Earthquakes of the Indian Subcontinent*,
GeoPlanet: Earth and Planetary Sciences,
https://doi.org/10.1007/978-981-16-4748-2_3

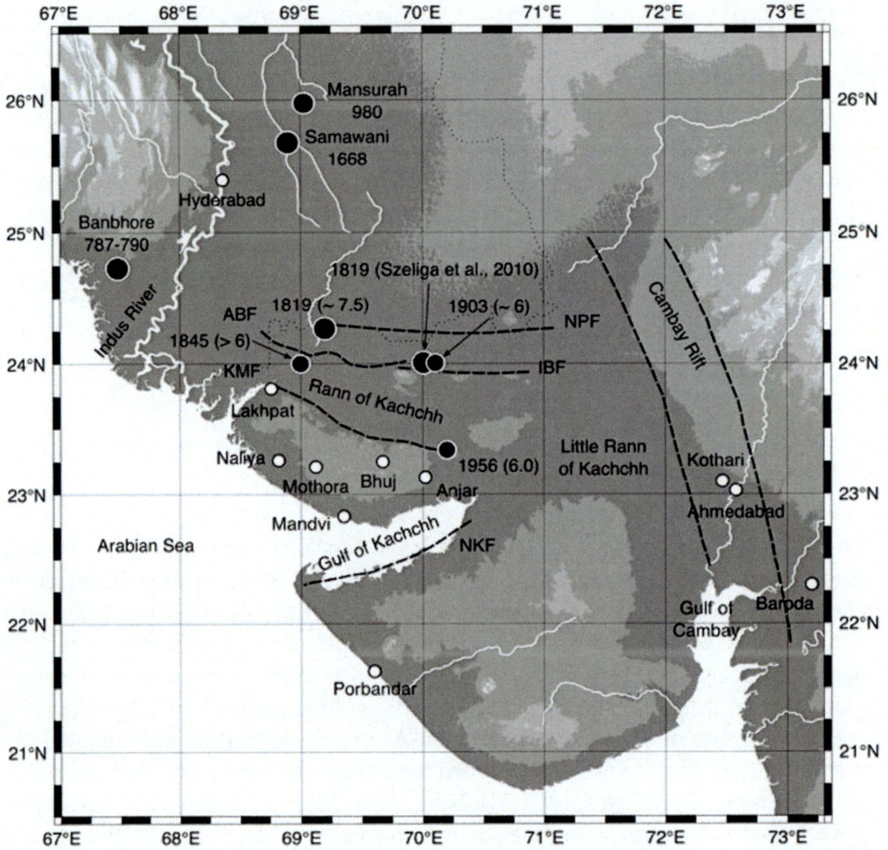

Fig. 3.1 Location of the 1819 earthquake and other historical earthquakes (M > 6) in the Kachchh region. Alternate locations for the1819 earthquake by Szeliga et al. (2010) are also shown. ABF: Allah Bund fault; NPF: Nagar Parkar Fault; IBF: Island Belt fault; KMF: Kachchh Mainland fault; NKF: North Kathiawar Fault (structures are after Biswas, 1987)

Despite the initial efforts in reporting the co-seismic morphological changes and the observed damage near and distant sites, it took more than a decade and a half for initiating surveys of the Allah Bund. The remoteness of the area, lack of logistics, and distance from settlement areas could have acted as a deterrent for undertaking any investigations. Another grave disaster to strike this area was the flood in 1826 that had not only destroyed several of the local artificial dams built across the upstream rivulets in the southern part of the Indus delta but also breached the barrier formed by the Allah Bund (e.g., Syvitski et al., 2013). Probably, this regional flooding event that had disrupted the then existing water management systems in this arid region provided a stimulus to explore and survey the region. In March 1827, Alexander Burnes, a British explorer and geographer, who was serving as an officer in Bhuj, proceeded to explore the Allah Bund and the neighbouring areas. He published his account of the landscape changes brought about by the earthquake in the Transactions of the Royal Asiatic Society (Burnes, 1835). This work was followed by several other explorations, as discussed in the report by Oldham (1926). Since those early explorations, the Allah Bund and the nearby regions had remained unattended by geologists for a long time. Geological studies in this Kachchh basin picked up in the 1980s, when exploratory drilling was carried out by oil companies. The data generated through these surveys were helpful in mapping the deep-time stratigraphy and tectonic architecture of this Mesozoic-Tertiary rift basin (Biswas, 1987). These details have provided the basic tectonic framework of the region.

3.2 Felt Reports and Surface Deformation

The mainshock struck Kachchh on June 16, at 6:50 pm, when Captain Mac Murdo, who was on a visit to Anjar, located ~40 km east of Bhuj and was getting ready for dinner. He described the initial shaking as vertical, and later ground motions as undulatory, and perfectly perceptible. In an article on the earthquake effects presented at the Literary Society of Bombay on April 28, 1820, Mac Murdo says: "*the shock lasted from two to three minutes, and during that short period the city of Bhooj was almost levelled with the ground*" (Mac Murdo, 1824). There is no reliable information of any foreshocks, but numerous aftershocks were felt every day until the end of November 1819. Felt effects of the earthquakes were documented by several others like Frere (1870), the chief commissioner of Sindh.

From various reports, it can be surmised that the earthquake was strongly felt over a large area. It damaged nearly 7000 houses and caused more than 1500 deaths, mostly in and around the towns of Bhuj and Anjar. Several other smaller towns like Kothari, Lakhpat, Mandvi, Mothora, and Naliya in the Kachchh Mainland also suffered severe damage. Strong shaking was felt at far away cities such as Ahmedabad and Baroda, where buildings suffered minor structural damage. In Ahmedabad, ~300 km southeast of the epicenter, a 400-year-old mosque was damaged and this was the case of the farthest reported structural failure. Damage was also reported from Hyderabad in the Sind Province of Pakistan ~200 km northwest of the epicenter (see Fig. 3.1 for locations). Oldham reported that some parts of Jaisalmer town in Rajasthan (26.9°N, 70.90°E, outside the frame of Fig. 3.1) were reduced to ruins. There was a specific mention of a collapsed fort at this location, killing about 500 people who had assembled there for a social function during the time of the earthquake.

Although the earthquake took place during a dry period of the year, it caused widespread soil-liquefaction and out-pouring of water, flooding the area. The phenomenon was widespread and was facilitated by the local geological conditions including the shallow water table (Mac Murdo, 1824; Oldham, 1926). Liquefaction was reported from far away locations, as from Porbandar, 250 km south of the epicenter (see Fig. 3.1 for location). Mac Murdo (1824) described it, saying, that almost all of the rivers have been "*filled to their banks for a period of a few minutes.... The rivers in the valleys, and those with sandy beds, were alone affected.*

Wells everywhere overflowed, many gave way and fell in, and in numerous places spots of ground in circles of twelve to twenty feet diameter threw out water to a considerable height, subsided into a slough". The report mentions how jets of black, muddy water ejected from fissures to a height of 1.5–2.5 m had carried pieces of nails and iron, presumably remnants of sunken boats buried under sediments. Large-scale slumping was also reported from most parts of the Rann (Oldham, 1926).

The earthquake had considerable effects on the fluvial systems around the Great Rann of Kachchh, mostly attributed to the formation of a linear fault scarp, the "*Allah Bund*", or "*the mound of God*" (Baker, 1846; Burnes, 1835; Mac Murdo, 1824; Oldham, 1926; Wynne, 1872). Near the town of Veego-Gud (modern Vigakot), north of the Allah Bund, the level changes hindered the southward flow of the Nara River, resulting in a local pond where it was dammed, and subsequently, the downstream part of the river dried up. The southern edge of the main river channel (the Nara River) was terminated at the northern flank of the Allah Bund, following the coseismic change in elevation. There are visible evidence for the poorly organized drainage system that features several abandoned channels, beheaded streams, and incipient drainage along this segment of the bund (Fig. 3.2a). The original extension of the river as a 10-m-wide and 3-m-deep entrenched river valley cutting through the central segment of the bund has also been traced. Rajendran and Rajendran (2001) interpreted this currently defunct channel, which severed the southern extension of the river, as the outcome of coseismic vertical uplift (Fig. 3.2b). An extension of this channel would lead towards the Sindri Lake, as it would have existed during the pre-earthquake conditions. Another distributary of the Indus River that used to flow into the Nara from the west is also reported to have dried up after the earthquake (Glennie & Evans, 1976), suggesting that the earthquake-related land-level changes had a regional impact.

The Sindri Fort, a revenue outpost in the downstream part of the Nara River and an important landmark in the region, has been used as a proxy to describe coseismic

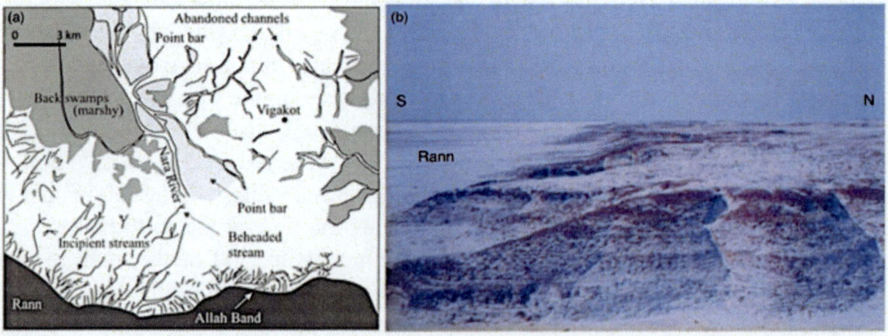

Fig. 3.2 **a** Drainage pattern in and around Vigakot and the swampy areas impounded by the Nara River (after Roy & Merh, 1977). **b** transverse section of the Allah Bund as seen in the central segment, viewed from the floor of an entrenched river (Nara) valley, facing west (after Rajendran & Rajendran, 2001)

subsidence. Located ~8 km south of the Allah Bund, the fort was damaged and had partly submerged during the earthquake. As Lyell (1857) writes in his treatise, "*one of the four towers, the north-western, still continuing to stand the day after the earthquake, the inhabitants who had ascended to the top of this tower, saved themselves in boats.*" Subsidence of the land is believed to have resulted in the submergence of the Sindri Fort at the foundation level and such large-scale inundation is believed to have led to the formation of a lake several hours after the earthquake (Baker, 1846; Burnes, 1835). Oldham (1926) used these reports to estimate the coseismic subsidence as ~1 to 3 m, which was attributed to downfaulting during the earthquake. However, subsequent investigations suggested that the foundation of the Sindri Fort had remained intact (Wynne, 1872), although it does not mean that the fort had not been lowered relative to the mean sea level. The condition of the Sindri Fort, as interpreted by various investigators in relation to the earthquake, will be discussed in some detail later.

3.3 Intensity, Location and Magnitude

Oldham (1926) assigned a maximum intensity of XI for the Kachchh earthquake, which was also the basis for the estimation of magnitude as M = 8, by the India Meteorological Department (IMD). A revision of the magnitude was done by Ambraseys and Douglas (2004) who re-calibrated the intensity estimates of significant historically documented Indian earthquakes. Their analyses based on a re-evaluation of the Indian and Tibetan earthquakes during the last 200 years, used the revised MSK scales. An important observation was that, in general, the upper range of the MSK scale is effectively the same for all earthquakes, as it saturates at VII–VIII. At that level of the intensity scale, all adobe, rubble stone masonry houses are damaged beyond repair or destroyed. The revised intensity map of the Kachchh earthquake, based on 30 intensity points shows the maximum intensity of VIII (Fig. 3.3a). The later study by Szeliga et al. (2010), that relied on empirically derived intensity attenuation relationship for the 1819 earthquake, also yielded similar results. The algorithms they used relate the intensity with the moment magnitude, hypocentral distances and other parameters, to account for attenuation and geometrical spreading.

Revised intensity estimates for the earthquake based on conventional methods are also available. Pande (2011) used data from 56 nearby localities and reported the maximum intensity of X at Bhuj, followed by IX at Anjar (Fig. 3.3b). Naik et al. (2019) used the newly developed Environmental Intensity Scale (ESI-07), and the macroseismic observations of the 2001 Bhuj earthquake to re-estimate the intensity of the 1819 event. Their study indicated maximum intensity of XI at locations in close proximity to the epicentre, an observation that was validated using the effects of the 2001 Bhuj earthquake. The estimates by Pande (2011) as well as Naik et al. (2019) are on the higher side, possibly because of the way the attenuation relations are factored in. The intensity estimates of Szeliga et al. (2010) are consistent with that of Ambraseys and Douglas (2004). However, they place the source of the 1819

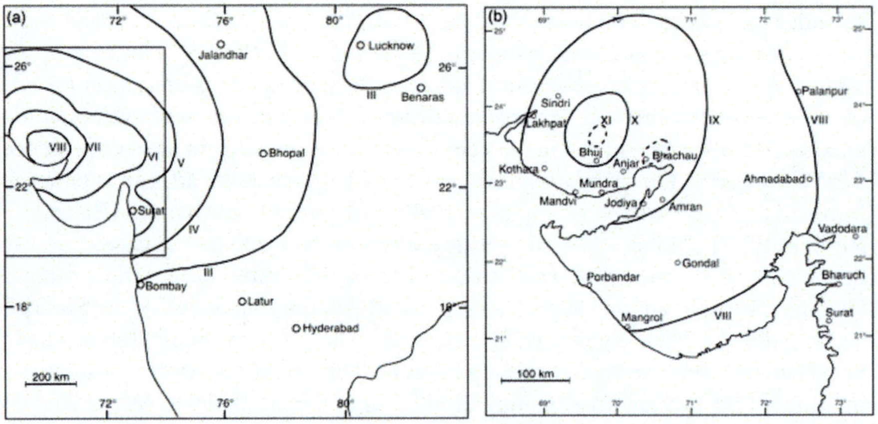

Fig. 3.3 **a** Isoseismal map of the 1819 earthquake showing intensity values MSK-64 scale (Ambraseys & Douglas, 2004); **b** Isoseismal map for a closer region, following the same scale (after Pande, 2011). The difference between the two estimations is attributed to the effects of attenuation

earthquake ~100 km east of Vigakot and close to the Island Belt fault, quite different from all the previous estimates (Fig. 3.1), as discussed next.

The epicentral location of the 1819 earthquake had remained uncertain for a long time. and most workers had placed it to the south of the Allah Bund (Chung & Gao, 1995). Using geometric models for a range of values of fault slip and dip attitudes, Bilham (1998) located it 5–15 km north or northeast of the Allah Bund. For a fault plane dipping 45°N and a focal depth of 15 km and general strike of N290° for the bund, this would fall about 10 km north of the kink where the bund takes a south-westerly swing. The frequency and dimensions of sandblows, maximum scarp height, and coseismic surface effects also favoured this location (Fig. 3.1). Szeliga et al. (2010) used recalibrated intensity estimates to suggest a range of values (23.67°N to 24.12°N and 70.21°N to 70.58°E), and a representative location is shown in Fig. 3.1. However, this location is inconsistent with the geological and geodetic evidence that favors the western part of the Allah Bund as the likely location. The morphological changes caused by the earthquake, such as the damming of the Nara River, were most pervasive in the western segment. Shaking effects were also quite severe near Vigakot, the prominent upstream trading post located north of the Bund. Remains of the fort whose existence and collapse during the earthquake are well-documented. Trenches excavated at the site of the ruined fort at Vigakot exposed larger sandblow craters compared to other sites (Rajendran & Rajendran, 2001). In fact, closer to the locations suggested by Szeliga et al. (2010), the sandblows were sparse, and much smaller in dimensions.

The magnitude of the earthquake has been estimated using various methods. Using the along-strike length of the Allah Bund as 80–150 km, Bilham (1998) estimated the geometric seismic moment and estimated the local magnitude as ML 7.7 $\pm$ 0:2. Using the rupture length of 90 km and width of 15 km, and the empirical relation

Table 3.1 Magnitude estimates of the 1819 earthquake

Magnitude	Method	References
7.5–8.0	Empirical relation between the intensity and the area of shaking	Johnston (1996)
$M_W 7.7 \pm 0.2$	Based on slip parameters and intensity distribution	Bilham (1998)
M_W 8.3	Felt area estimates	Dunbar et al. (1997)
M_W 8.2	Recalibration of felt area reports	Ambraseys and Douglas (2004)
$8.0 \leq M_W \leq 8.2$	Intensity-attenuation relationship for cratonic regions	Szeliga et al. (2010)
M_W 7.37 to 7.5	Inferred rupture dimensions and scaling relations; distance to farthest liquefaction	Rajendran et al. (2001)
M_W 7.6	Comparative studies of 1819 and 2001 earthquake effects	Hough et al. (2002)

between magnitude and surface rupture length (Wells & Coppersmith, 1994), the magnitude was estimated as M_W 7.37. Relation between magnitude and the distance to the farthest liquefaction also suggested a similar value, at M_W 7.5 (Rajendran & Rajendran, 2001). Post-2001 Bhuj earthquake, some researchers have used the intensity reports of the Bhuj and the Kachchh earthquakes to calibrate magnitudes based on their respective intensity values. Thus, by using denser sampling and the similarities between the felt reports, Hough et al. (2002) concluded that both earthquakes are of similar magnitude, at about M_W 7.6. Ambraseys and Douglas (2004) estimated the magnitude to be M_W 8.2, but they admit the possibility for errors due to the lack of detailed re-evaluation of the felt reports. Szeliga et al. (2010) arrive at magnitude values ranging from M_W 8.0 to 8.2. The magnitude ranges for these earthquakes, following various methodologies are summarized in Table 3.1.

3.4 Regional Tectonics and Morphology

The 1819 earthquake occurred within a Mesozoic basin on the north-western margin of the Indian craton, known in the geological literature as the 'Kachchh' rift (Fig. 3.4). Part of an aulocogen (the failed arm of a triple junction), this rift owes its origin to the Mesozoic tectonics associated with the break-up of the supercontinent called "Gondwanaland" and the subsequent northward drift of the Indian plate. The rift basin is riddled with numerous parallel normal faults expressed as horsts and grabens that are either exposed or remain blind under the sedimentary cover (Biswas, 1987). Compressional stress regime ensued around 40 Ma, following the collision of India with Eurasia expressed in the generation of reverse faulting events, as evident from the focal mechanisms (Chung & Gao, 1995). The tectonic setting of the Kachchh rift is outlined in the next chapter, as part of the introductory discussions on the 2001

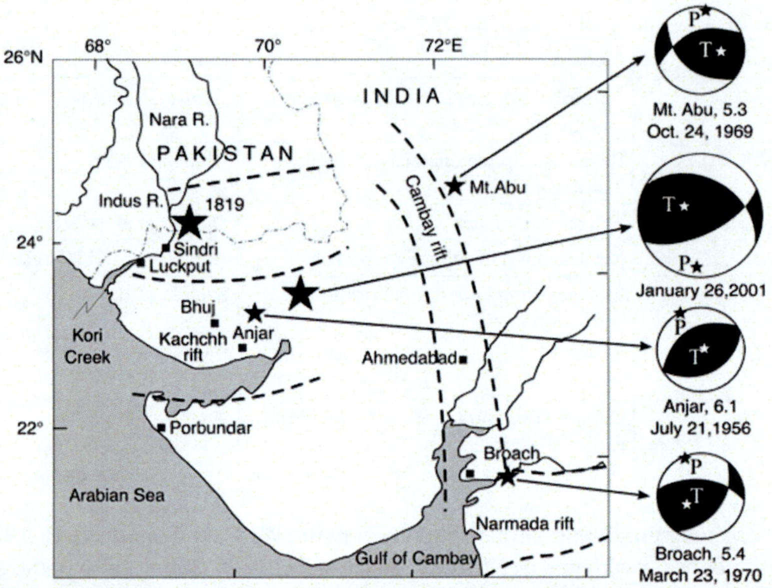

Fig. 3.4 Rift-related structures in the Rann of Kachchh and adjoining areas of Sind Province (Pakistan). Dashed lines show boundaries of the major Mesozoic rifts. Large stars: locations of the 1819 and the 2001 earthquakes. Smaller stars: locations of instrumentally recorded pre-2001 moderate events and their focal plane solutions from published sources. Harvard CMT for the 2001 event is also shown (modified after Rajendran & Rajendran, 2001)

Bhuj earthquake, which belongs to the same tectonic environment. We focus here on the morphological evolution of the terrain that is well-represented in the northwest part of the Great Rann of Kachchh—the source region of the 1819 earthquake and role of tectonism in its transformation from a sea-inlet to a tidal flat.

Geomorphologically, the Rann of Kachchh is a dynamic region subject to rapid terrain transformations reflected in its various stages of landform evolution. Lyell (1857) sums up the uniqueness of the Rann as a "singular flat region... It is neither land nor sea but is dry during a part of every year and again covered by saltwater during monsoons". The historical documents testify to the rapid landform changes experienced in this region. For example, it is suggested that the Rann was widely navigable at least until 325 BCE, at the time of the military campaign of Alexander of Macedonia. The Arabs invaded the Sind Province, located to the west of the Great Rann, in 712 CE, and their chronicles provide many details about the growth of the delta. A map during that period developed by Siveright (1907) using the Arab records conveys a spectacular image of the shallow marine setting of the region that existed up until 1000 years ago (Fig. 3.5).

Based on the Arab chronicles, flourishing harbour towns had existed along the banks of the Rann during the time of their invasion. The Indus used to feed the Nara River that flowed into the Arabian Sea through the Kori Creek, and it was this passage that connected the riverine outposts like the Sindree/Sindri and Veego-Gud/Vigakot

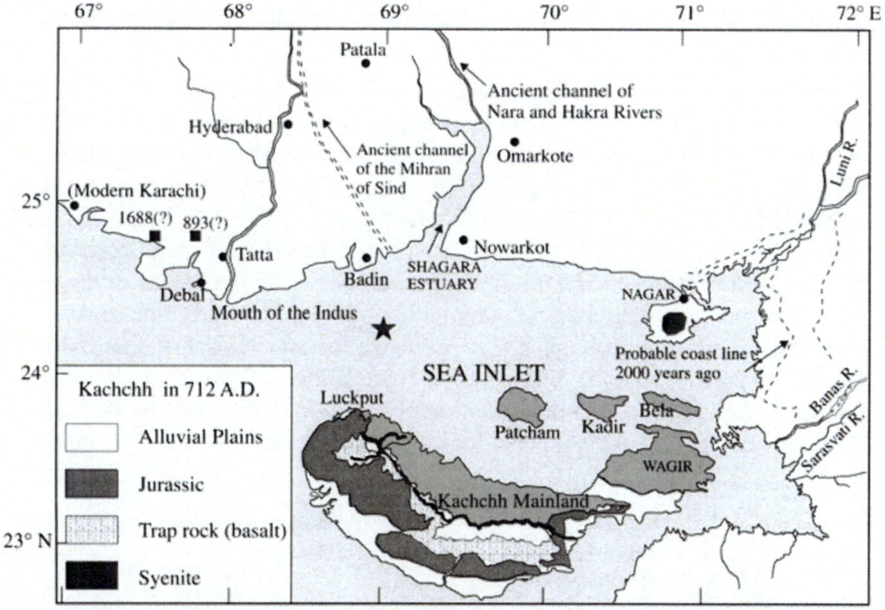

Fig. 3.5 Morphology of the Kachchh region as it existed in 712 CE based on the Arab chronicles. The area shown as sea inlet is the present-day Rann. The previously inferred locations of the 893 and 1668 CE earthquakes are shown. Reproduced from Siveright (1907)

(Oldham, 1926). Over the past thousand years or so, the Rann of Kachchh has undergone drastic topographical changes that have progressively rendered it less navigable. The Indus River itself was diverted far to the west, as the level of the Rann was gradually raised. During the 1000 years following Alexander's campaign, the coastline must have gradually migrated southward, but the shallow inlet remained somewhat navigable. Between 712 and 1361 CE, the inland sea transformed into a tidal marsh, and as noted by some Arab historians, the region had turned into a "howling desert" by 1361 CE (Siveright, 1907). Historians also note that the surface deformation from a large earthquake that occurred on the northern shores of the Rann of Kutch around this time must have contributed to these spectacular alterations in the landscape (Williams, 1958). The relative roles of both natural and anthropogenic forces in the evolution of the drainage network in the lower part of the Indus delta system are elucidated by Syvitski et al. (2013). Thus, it can be surmised that the morphological transformations of the Indus deltaic region must have been a continuing process, and large earthquakes seem to be one of the factors that provide an occasional, sudden stimulus.

3.5 Morphology of the Allah Bund: Various Interpretations

An elongated ridge of emergent land, named the "Allah Bund", bordered by the salt-encrusted surface of the Rann to its south, is the most dramatic morphological outcome of the 1819 earthquake (Fig. 3.2b). The actual spatial dimension of the bund has been a matter of conjecture for a long time. Burnes (1835) was the first to provide an eyewitness account of the bund and estimated its elevation, based on surveys conducted in 1827–1828. These surveys initially indicated its length to be 50 km, and later revised as 80 km. Burnes also reported the height of the bund as about 3 m and its width to be >25 km, with a gentle northward slope. Baker (1846) measured its maximum height as 6.2 m above the water level of the Sindri Lake and with a northward extending width of 6 km. Bilham (1998) has reproduced the profile by Baker, reflecting the elevation of the Allah Bund and the subsidence of the Sindri Fort (Fig. 3.6). Neither Burnes nor Baker traversed the entire length of the

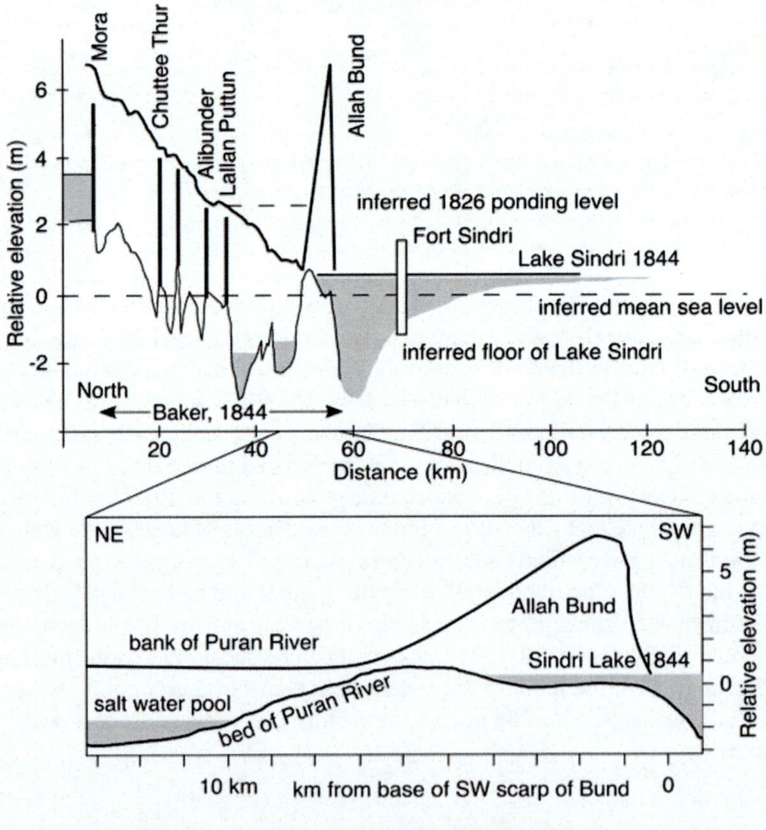

Fig. 3.6 Baker's 1844 profile of bank and bed levels projected on a north–south section and close-up view of a section through the Allah Bund. Artificial dams are shown as vertical lines. The lake level at Sindri approximates as the high-tide level (after Bilham, 1998)

Table 3.2 Dimensions of the Allah Bund reported by various surveys

Source	Elevation (m)	Subsidence (m)	Width (km)	Length (km)
Burnes (1835)	3.0	0.6–0.9	25	80
Baker (1846)	6.0	3	16	–
Raikes (1855)	5.4	–	–	–
Wynne (1872)	No uplift	1.5	–	–
Survey of India (1880–1884)	3.0	–	–	80
Rajendran and Rajendran (2001)	4.3 (coseismic uplift)	~1	15	80
-do-	5.3 (cumulative height of the scarp)			

bund to establish its lateral extent. In his monograph published 26 years later, Wynne (1872) expressed reservations about the reported height of 6.2 m, but the region was not resurveyed. In an account compiled by Raikes (1855), who was a political agent of the British Government, the elevation of the bund is given as 5.4 m. The Survey of India (1880–1884) estimated its height as 3 m and the length as 80 km, as also referred to by Oldham (1926). These estimates are summarized in Table 3.2.

Efforts to map the morphology of the bund and develop the history of earthquakes were initiated in the late 1990s (Rajendran & Rajendran, 2001). Using satellite images, field data and topographic maps, the length of the Allah Bund was estimated as 80 km with a general east–west trend and a northerly swing from N290° to N315°. Further, it was noted that tidal scouring was active in its central segment, along the southern margin, which probably resulted in the development of a local kink (Fig. 3.7). The crest of the scarp was eroded, resulting in the decline of its elevation in some places. Although the overall topography of the scarp remained intact, it has been dissected in numerous locations by storm discharges and gully erosion.

The ground surveys along 14 profiles, varying in length between 0.5 to 14 km, were used to describe the morphology of the Allah Bund in greater detail (Rajendran & Rajendran, 2001; Fig. 3.7). The surveys also documented how features like incipient streams and associated gully erosion define its lowermost surface. A composite of the profiles suggests three major breaks in slope that are most prominent across the central part of the mound, with a steep southern face and a gentle northward slope (Fig. 3.8). The lowermost surface of the bund and the southern limits of the Bet zone are separated by 0.5–1.0 km in the central part of the bund, which was interpreted as the evidence of degradation and retreat of older scarps. The breaks in slope represented by profiles II and III were interpreted as elevated tracts, but the top part was considered as a relic of a pre-1819 surface that is poorly preserved and could be traced only in the central part of the bund (shaded grey in Fig. 3.8). The maximum height of the bund was s 5.3 m, with an average value of 3.7 m, and the

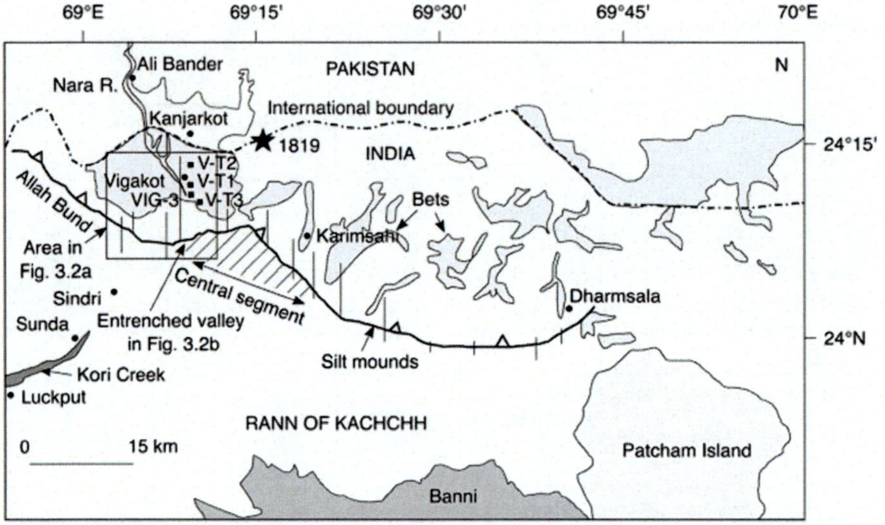

Fig. 3.7 Region around the Allah Bund. Transverse lines (scaled to length) are topographic profiles (1 to 14 from the right, as in Fig. 3.8). Star shows the inferred epicenter of the 1819 earthquake. Hatched lines demarcate the central segment, a region of tidal scouring. Black squares show locations of trenches. The box around Vigakot and the region marked central segment are shown in detail in Fig. 3.2a, b. (modified from Rajendran and Rajendran, 2001)

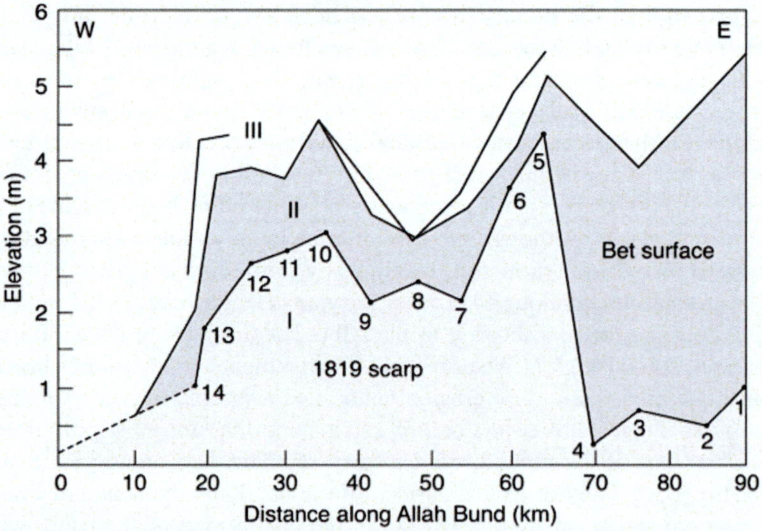

Fig. 3.8 A composite of 14 profiles showing scarp heights along the strike of the uplifted zone. The area between profile numbers 7 and 12 (central segment) is more susceptible to erosion, possibly explaining its lower elevation. The western end of the bund shown by dashed line was not surveyed, and the decrease in scarp height here was inferred from visual assessment and satellite imagery (after Rajendran & Rajendran, 2001)

structure was interpreted as a compound scarp with three major breaks in slope that preserves evidence for at least three uplift events, the youngest and the lowermost surface representing the 1819 uplift. Bilham (2004) has put forth several explanations to reconcile the modern estimates of levelling with that of Baker's (Table 3.2). The possibility that the crest of the bund has been eroded, resulting in a reduction from its original height is one of the possible explanations for the discrepancies in the estimates. That Baker (1846) may have used a much lower vertical datum than what is used in the 1990's surveys is pointed out as another cause for such differences.

3.6 Sindri Fort as a Marker of Subsidence

The Sindri Fort, located 8 km south of the Allah Bund, an important landmark in the Rann of Kachchh, had partly submerged during the 1819 earthquake, forming a lake (see Fig. 3.6). Submergence of the fort during the earthquake was explained as being due to the tectonic subsidence of the lake area, accompanied by a small tsunami (Oldham, 1926). When Alexander Burnes visited the newly formed lake (known as Lake Sindri) in 1828, the fort's location was marked by a single tower rising above the water. A sketch drawn in 1838 by Captain Grant (reproduced in Fig. 3.9a) has been widely used to demonstrate the magnitude of submergence (Lyell, 1857). The picture of the fort, as it appeared eleven years before the earthquake, shows a rectangular brick structure with towers on four corners and a viewer's gallery in the north-western corner. The fort reportedly submerged a few hours after the earthquake, and the area was converted to an inland lake that extended for 25 km. The lone tower that was reported to have been standing, as represented in Fig. 3.9b is where people took asylum, before they were rescued in boats, the next day (Lyell, 1857).

According Burnes (1835), the coseismic subsidence was estimated to be 1 to 3 m. However, Wynne (1872) mentions that the 4.5 m tall fort was built on a mound, 1.5 m above the level of the dry Rann, next to the Nara channel. The western half of the tower is reported to have crumbled and fallen, with the resulting debris lying 3 m above the level of the Rann (Fig. 3.9c). The reliability of these estimates is open to questions, as they were made long after the earthquake occurred and without due considerations to potential intervening processes. For example, the sediment eroded

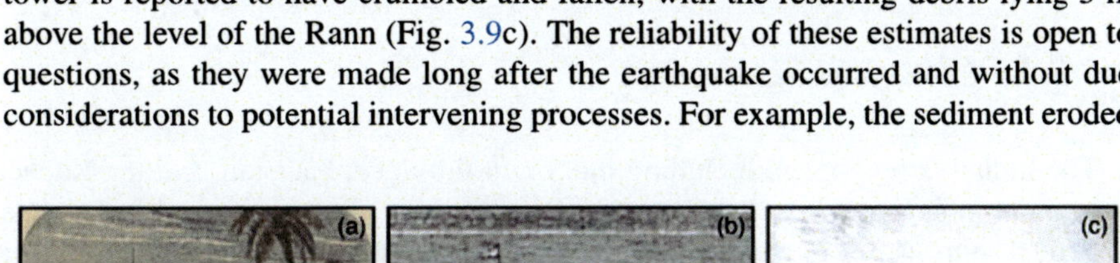

Fig. 3.9 Sketches of the Sindri Fort as it appeared **a** in 1808; **b** in 1868 (reproduced by Lyell, 1857) and **c** as reproduced by Wynne (1872)

from the southern edge of the Allah Bund and the contribution of the Nara River during floods, could have filled the Sindri depression over the years, raising the level of the Rann, masking the actual coseismic subsidence (Bilham, 2004). The 1826 flood, which destroyed all the man-made barrages on the south-eastern part of the Indus drainage and also breached the Allah Bund, is possible source for the sediment load (Syvitski et al., 2013).

Bilham (1998) used the height estimate (6.2 m) from the original survey by Baker (1846) and suggested that the earthquake may have occurred on a north-dipping, buried reverse fault, with its source located 5–15 km north or northeast of the Allah Bund. The fault dislocation model invokes a near-surface reverse fault that slipped more than 11 m locally, with a rupture that extended at least 80 km along its strike. However, this model required a dip of 50°–70° that is unfavourably steep for reverse faulting, with its down-dip width (6–10 km), much smaller than what is expected for >10 m slip. This geometric incompatibility has been addressed using a listric fault geometry at depth that would result in the observed surface deformation.

In an alternate conceptual model, Rajendran and Rajendran (2003) suggested that the surface deformation corresponds to a broad zone of up-warping caused by a buried north-dipping low-angle thrust fault. The E-W trending parallel normal faults that flatten at depth, could act as detachment surfaces in the current post-rift compressional stress field. The folding/flexuring of rock formations exposed in parts of Luckpat, southwest of the bund are suggested as field evidence for the proposed style of rock deformation (Fig. 3.10a). Thakkar et al. (2012) concur with the idea of folding and present some supporting morphological evidence from the southwestern part of the Allah Bund scarp between Sindri and the Kori Creek (Figs. 3.10b). They used the presence of intervening subsiding areas (e.g., the Sindri depression) separated by marginally elevated land as indirect indications of flexural folding of the footwall during the 1819 earthquake.

3.7 Predecessors of the 1819 Earthquake

3.7.1 Historical Seismicity

The Indus Valley region, including the Sindh Province, Pakistan, and the Kachchh basin in India, has a history of damaging earthquakes. From historical and archaeological sources, three earthquakes are known to have occurred in the past 1300 years (Table 3.3). There is evidence for their occurrence because they destroyed cities on the banks of the Indus River, where towns of historical importance were located. The earliest historically documented event in the region is believed to have occurred between 787 and 790 CE (Fredunbeg, 1902). Another two earthquakes are believed to have occurred NW of Hyderabad between 980 and 1200 CE and in 1668 CE.

Before further discussions, we need to dispel a myth concerning an event often cited as having occurred in the Indus delta region, referred to as the 893 CE

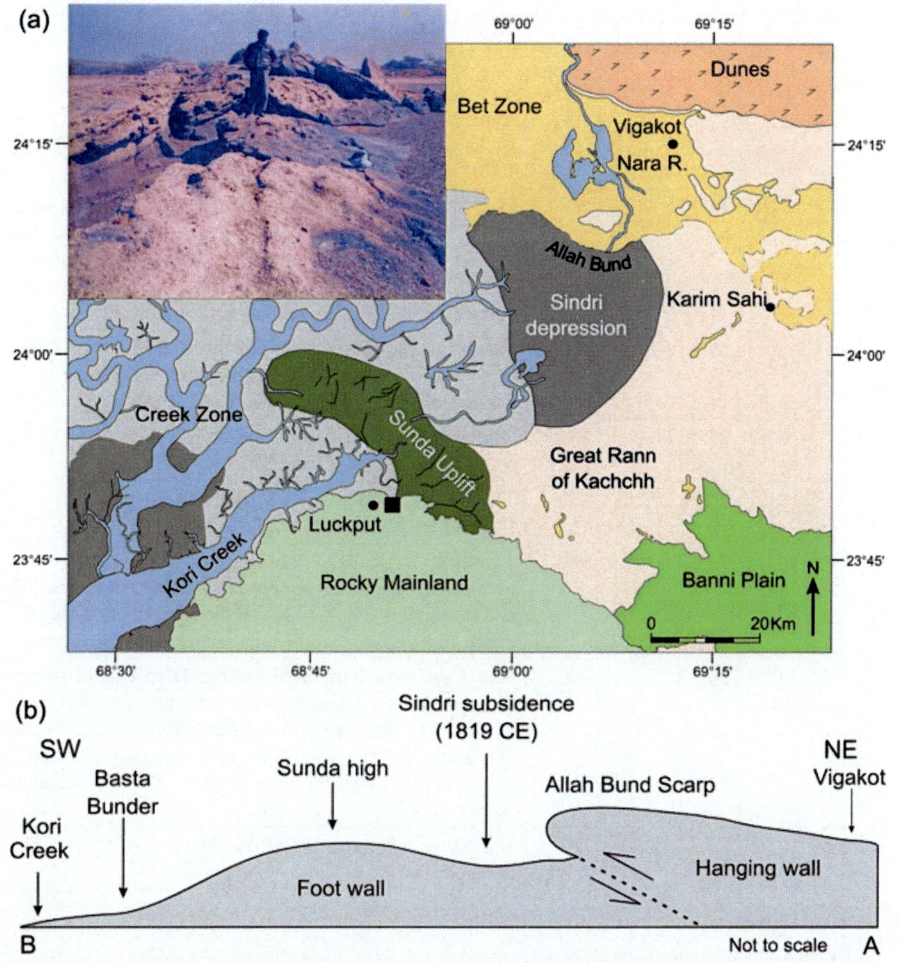

Fig. 3.10 **a** Physiography of the western Rann of Kachchh that was affected by the 1819 Kachchh earthquake. Note how the Nara River abuts at the coseismic uplift of Allah Bund. Sunda is an upwarped region between two coseismic depressions in Sindri in the northeast and Basta Bunder in the southwest. Inset: Surface folding exposed near Luckput **b** Schematic cross-section from Allah Bund to Kori Creek showing a major coseismic uplift of Allah Bund accompanied by subsidence in Sindri and Basta Bundar, separated by the Sunda high (modified after Thakkar et al., 2012)

Daibul/Debal/Dabil earthquake. There was some confusion about the name of the location as this earthquake was originally believed to have occurred at Debal. Le Gentil's 1770 map shows Dabil, as a destroyed village near Thatta, in the Indus Delta as cited by Bilham et al. (2007). However, Ambraseys (2004) equates this earthquake as being conflated with one that occurred on the night of December 28, 893 CE at Daibul in Armenia (also called Dvin earthquake). It is believed that the confusion in reporting the location of this event arose from early explorers' lack

Table 3.3 Significant historical earthquakes (>6) in the Kachchh region

Date	Lat/long°N/E	M_w	References/Comments/source of data
325 BCE	–	>7 (?)	Source could be in the Gulf of Kachchh or Makran; tsunami waves partly destroyed Alexander's army anchored at the ancient mouth of the Indus River (Lisitzin, 1974)
780–790 CE	24.75 67.52	?	Evidence is from Banbhore, an archaeological site 64 km east of Karachi. One of oldest mosques constructed during 715–750 CE was reportedly damaged in an earthquake and historians record that the event could have occurred between 780–790 CE (Fredunbeg, 1902)
980 CE	25.88 68.77	~7 (?)	Mansurah (Brahminabad). The location of this earthquake was confirmed based on archeological evidence and excavations of ruins (Bilham & Lodi, 2010)
06-05-1668	25.70 68.9	~7 (?)	Samawani (Indus Delta) relocated based on a review of original reports (Ambraseys, 2004)
16-06-1819	24.25 69.25	~7.5	The Allah Bund (Rann of Kachchh) earthquake (Rajendran & Rajendran, 2001)
19-06-1845	24.69	>6	1-m-uplift at Sunda, 40 km south of Allah Bund; numerous aftershocks for a week starting from June 19, and flooding of the Kori River. Nelson (1846); (Malik et al., 1999)
14-01-1903	24.70	~6	Felt in the area south of Hyderabad District (Pakistan). Fissures and sprouting of water and mud occurred in parts of Badin Taluka. Sandblow craters with various sizes with maximum size ranging from 4 to 6 m in diameter and 2 to 4 m in depth. Location from Malik et al., (1999). See also Aitken (1907)
21-07-1956	23.34 70.20	6.0	M_w 6.0; source: near Anjar, 50 km south of Bhuj (Chung & Gao, 1995)

of familiarity with the geography of the Middle East. The earthquake was listed in Oldham's 1893 catalogue, and numerous authors have mistakenly repeated the error, speculating both about its location on the Indus delta, and its magnitude. The articles of Ambraseys (2004) and Kovach et al. (2010) provide a wealth of background information for the interested reader. From these discussions one may conclude that the 893 CE earthquake needs to be removed from the catalog of Indian earthquakes. It must be recognized that two major earthquakes did occur within an interval of few hundred years in the Sindh Province. A century earlier, an earthquake had destroyed the abandoned port city of Banbhore on the now silted distributary of the Indus delta near Karachi. Later on, an 8th-century earthquake destroyed the ancient Arabic capital of Mansurah, near the city of Brahmanabad, now known as Brahminabad.

3.7.2 *Banbhore Earthquake (787–790 CE; 24.75°N, 67.52°E)*

A Kufic (a style of Arabic script) inscribed tablet unearthed during archaeological excavations of the port city of Banbhore indicates that it was rebuilt around 906 CE after a ruinous earthquake speculated to have occurred during the caliphate that destroyed the ancient city between 787 and 780 CE (see Kovach et al., 2010 and references therein). Banbhore was once a thriving port located partly on bedrock at the extreme western edge of the Indus Delta. Its battlements, sea walls, and stone steps leading to what must have been an important harbour in medieval times, have survived, but its buildings are now lost. It was founded c.750 CE and was abandoned by about 1250 CE, presumably due to avulsions in the upper courses of the Indus, rendering sea access to the port becoming no longer viable. The largest of the structures at the archaeological site is a mosque that was evidently partly reconstructed from damaged stonework following the 8th-century earthquake. It was here that the inscribed tablet was found. In the absence of additional historical data, it is impossible to estimate the magnitude of the causal earthquake. Collapse of masonry structures and the damage characteristics concur with that of a M_w 6.5 earthquake.

3.7.3 *Mansurah Earthquake (Tenth Century; 25.88°N, 68.77°E)*

In a series of articles, Bellasis described his 1854 excavations of the ancient archaeological site of Mansurah, 60 km NE of Hyderabad, as the "Pompeii of the East" (Bellasis, 1857a, 1857b; Sykes, 1857a, 1857b). Bellasis noted that the abandonment of the fortified city appeared to have happened in great haste, with bodies, coins, and other items strewn around the passages, beneath construction debris. No attempt had been made to bury people, or to retrieve valuables.

> The human bones were chiefly found in doorways, as if the people had been attempting to escape, and others in the corner of rooms. Many of the skeletons were in sufficiently perfect state to show the position the body had assumed; some were upright, some recumbent with their faces down, and some crouched in a sitting posture. One in particular I remember finding in a doorway; the man had evidently been rushing out of his house, when a mass of brickwork had in its fall crushed him to the ground, and there his bones were lying extended full length and the face downwards (Bellasis, 1857a, p. 417).

The account has been interpreted by some as the result of a punitive army attacking the city (Cousens, 1905). However, Bellasis (1857a, 1857b), Kovach et al. (2010) and Bilham and Lodi (2010) argue that the type of damage makes a compelling case for an earthquake near the city. In particular, the archaeological excavation of four 50-cm-diameter door knockers that were once attached to the entry doors of a large civic structure would most certainly have been plundered had they not been buried deep beneath the rubble of the building they once adorned.

The date of the earthquake was ascribed to postdate the last minted coins in the city 975 CE (Bilham & Lodi, 2010), but a more precise date is rendered difficult because the city survived only until 1020 CE. Kovach et al. (2010) bracket the earthquake conservatively in the early eleventh century. Dating the earthquake debris or the skeletons exhumed would have provided an unequivocal constraint on the time of the earthquake. For this earthquake too, it is possible only to assess a minimum magnitude (M_w 6) capable of reducing a city of sun-dried bricks to ruins. One tough mortar and brick tower known as the Thul had survived amidst the ruins of the city.

3.7.4 Samawani Earthquake (1668 CE; 25.53°N, 68.62°E)

Revenue reports dispatched to the Mughal court in 1724 relate the destruction of the town of Samawani in an earthquake in June 1668. "*It was reported from the province of Thatta, that the village of Samawani, in the jurisdiction of Bandar Lahori, had sunk down with 30,000 residents, owing to an earthquake...*". (Musta'id Khan, Saqi, d. 1724). Some 60 years earlier, Samawani was the fifth largest town in the province of Thatta, the largest being the port city of Bandar Lahori, which had by then replaced Banbhore as the principal Indus estuary port. The wording of the text in various translations (Bilham et al., 2007) has been interpreted to signify liquefaction, which must have been quite local since none of the other towns of Thatta Province are mentioned in the report.

The location of Samawani (with various similar spellings) was until quite recently a source of confusion to those who have discussed it (e.g., Burnes, 1835; Oldham 1883; Rajendran and Rajendran, 2002; Ambraseys, 2004; Bilham & Lodi, 2010). Following the earthquake, the course of the Indus shifted westwards causing new towns to spring up in its wake and former villages to be abandoned (Syvitski et al., 2013). The ruined village of Samawani was apparently one of these abandoned villages and in 1939 it was described as "a poor place with fewer than 500 houses" by Hodivala (1939), who placed it near the current village of Nasirpur. A visit to the village by Bilham and Lodi (2010), 1.5 km NW of Nasirpur, revealed an oral tradition of its name. The folklore also suggests that the damaged structures, now surrounded by farming activities, are relicts of an ancient earthquake.

The date of the earthquake is not known precisely because the A'in Akbari damage entry is itself undated, although the preceding and following entries bracket it to have occurred between May 2 and May 11, 1668 (Ambraseys, 2004). As with all ancient earthquakes in the region with sparse historical data, the magnitude is difficult to be estimated. Liquefaction on the banks of the Indus could have accompanied an earthquake, as small as M_w 6. Thus it is assumed that this could not exceeded the magnitude of a moderate earthquake as there is no evidence of damage from the neighbouring areas of Thatta and Lahori (Ambraseys, 2004). No causal fault is known, either for the Samawani or the Mansura earthquake. However, their proximity within 100 km in a north–south direction might imply a common source.

Another earthquake in 1844/1845 (the exact year uncertain) is supposed to have caused a 1-m uplift at Sunda, five km south of Sindri (Nelson, 1846), but according to Bilham (1998), these descriptions resemble the wording of the 1819 earthquake accounts. As reported by Nelson (1846), the uplift reported from Sunda, the downstream part of the Kori Creek could very well be the surface deformation associated with the 1819 earthquake, as marked in Fig. 3.10a, b. Among the historical earthquakes, the one that occurred in the year 1903 in the southern parts of Hyderabad District is rarely mentioned in the literature. But it appears to be a strong one that generated widespread soil liquefaction in the western parts Allah Bund (Aitken, 1907). The minimum magnitude of the 1903 earthquake is likely to be close to M_w 6.0. The M_w 6.1, 1956 Anjar event that killed about 115 people and injured hundreds was a significant modern day event that had some instrumental constraints. No surface deformation was reported for the 1956 earthquake, but nodal planes suggested reverse faulting on a plane dipping 45°N and striking in the NE–SW direction (Chung & Gao, 1995). These authors postulate that both these earthquakes have occurred on the boundary faults of the Kachchh rift in a reversed stress regime. They have used the thrust fault mechanisms of three regional earthquakes to suggest that compressional stress regime in response to the convergence along the plate boundary is prevalent in this region (see Fig. 3.4).

3.8 Paleoliquefaction Features

Although extensive liquefaction is reported to have occurred during 1819, no surface evidence is preserved probably due to the action of floods and storms that frequent this region, as discussed earlier. Vigakot was a major settlement and a fortified township with houses, government offices, and a fort that worked as the region's revenue and trading outpost (Wynne, 1872). The red brick fort, situated on the elevated ground in Vigakot, was destroyed in 1819 with its ruins spread over an area of ~500 m^2. With the cultural heritage of this location, the trenches excavated here were expected to provide better age constraints (see Fig. 3.7 for trench locations). The stratigraphy depicted layers of thick brownish-black silty clay interspersed with layers of fine-grained sand, representing phases of transgression and regression of tidal conditions. The vertical conduits of sand were punctuated with rip-up clasts mobilized from the underlying layer of brownish clay, a feature that is typically observed in seismically induced liquefaction features. Some of the trenches also exposed alternating thin bands of fine sand mixed with bricks and pottery, which provided approximate ages of destruction to occupation levels. Excavations within and outside the fort at Vigakot revealed multiple liquefaction events, which crosscut two occupation levels. The older horizon yielded calibrated ages of 885–1035 and 875–1025 CE that represent the maximum age for the older liquefaction event. The penultimate earthquake in the vicinity of the 1819 source is suggested to have occurred between 875 and 1035 CE (Fig. 3.11a, b). Interestingly, these sandblow dates overlap with the historical earthquake of the

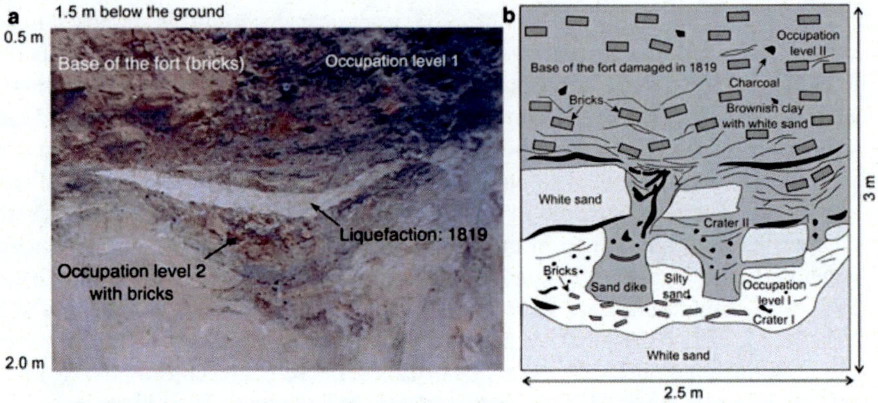

Fig. 3.11 **a** Multiple phases of liquefaction from a section near Vigakot (VIG-3; western wall of he trench). A sill formed from a younger liquefaction event, composed of clean white sand has intruded into an existing sandblow crater, which contains bricks, bones, pottery, and charcoal. The base of the Vigakot Fort, destroyed in 1819, is exposed 1.5 m below the surface. **b** Log of a trench section (VIG-3; southern wall of the trench). See Fig. 3.7 for the location of the trench. (after Rajendran & Rajendran, 2001)

eleventh century earthquake that ruined the settlement in Brahminabad, mentioned earlier.

3.9 Significance of the 1819 Earthquake

The 1819 Kachchh earthquake is widely referred to in the early geological literature for the perceptible land-level changes that it brought about. This became an early global example to argue for the fault movements and that the earthquakes are genuine agents of land level changes. The application of modern tools and extensive field surveys added much insight to our understanding of this earthquake. The occurrence of the 2001 Bhuj earthquake, provided a fortuitous opportunity to revisit its well-known predecessor, and to compare data and recalibrate some observations.

There are different proposals on the style of deformation of the 1819 earthquake. Slip on a steeply dipping fault resulting in the observed surface feature is a widely regarded view. An alternate explanation, proposed based on field observations, is that the Allah Bund could be considered as a compound scarp evolved through repetitive faulting. It has also been proposed that the overall morphological characteristics of the deformation zone may be that of a surface fold formed in response to a reverse movement on a north dipping shallow fault at depth. The observation that at least one large pre-1819 earthquake has occurred here, as evident from geologic and historical data, suggests that the western segment of the bund appears to be more productive in terms of large earthquakes. It could be argued that the segmented structure extending further to the west and east of the 1819 source may host potential seismic sources whose histories are unknown. The eighth century Banbhore and seventeenth century Samawani earthquakes imply that the western part of the Indus Delta (near the megacity of Karachi in Pakistan) may host unknown seismic sources with potential threat to a densely populated coastal urban center.

The occurrence of the 2001 Bhuj earthquake, not associated with the Allah Bund but sourced on a hitherto unmapped blind fault within the rift (see Chap. 4), proves the point of the underestimated hazard posed by hidden faults. While the Bhuj earthquake was associated with a deep-seated and steep south-dipping blind fault within the basin, the 1819 earthquake is believed to have been sourced on a shallow north-dipping fault along the boundary of the rift. While the predecessor event of the 1819 earthquake might be 1000 years old, the 2001 source, has a much longer interseismic interval as discussed in the next chapter. From these two examples, it become clear that the deformational characteristics, style of faulting, and the mode of recurrence of major earthquakes may differ between the faults within the same rift.

How well justified is the classification of the Kachchh seismic zone as a typical Stable Continental Region (SCR) source, is a question that has been frequently asked, especially after the 2001 earthquake. The Kachchh rift basin has been compared with the 1811–1812 Reelfoot Rift (the New Madrid seismic zone) within the Mississippi embayment (Johnston, 1989). Some researchers like to consider the Kachchh rift as part of an extended diffused plate boundary, proximal to the India-Eurasia collision boundary, unlike the New Madrid region, which is 1000 km away from the nearest plate boundary (e.g., Li et al., 2002; Stein et al., 2002). The occurrence of two large earthquakes within a short span of 182 years although associated with two distinct sources, makes this question even more relevant. These aspects are further discussed in the next chapter.

References

Aitken, E. H. (1907). *Gazetteer of the Province of Sind* (p. 519). Printed for Government at the Mercnatile P Steam Press.

Ambraseys, N. N. (2004). Three little known earthquakes in India. *Current Science, 86*(44), 506–508.

Ambraseys, N., & Douglas, J. J. (2004). Magnitude calibration of north Indian earthquakes. *Geophysical Journal International, 159*, 165–206.

Baker, W. E. (1846). Remarks on the Allah Bund and on the drainage of the eastern part of the Sind basin. *Transactions of the Bombay Geographical Society, 7*, 186–188.

Bellasis, A.F. (1857a). An account of the ancient and ruined city of Brahminabad, in Sind. *Journal of the Bombay Branch of the Royal Asiatic Society, 5*, 413–425.

Bellasis, A.F. (1857b). Further observations of the ruined city of Brahminabad, in Sind. *Journal of the Bombay Branch of the Royal Asiatic Society, 5*, 467–477.

Bilham, R., Lodi, S., Hough, S., Bukhary, S., Khan, A.M., Rafeeqi, S. F. A. (2007). Seismic Hazard in Karachi, Pakistan: Uncertain Past, Uncertain Future. *Seismological Research Letters, 78*(6), 601–613.

Bilham, R., Lodi, S. (2010) The door knockers of Mansurah: Strong shaking in a region of low perceived seismic risk, Sindh, Pakistan, in Sintubin, M., Stewart, I.S., Niemi, T.M., and Altunel, E., eds., *Ancient Earthquakes: Geological Society of America Special Paper* 471, p. 29–37.

Bilham, R. (2004). Earthquakes in India and the Himalaya: Tectonics, geodesy and history. *Annals of Geophysics, 47*, 839–858.

Bilham, R. (1998). Slip parameters for the Rann of Kachchh, India, 16 June 1819, earthquake, quantified from contemporary accounts. in Stewart, I.S., Vita-Finzi, C., eds., *Coastal Tectonics*(Special Publications, Vol. 146, Geological Society, London) pp. 295–319.

Biswas, S. K. (1987). Regional tectonic framework, structure and evolution of the western marginal basins of India. *Tectonophysics, 135*, 307–327.

Burnes, A. (1835). Memoir on the eastern Branch of the River Indus, giving an account of the alterations produced on it by an earthquake, also a theory of the formation of the Runn and some conjectures on the route of Alexander the Great; drawn up in the years 1827–1828. *R. Asiatic Soc. Trans., 3*, 550–588.

Chung, W.-P., & Gao, H. (1995). Source parameters of the Anjar earthquake of July 21, 1956, India, and its seismotectonic implications for the Kutch rift basin. *Tectonophysics, 242*, 281–292.

Cousens, H., 1905, Conservation of ancient monuments in the Bombay Presidency, in Scott, R., ed., J. Bombay Branch R. Asiatic Society, Extra Number, Centenary Memorial Volume: London, Kegan Paul, Trench, Trübner & Co., p. 149–162.

Dasgupta, S., & Mukhopadhyay, B. (2014). Historiography and commentary on the 16 June 1819 Kutch Earthquake. *Gujarat, India, Indian Journal of Geosciences, 68*(1), 57–126.

Dunbar, P. K., P. A. Lockridge, Whiteside, L.S (1997). Catalog of significant earthquakes 2150 B.C.–1991 A.D. (with addendum through 1997) including quantitative casualties and damage. National Geophysical Data Center, Boulder, Colorado

Fredunbeg, M. K (1902) The History of Sind, Volume 2: Karachi, Printed at the Commissioner's Press, 346 p.

Frere, H. B. E. (1870). Notes on the Runn of Cutch and neighboring region. *J. r. Geograph. Soc. London, 40*, 181–207.

Glennie, K. W., & Evans, G. (1976). A reconnaissance of the recent sediments of Ranns of Kutch. *India, Sedimentology, 23*, 625–647.

Hodivala, S.H., 1939, Studies in Indo-Muslim History: A Critical Commentary on Elliot and Dowson's History of India as Told by Its Own Historians, Volume 1: Bombay, Kokil and Co., 712 p.

Hough, S. E., S. Martin, R. Bilham, and G. M. Atkinson (2002). The 26 January 2001 M 7.6 Bhuj, India, earthquake observed and predicted ground motion, Bull. Seismol. Soc. Am., 92, no. 6, 2061–2079.

Johnston, A. C. (1989). Seismicity of stable continental interiors. In S. Gregersen & P. W. Basham (Eds.), *Earthquakes at North-Atlantic Passive Margins: Neotectonics and Postglacial Rebound* (pp. 299–327). Kluwer Academic Publishers.

Johnston, A. C. (1996). Seismic moment assessment of earthquakes in stable continental regions. *I. Geophys. J. Int., 124*, 381–414.

Kovach, R.L., Grijalva, K., and Nur, A., 2010, Earthquakes and civilizations of the Indus Valley: A challenge for archaeoseismology, in Sintubin, M., Stewart, I.S., Niemi, T.M., and Altunel, E., eds., Ancient Earthquakes: Geological Society of America Special Paper 471, p. 119–127.

Li, Q., Liu, M., & Yang, Y. (2002). The 01/26/2001 Bhuj earthquake: Intraplate or interplate? In S. Stein & J. Freymueller (Eds.), *Plate Boundary Zones: Washington* (pp. 255–264). American Geophysical Union.

Lisitzin, E. (1974). Sea-level Changes, Oceanography Series No.8, Elsevier Publishing Company, Amsterdam. 286 pp.

Lyell, C. (1857). *Principles of Geology* (Eleventh, p. 834). Appleton & Co.

Mac Murdo, J. (1824). Papers relating to the earthquake which occurred in India in 1819. *Philosophical Magazine, 63*, 105–177.

Malik, J. N., Sohoni, P. S., Karanth, R. Y., & Merh, S. S. (1999). Modern and Historic Seismicity of Kachchh Peninsula. *Western India. Jour. Geol. Soc. India, 54*, 545–550.

Musta'id Khan, Saqi, (d. 1724) Maasir-i- 'Alamgiri, and tr. J-N Sarakar, Bibliotheca Indica, Calcutta, 1947, pp. 73–74.

Naik, SP, Mohanty, A, Porfido, S Sabina, Tuttle, M, Ohsang , G., Kim Young-Seog. (2019). Intensity estimation for the 2001 Bhuj earthquake, India on ESI-07 scale and comparison with historical 16th June 1819 Allah Bund earthquake: A test of ESI-07 application for intraplate earthquakes. Quaternary International. 536. https://doi.org/10.1016/j.quaint.2019.12.024.

Nelson, R. E. (1846). Notice of an earthquake and a probable subsidence of the land in the district of Cutch, near the mouth of the Koree, or eastern branch of the Indus, in June 1845. *The Quarterly Journal of the Geological Society of London, 2*, 103.

Oldham, T. (1883) A catalogue of Indian earthquakes from the earliest to the end of 1869. Mem. *Geol Surv India, 19* 1–53.

Oldham, R. D. (1898). A note on the Allah Bund in the northwest of the Runn of Cutch. *Geological Survey of India Memoirs, 28*, 27–30.

Oldham, R. D. (1926). The Cutch (Kachh) earthquake of the 16th June, 1819 with a revision of the great earthquake of the 12th June, 1897. *India Geological Survey Memoir, 46*, 71–147.

Pande, P. (2011). 16 June Kachchh earthquake, Gujarat, India and its isoseismals. *Indian Jour. Geosci, 652*(2), 97–106.

Raikes, SN. (1855). Bombay Government Records, New series XV: p. 46

Rajendran, C. P., & Rajendran, K. (2001). Characteristics of deformation and past seismicity associated with the 1819 Kutch earthquake, northwestern India. *Bulletin of the Seismological Society of America, 91*(3), 407–426.

Rajendran C. P., & Rajendran K. (2002). Historical constraints on previous seismic activity and morphologic changes near the source zone of the 1819 Rann of Kachchh earthquake: Further light on the penultimate event Seismol Res Lett. *73*, 470–479.

Rajendran, C. P., Rajendran, K., Vora, K. H., & Gaur, A. S. (2003). The odds of a seismic source near Dwarka, NW Gujarat: An evaluation based on proxies. *Current Science, 84*, 695–701.

Roy, B., & Merh, S. S. (1977). Geomorphology of the Rann of Kutch and climatic changes. In D. P. Agrawal & B. M. Pande (Eds.), *Ecology and Archaeology of Western India* (pp. 195–200). Concept Publishing Company.

Siveright, R. (1907). Cutch and the Ran, The Geographical Journal XXIX, 519–539.

Stein, S., Sella, G. F., & Okal, E. A. (2002). The January 26, 2001, Bhuj earthquake and the diffuse western boundary of the Indian plate. In S. Stein & J. Freymueller (Eds.), *Plate Boundary Zones: Washington* (pp. 243–254). American Geophysical Union.

Sykes, W. H., (1857b). The ancient and ruined city of Brahmunabad in Sind: Illustrated London News, 847, 28 February 1857, p. 187–189

Sykes, W. H. (1857a) Relics from the buried city of Brahmunabad: IllustratedLondon News, 846, 21 February 1857, p. 166–167.

Syvitski, J. M. P., Kettner, A. J., Overeem, I., Giosan, L. B., & GR, Mark Hannon, M and Bilham, R,. (2013). Anthropocene metamorphosis of the Indus Delta and lower food plain. *Anthropocene, 3*, 24–35.

Szeliga, W., Hough, S., Martin, S., & Bilham, R. (2010). Intensity, Magnitude, Location, and Attenuation in India for Felt Earthquakes since 1762. *Bulletin of the Seismological Society of America, 100*, 570–584. https://doi.org/10.1785/0120080329

Thakkar, M.G., Mamata Ngangom, P. S. Thakker and N. Juyal (2012). Terrain response to the 1819 Allah Bund earthquake in western Great Rann of Kachchh, Gujarat, India Author(s): Current Science, Vol. 103, No. 2 (25 July 2012), pp. 208–212

Wells, L. D., & Coppersmith, K. J. (1994). New empirical relationships among magnitude, rupture length, rupture width, rupture area, and surface displacement. *Bull. Seism. Soc. Am., 84*, 974–1002.

Williams, R. L. F. (1958). *The Black Hills-Kutch in history and legend* (p. 276). Weidenfeld and Nicolson.

Wynne, A. B. (1872). Memoir on the geology of Kutch. *Indian Geological Survey Memoir, 9*, 29–47.

Chapter 4
Bhuj 2001

The January 26, 2001, M_w 7.7 earthquake located near Bachau, in the state of Gujarat in western India, referred to as the Bhuj earthquake, after the major city in the region, is the largest to have occurred in the Kachchh region since the 1819 earthquake (Fig. 4.1). Although located ~70 km southwest of the epicenter, the semi-urban city of Bhuj took the major brunt, in terms of loss of life and damage to structures. According to the estimate by the official Indian agencies, the earthquake caused ~20,000 deaths; destroyed ~400,000 houses and partially damaged another 800,000. The direct economic losses were calculated at US $1.3 billion, which was re-estimated as US $>5 billion. This earthquake attracted the global attention from seismologists because of its association with a failed rift and its potential similarity with events such as those in the New Madrid seismic zone, Missouri, the site of a series of enigmatic earthquakes in 1811–1812 (see Chap. 2). It is also the first large SCR earthquake to be investigated using modern tools of seismology and earthquake geology. Thus, together with its 1819 predecessor, the Bhuj earthquake provided an unprecedented opportunity to learn more about large, mid-continental earthquakes associated with failed rifts.

Seismic networks established by the National Geophysical Research Institute (NGRI, India), Center for Earthquake Research and Information (CERI, University of Memphis), and the Japanese seismologists, provided a large set of aftershock data. This data facilitated studies on their spatial and temporal distribution, development of source models, and tomographic imaging. Coseismic deformation leading to lateral spreads, sandblows, and structural features including secondary faults served as proxies for the paleoseismologic exploration. As the largest among the digitally recorded intra-continental earthquakes, and as an intraplate event analyzed using GPS data, geophysical and geological tools, investigations on the Bhuj event have added many new insights. In this chapter, we review the results of these investigations and highlight how these studies have furthered the understanding of seismogenesis in the Kachchh region.

C. P. Rajendran and K. Rajendran, *Earthquakes of the Indian Subcontinent*, GeoPlanet: Earth and Planetary Sciences,
https://doi.org/10.1007/978-981-16-4748-2_4

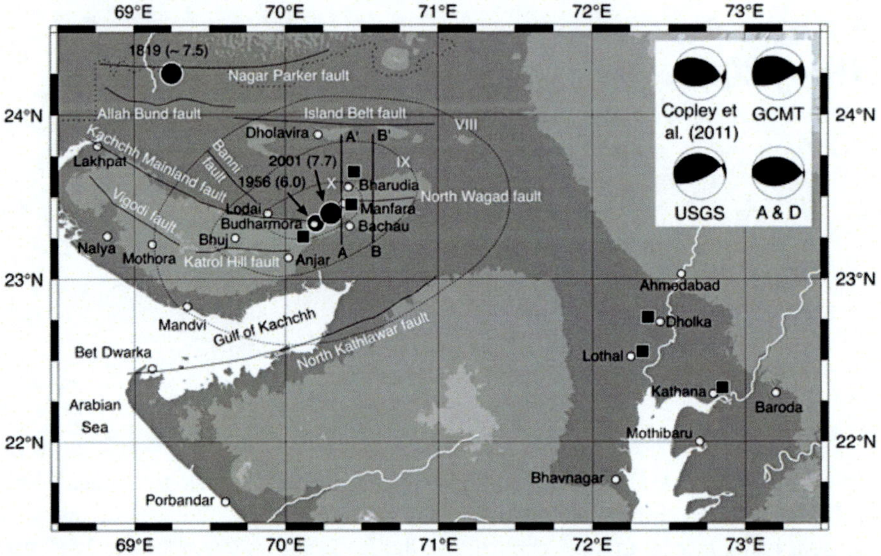

Fig. 4.1 Map of the Kachchh region showing major faults (Biswas, 1987); location of the 2001 earthquake (Mandal, 2013) and isoseismals (Narula et al., 2002). Inset: fault plane solutions for the 2001 earthquake; A&D: Antolik and Dreger (2003). Filled squares are the sites of paleoseismological investigations by Rajendran et al. (2008). Lines AA′ and BB′ are referred to in Fig. 4.4

4.1 Tectonic Setting of the Kachchh Basin

The pericratonic Kachchh basin in northwest India is part of an intra-continental rift system within the Precambrian shield region, believed to have developed during the Mesozoic times. The present landscape of the Kachchh basin is quite striking, showcasing how the interplay of tectonism and climate could have shaped its morphology, as discussed in Chap. 3. The basin is bounded by two major faults, the Nagar Parker Fault (NPF) to the north, and the North Kathiawar Fault (NKF) to the south. In addition, there are several E-W striking normal faults within the basin, which were formed during the rifting phase (Fig. 4.2). The basin is believed to have opened along the major Precambrian trends in the early Jurassic Period and was later filled with sediments from the middle Jurassic through the Holocene. Precambrian granitic basement exposed in the Nagar Parker Hills bordering the northern flank of the rift graben can be considered as relics of uplift within the rift system (Biswas, 1987).

A second phase of deformation is believed to have commenced from the Late Cretaceous–Early Paleocene, when the Deccan/Reunion hotspot erupted, and the western margin of India passed over it around 66 Ma. The onset of the collision of India with the southern margin of Eurasia (~50 Ma) changed the stress regime from extensional to compressional. The deformation within the rift was significant in the early phases, resulting in folding and faulting of the Mesozoic strata. However, the

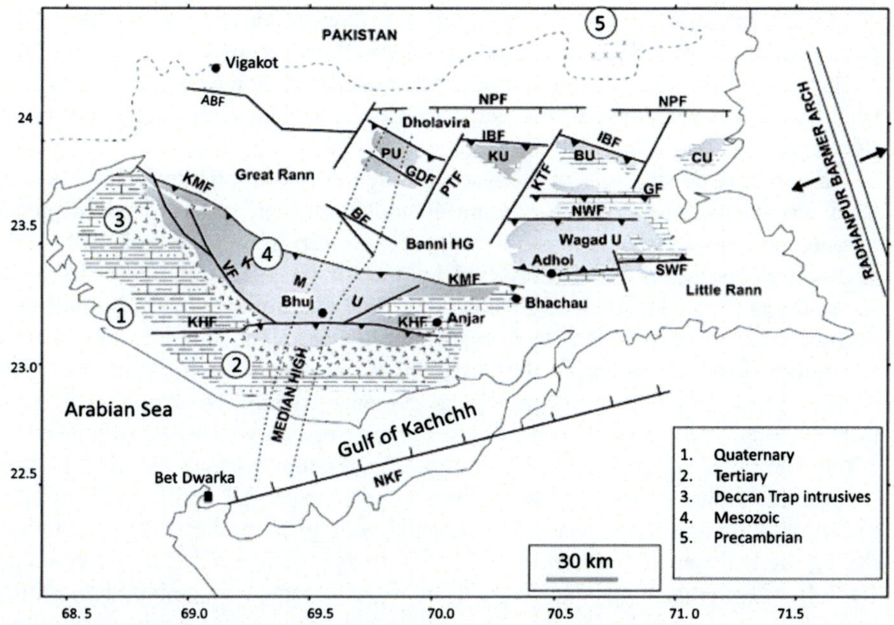

Fig. 4.2 Distribution of various tectonic structures in the Kachchh region. Faults: ABF: Allah Bund; IBF—Island Belt; KMF—Kachchh Mainland; KHF—Katrol Hill; VF—Vigodi NPF- Nagar Parker; KMF—Kachchh Mainland; BF—Banni, GF—Geddi, GDF—Gora Dunga; NWF- North Wagad; SWF—South Wagad. Uplifts: PU—Pachham; KU—Kadir; BU—Bela; CU—Chorar; WU—Wagad; KMU—Kachchh Mainland; HG—High grounds; PTF, KTF—Transverse faults (figure redrawn from Biswas & Grasemann, 2005)

deformational spree slowed down with time, as evident from the gentle dips of the Tertiary strata exposed here (Biswas, 1982). By the Late Miocene (~20 Ma), the east–west-trending Kachchh Rift Basin was already being subjected to an approximately north–south compressive stress field in response to the plate boundary forces that continue to the present day, as reflected in the earthquake mechanisms (Chung & Gao, 1995; Talwani & Gangopadhyay, 2001 and references therein).

The Kachchh rift basin is traversed by major E-W-trending faults, such as the Nagar Parkar Fault (NPF), Kachchh Mainland Fault (KMF), Katrol Hill Fault (KHF), South Wagad Fault (SWF) and Island Belt Fault (IBF) (Biswas & Grasemann, 2005; Maurya et al., 2017a) (Fig. 4.2). Of these, the Island Belt (IBF), Kachchh Mainland (KMF), and South Wagad (SWF) faults are expressed as distinct topographic features. Based on paleoseismological evidence, a minimum average slip rate of 0.23 mm/yr has been inferred on the KHF, with long-term average uplift rate of 0.8 mm/yr (Das et al., 2016). The much-faulted southern part of the WU displays several converging and diverging faults, collectively called the South Wagad Fault (SWF) system (Kothyari et al., 2016). The geomorphic signature of the SWF is characterized by tectonically governed landforms, like strath terraces along river valleys, truncation of alluvial fans, offsets, and incision of rivers (Maurya et al.,

2017a, 2017b). Some strike-slip faults are also present and the near vertical Gedi Fault inferred from seismological and geophysical data being one of them (Kumar et al., 2017). Since the post-collisional stress inversion, some of these faults have been associated with the seismic activity in the Kachchh region (e.g., Biswas & Khatri, 2002).

As best-studied structure in the Kachchh area, the E-W oriented ~150 km-long KMF that forms steeply north-dipping 40 to 100-m-high scarps, needs a special mention. It is expressed as a north-facing steep scarp in the Northern Hill Range of Kachchh mainland and is dissected by several younger, transverse strike-slip faults (Maurya et al., 2003; Mohan et al., 2018). It originated as a normal fault during the rift stage when the extensional stress was dominant, until it was subjected to compressional stress during the post-collisional inversion stage (Biswas, 2016; Maurya et al., 2003). Paleoseismological evidence for mid-Holocene reactivation has been reported from the central and eastern segments of KMF (Malik et al., 2008; Morino et al., 2008; Rajendran et al., 2020 and references therein). Despite its prominent surface expression, the seismogenic potential of this fault remains uncertain, although some recently conducted geological studies suggest activities during the Holocene (Kothyari et al., 2020). Prizomwala et al. (2016) divided the KMF into segments, with a westward decrease in tectonic activity based on limited quantitative geomorphic analysis. These observations come as a reiteration of the deterministic seismic hazard analysis implying the potential for a large earthquake on the KMF (Chopra et al., 2012; Mohan, 2014). Despite the striking surface expressions of the various faults and their geomorphic prominence, surprisingly none of them could be attributed to the 2001 earthquake. In fact, a hitherto unmapped blind fault was eventually identified as its causative fault.

Geomorphology of the Kachchh Basin characterized by highlands (the tilted uplifts along the E-W trending faults) surrounded by lowlands, with intervening sediment-filled half-grabens, reflects its underlying tectonic architecture (Figs. 3.10a and 4.2). These sub-basins comprise unconsolidated near-shore marine and fluvial sediments from the Tertiary and Holocene periods. The Great Rann of Kachchh and the Little Rann, regarded as a salt wasteland at the sea level, are inundated during the monsoon with marine waters and terrestrial water runoff. With such dramatic seasonal changes, this region is endowed with unique ecological characteristics and biological diversity. As mentioned in Chap. 3, several hundreds of years ago, the Great Rann was apparently inundated by the sea deeply enough that large seagoing vessels could sail far into the inland regions. The southern and northern margins of the Rann have slightly uplifted landscapes called the Banni and the Bet, respectively. Composed of sandy to silty loam with intervening lenses of clay, suggestive of a fluvial origin, and these are elevated to about 3–10 m from the sea level, the Banni supports rich vegetation in an area of about 3000 km^2. The Bet zone with slightly elevated patches of grasslands are composed of fine micaceous sand and silt with clay intercalation, and they are considered as remnants of the abandoned river channels or patches of an ancient delta formed by the Indus River (Snelgrove, 1979).

4.2 Source Characteristics of the Bhuj Earthquake

4.2.1 Fault Geometry

While there was a consensus among the researchers that the 2001 Bhuj earthquake occurred by the reactivation of faults within the paleo-rift, a lot more had to be learned about the nature of its source. This was possible from the large amount of aftershock data, which helped to map the geometry of the fault and constrain the shape of the rupture volume. The focal mechanism suggests thrust faulting on a near E-W, south-dipping fault (Fig. 4.1). Negishi et al. (2002) were among the first to report the source characteristics using the spatial distribution of aftershocks from the Japanese array. They suggested that the Bhuj earthquake was sourced on a 50° south-dipping fault plane, but the surface projection of this plane did not match with any of the mapped faults in the area. Depths of aftershocks ranged from 10 to 35 km, deeper than usual for crustal earthquakes and covered an area of about 1260 to 1960 km^2, regarded as small for its magnitude. Studies in the later years have provided additional constraints on the source characteristics and relation to causative structures.

The National Geophysical Research Institute (NGRI), Hyderabad (India) established several seismic stations in the area, which facilitated the study of aftershocks, seismic imaging, and quantification of source parameters. Mandal et al. (2004) used hypocenters of 600 well-located aftershocks (M_w 2.0–5.3) from 1-D simultaneous inversion to constrain the geometry of the causative fault. Their analysis suggested that the earthquake occurred on an east–west trending, south-dipping (~45°) blind thrust fault, extending to a depth of 10–45 km, which they named as the North Wagad fault. Seismic imaging of the source region helped to demonstrate that a high-velocity intrusive body exists at 10 to 30 km depth under the Wagad uplift, and that the mainshock hypocenter coincides with the high-velocity body (Mandal & Pujol, 2006). The analyses of the relocated aftershocks suggested a distinct, south-dipping plane extending up to 35 km depth, and a near-north–south orientation of the P-axis, consistent with the N-S compression prevalent over the Indian plate (Mandal and Horton (2007). The interpretation of aftershock data has also reiterated the tectonic link with the ENE–WSW trending, south-dipping blind reverse fault (North Wagad fault), which almost coincides with the northern limit of the Wagad uplift.

Additional data was collected from the seismic network installed by the Center for Earthquake Research and Information (CERI at the University of Memphis) and the Institute of Science and Technology for Advanced Studies and Research (ISTAR), two weeks after the mainshock. The network recorded more than 2000 events during its 18-day-long operation. The data pointed to a blind, and ~25-km-long and ~35-km-deep reverse thrust (Bodin & Horton, 2004). The inferred trapezoidal-shaped aftershock zone tapers from ~45 km along strike to ~25 km at a depth of 35 km, with total rupture area of 1300 km^2. The thrust faulting mechanism and the geometry of the inferred fault indicate that the earthquake occurred by the reactivation of a fault in response to the current north–south compressive stress. The potential role of crustal fluids in weakening the faults in triggering the failure was another upshot of these studies.

Chandrasekhar et al. (2004) used the pre- and post-2001 elevation and gravity anomaly values in the epicentral area (1.57 ± 0.5 m and –393 ± 18 m Gal respectively), to develop a slip model. The best-fit uniform-slip dislocation model suggests a 23-km-long, 12-km wide fault plane that dips southward (51°dip angle) and extend from 12 to 22 km depth. A dip slip of 10.8 ± 0.5 m with a small right lateral strike-slip component of 0.7 ± 1.1 m was reported. An InSAR-based model by Schmidt and Burgmann (2006) proposed a relatively deep, east–west striking, south-dipping fault with its plane entirely within the cloud of aftershocks. The slip within a 20 × 20 km^2 region was centered at a depth of ~20 km, with the up-dip edge of the fault plane at a depth of 9–15 km. The InSAR-based model is consistent with that of Antolik and Dreger (2003), who found a deep slip patch from the inversion of teleseismic waveform data with much of the moment release between 12 and 25 km depth. The teleseismic body wave modeling by Copley et al. (2011) suggest that the shallow deformation may have been distributed within the thick sediments of the basin rather than localized onto a single discrete surface.

4.2.2 Stress Drop

The static stress drop of the Bhuj earthquake is anomalously high (at ~20 MPa). Many researchers believe that the rapid onset of moment release must have caused the unusually large high-frequency ground motions and hence the severity of damage (e.g., Hough et al., 2002). The estimate of stress drop is considered as a useful indicator to assign interplate/intraplate status to the 2001 earthquake (Ellis et al., 2001; Stein et al., 2002). The high stress-drop of the Bhuj earthquake places it well within the class of intraplate events and the value of 20 MPa, from most estimates, is close to the suggested average value (Scholz et al., 1986).

A wide range of stress drop values for 2001 earthquake has been reported. Negishi et al. (2002) had estimated the area of the fault as about 40 × 40 km^2, and estimated the static stress drop between 13 and 25 MPa. From inversion of teleseismic body waves, Antolik and Dreger (2003) suggested ~20 MPa, slightly higher than 16 ± 2 MPa by Bodin and Horton (2004). Copley et al. (2011) used teleseismic waveform inversion along with surface displacements from interferometric synthetic aperture radar (InSAR), to suggest a value of ~35 MPa. A later work by Wang et al. 2015 used Synthetic Aperture Radar (SAR) to suggest a stress drop of ~20 MPa and a rise time of < 3 s. A recent estimate by Silpa and Earnest (2021), that incorporated the spatial heterogeneity of slip in the computation, suggested an average of ~27 MPa.

The source model developed by Copley et al. (2011), with its highest stress drop estimate, is compact and consistent with other models, showing a maximum slip of ~14 m within the ∼30-km wide zone. The SPOT satellite images also point to the existence of a ∼15 km long, NW–SE trending displacement discontinuity, which corresponds to what Wesnousky et al. (2001) describe as 'tear-faults' on the hanging wall near Manfara (see Fig. 4.1 for location). However, the north–south displacement (<0.4 m) and the eastward uplift component (~0.6 m) inferred from the satellite image is larger than what was reported from the field investigations, the discrepancy attributed to the field measurements, in the large zone of diffuse deformation.

4.3 Seismic Images and Proposed Mechanisms

Tomographic imaging suggested that the source zone of the Bhuj earthquake is characterized by anomalously high-Vp, low-Vs, and high Poisson's ratio (σ), related to the fluid-filled, fractured rock matrix in the source zone (Kayal et al., 2002; Fig. 4.3). The presence of a zone of relatively high velocity (Vp and Vs) intrusive body extending from 10 to 35 km depth in the hypocentral region was also evident in a later work (Mandal & Pujol, 2006). The inferred high-velocity zone was supported also by gravity and magnetic data that indicate gabbroic crustal intrusive bodies associated with the Patcham and Wagad uplifts and presence of E-W trending dykes and patches of Deccan basalts in the area (Chandrasekhar & Mishra, 2002).

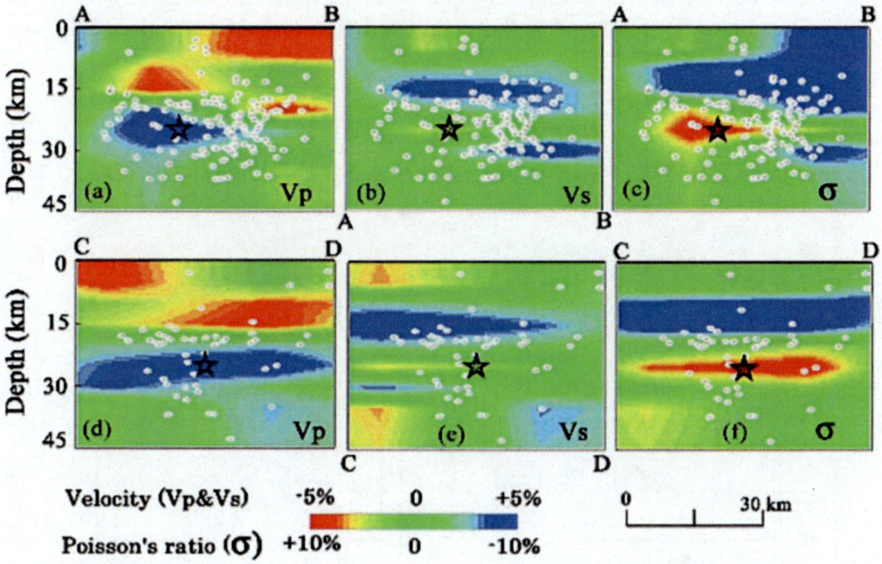

Fig. 4.3 Top panel: (a–c): Vertical West–East (A–B) cross-section showing Vp, Vs and Poisson's ratio σ, respectively. Bottom panel (d–f): same as the top panel, along North–South (C–D) cross-section. Small white circles are aftershocks within 9-km width along the sections. The star denotes the Bhuj mainshock hypocenter; perturbation (in %) scales are shown at the bottom (after Kayal et al., 2002)

Two-dimensional modelling based on magneto-telluric data has revealed an assemblage of resistive and conductive features down to the upper mantle (Kumar et al., 2017). A combined interpretation of the 2-D geoelectric model and seismic tomography image points to a fluid reservoir at depths of 35–40 km. The near-vertical North Wagad fault and the South Wagad faults that almost reach the top of the mantle serve as conduits for fluid flow to the hypocentral depths. The aftershocks of magnitude $\geq$ 2.5 during 2006–2015 are largely confined to the middle to lower crustal depths and they coincide well with the high conductivity zone below the SWF and the transition zone along the NWF. The Bhuj earthquake is located at the junction of the two fault traces (Fig. 4.4) and the fluids originated from the mantle, forming a reservoir at upper mantle depths. The presence of high-pressure fluids can mechanically weaken the fault planes as well as the entire rock volume, providing conditions for failure.

Large lower crustal earthquakes in rift basins are often attributed to the reactivation of old rifts, presence of rift pillows, and/or changes in the brittle-ductile transition (e.g., Pollitz et al., 2001). Association with mafic plutons have been cited as causative factors for the 1811–12 New Madrid (USA), the 1963 and 1983 Amazonas and the 1997 Jabalpur (India) earthquakes (Rao et al., 2002; Zoback & Richardson, 1996). Lower crustal processes leading to the emplacement of high-density plutonic bodies, heterogeneous lithology, and deep-seated faults are attributed to rift tectonics. The E-W oriented North Wagad fault, considered as the source of the 2001 earthquake, is part of the imbricate fault system that originated as a normal fault within an active rift,

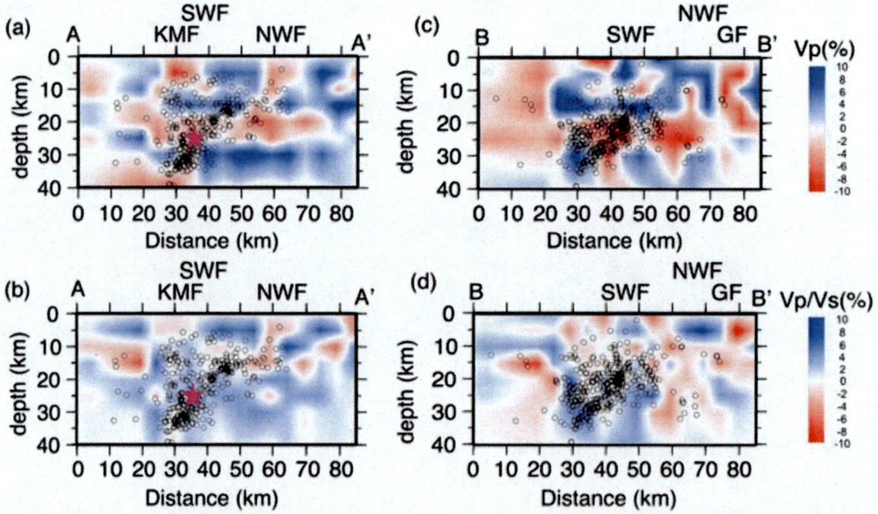

Fig. 4.4 Vp and Vp/Vs perturbations along the profiles AA′ (a and b) and BB′ (c and d) (profile locations given in Fig. 4.1). Open circles represent the seismicity ($M_w > 2.5$) within 15 km on either side of the profiles. Star marks the 2001 Bhuj mainshock. The dotted lines delimit the well-resolved portion of the tomographic models (after Kumar et al., 2017)

which evolved as a thrust fault in the post-collisional stress regime (e.g., Chung & Ghao, 1995). The tomographic images and geophysical models that indicate the presence of a high-density material at the lower crustal depths support the rift pillow hypothesis in the Kachchh rift.

The relatively high seismic productivity of the Kachchh rift is a point of debate, in particular the succession of two large earthquakes during a short span of less than 200 years. It is well known that the relaxation following an earthquake can change the stress in the surrounding crust and can enhance/delay the occurrence of earthquakes on the nearby faults. So, could the 1819 earthquake have stressed the near-by faults that led to the 2001 earthquake, is an interesting academic question. To et al. (2004) examined this possibility and explored the connection between the occurrence of the 1819 and the 2001 earthquakes. Results suggested that coseismic and post seismic changes in the Coulomb failure stress (ΔCFS) induced by the 1819 earthquake on the hosting fault of the 2001 event was nominal. The coseismic shear and normal-stress changes at the 2001 hypocenter due to the 1819 earthquake were 0.06 bar and -0.09 bars, respectively, which rose to 0.30 bar and −0.36 bars, after the 2001 event. Their model proposes that the earthquake of 1903 must have been influenced by the increased ΔCFS (+0.06 bar) from the 1819 event (see Fig. 3.1 for the location of the earthquake). Further, for a range of values of coefficient of friction (μ) between 0.2 and 0.8, they reported that ΔCFS was found to be positive at the location of the 2001 event, suggesting the possibility of continued seismicity for an extended period.

On a similar note, Rastogi et al. (2011) reported that aside from the strong aftershock activity for over a decade, seismicity has spread to the nearby faults in the Kachchh region and to several locations to the south, for about 200 km. Over 40 shocks of M_w ~3–5 have been recorded at twenty different sites and the spatial spread of seismicity is attributed to the stress perturbation by viscoelastic processes due to the 2001 Bhuj earthquake. Using RADAR interferometry, Kandregula et al. (2021) report an average surface deformation in the Kachchh mainland region of about 22 mm/yr during 2003 to 2005. The results indicate that the surface deformation attained a maximum by the year 2009 and started to decline further ahead. Based on the analysis of GPS data collected during 2009–2015, Dumka et al., (2019) also suggested low deformation in the Kachchh region, with the maximum value of 3.0 ± 0.5 mm/yr.

4.4 Recurrence History

The 2001 Bhuj earthquake caused wide-spread liquefaction, secondary faults, and other ground deformation features, such as lateral spreads and folds (Bendick et al., 2001; Rajendran et al., 2001; Wesnousky et al., 2001). Liquefaction features that developed in an extensive area near the epicenter as well as at distant locations served as useful proxies in the exploration for older events (Fig. 4.5; see Fig. 4.1). Various researchers have used the sandblow craters, lateral spreads, and secondary faults to locate sites for trenching excavations, and results are discussed here briefly.

Fig. 4.5 **a** A co-seismic sandblow feature formed near Vigakot **b** wall of a crater near Vigakot that exposes three generations of sand vents from 2001, 1819, and an older event; **c** view of a 3-m-wide sandblow crater near Lodai; **d** a cross-cutting sand dyke exposed in a trench near Dholka; (see Rajendran and Rajendran 2002 and Rajendran et al., 2002, for details)

McCalpin and Thakkar (2003) found the thrust scarp and the mole tracks north of Bharudiya Village to be parallel to the major plane of rupture of the main shock and linked them to the zones of dextral faulting near Manfara. If indeed these dextral fault zones define the boundary of the north-thrust block, they argued that the 2001 causative fault must project to their north, an argument that rule out KMF and the South Wagad fault as the source of the earthquake. The lateral spread, recorded north of the Budharmora Village, was interpreted to have preserved sedimentological evidence for a possible combination of an older fault and liquefaction features. Rajendran et al. (2008) examined the features at Manfara and Budharmora in detail and have provided additional constraints on the previous faulting episodes (see Fig. 4.1 for locations).

4.4.1 *Strike-Slip Fault at Manfara*

The 2001 earthquake generated an 11-km-long, northwest-striking fault referred to as the Manfara fault near the village of Manfara, which Wesnousky et al. (2001) described as a 'tear fault' on the hanging wall. The focal mechanisms of at least half a dozen aftershocks from this region had suggested strike-slip movement along near vertical planes, at focal depths of ~10 km (Mandal et al., 2004). The length of the

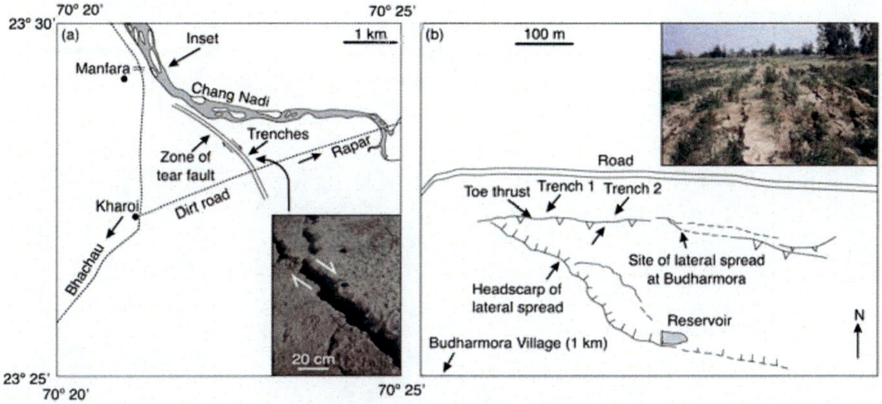

Fig. 4.6 **a** Location map of the fault at Manfara and the trench detailed in Fig. 4.7. Inset: A view of the tear fault. **b** Map of the lateral spread at Budharmora. Inset: a view of the lateral spread showing slumping (modified after Rajendran et al., 2008)

N 10° E to N 20° W-oriented fault mapped here is about 3 km, and it displays a right-lateral strike-slip, with a maximum lateral displacement of 32 cm. Part of the rupture is characterized by a series of right-stepping *en échelon* fractures showing right-lateral movement, with a maximum horizontal offset of ~16 cm and an oblique slip of ~10 cm (Fig. 4.6). An east–west flowing stream (Chang Nadi), north of this rupture zone, takes a northwesterly turn near the Manfara Village, but it is not clear if this is tectonically controlled.

The Bhuj earthquake had also generated several fissures on the ground, which were filled with loose material derived from the top layers. The 3-m-deep trench excavated across the Manfara fault revealed evidence for similar older features, which can generally be described as terminations of fault strands, vertical offsets of marker beds, colluvial wedges, and fissure fillings (Fig. 4.7). The thin colluvial wedge featuring a loose matrix with large clasts and poor stratification was considered characteristic of shaking derived debris. The optically stimulated luminescence (OSL) date of the sample taken from the gravel bed was estimated to be 5125 ± 840 years. A slip of 20 cm was observed on the gravel bed (GB), which was assumed to be cumulative, and a product of the previous and the 2001 events (Rajendran et al., 2008). The OSL date of 4424 ± 656 years was regarded as the minimum age of the penultimate event. Thus, based on the stratigraphic evidence combined with age data, the previous earthquake was dated at around 4000 years BP.

4.4.2 *Lateral Spread at Budharmora*

The lateral spread at Budharmora appeared as a disturbed zone, about 140 m wide and 400 m long, with a gentle (~1)° northerly slope, featuring 1-m wide extensional

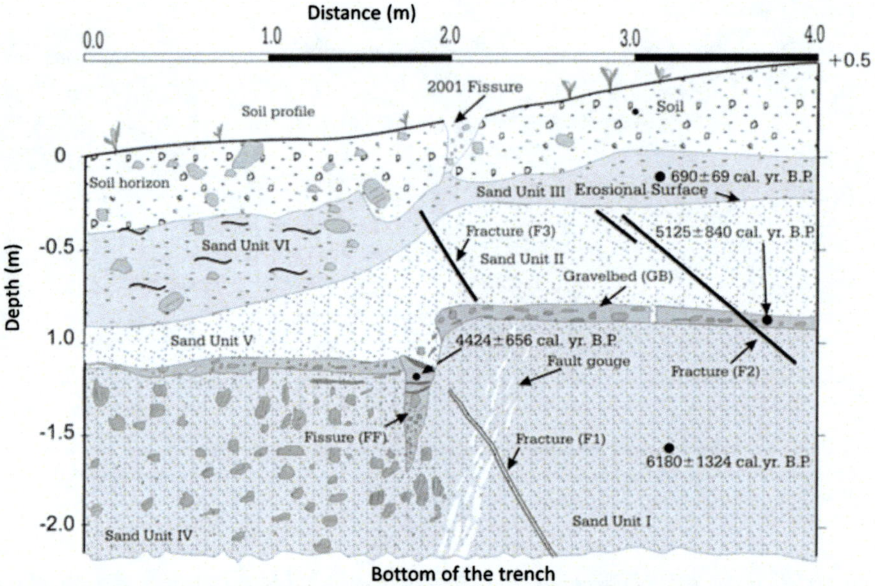

Fig. 4.7 Log of the western wall of a trench across the Manfara fault. The upthrown side is the footwall block; the downthrown side is the hanging wall (after Rajendran et al., 2008)

cracks and back-rotated soil blocks, combined with ejection of sand (Figs. 4.5d and 4.6b). The trenches did not reveal any fault-like features, but there were three generations of feeder dikes including one from the 2001 earthquake (Fig. 4.8). The earliest event and possibly the first-generation sand dikes observed in the trench are believed to have formed between ca. 10.1 ka and 13.8 ka (Rajendran et al., 2008). The fluidization seems to have had disturbed an occupation level that shows flow structures at ~1.5 m below the present surface. From the random distribution of rotated sandstone slabs, this event horizon is interpreted as an occupation level from 3850 to 4700 years BP (Joshi & Bisht, 1994).

McCalpin and Thakkar (2003) excavated the lateral spreads at Budharmora and reported that the sand dikes are overprinted by soil carbonate (6750–7180 cal years BP), which must have been formed in the mid-Holocene or earlier. According to the archeologists, the earliest human occupation in this region is dated to be between c. 2750 and 1900 BCE, and the disturbance, if attributed to an earthquake, post-dates this period. It has been documented that at the archaeological site of Dholavira, close to Khadir Island, a settlement had existed for about 1500 years (3450–4950 years BP), and an earthquake had caused minor damage sometime during 4150–4450 years B.P. (Joshi & Bisht, 1994; Fig. 4.2 for locations). Thus, the geological evidence together with the archaeological data point to an earthquake around 4000 years ago (Rajendran et al., 2008).

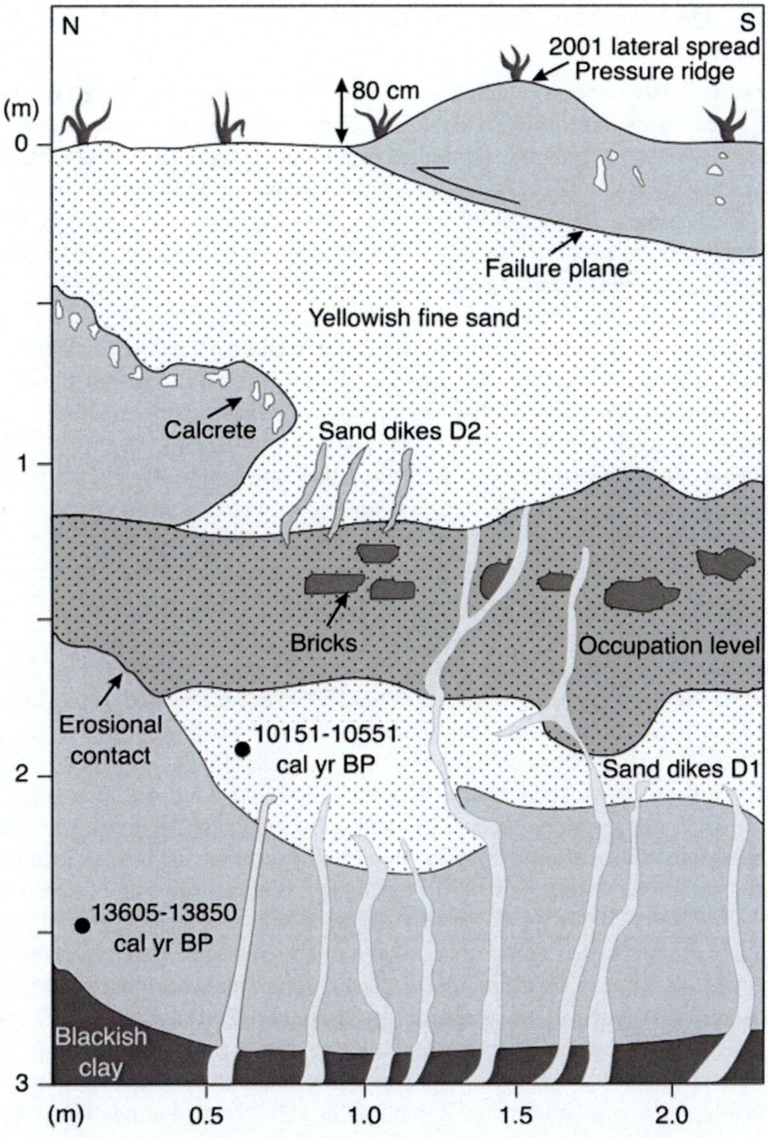

Fig. 4.8 Log of trench 1 excavated across the lateral spread at Budharmora, showing three generations of sand dikes. The second event has disturbed an occupation level (after Rajendran et al., 2008)

4.4.3 Observations from Other Locations

Excavations were also conducted at a few distant locations prompted by the widely distributed coseismic liquefaction features and references in historical documents pointing to possible previous earthquakes at some of these sites. These include sites in the Cambay Basin, Bet Dwarka, Vigakot, Dholka, and Dwarka (see Figs. 4.1 and 4.2, for locations). Large sandblow craters near Vigakot were useful to explore for past events. It appears that Vigakot has been affected by repeated events, and one of the trenches exposed a succession of three sandblow vents: 2001, 1819, a pre-1819 event (Fig. 4.5 a, b). The ~1000-year-old sandblow feature is from the penultimate earthquake as discussed in Chap. 3. The village of Dholka, near Ahmedabad, ~250 km from the source of the earthquake, was one of the farthest liquefaction features formed in 2001. Here, the small sand dike from recent sandblow was seen as cutting through a previous feature (see Fig. 4.5c). Luminescence dating of the source sand and the emplaced sand yielded dates of 5138 ± 514 years and 2945 ± 295 years. Thus, Rajendran et al. (2002) suggested that an event, whose source remains uncertain, must have affected this region about 3000 years ago. Several large sandblow craters, like the one near Lodai (Fig. 4.5d), could not be explored due to the shallow water table, even years after the earthquake.

At Bet Dwarka, there was no report of any coseismic liquefaction, but this site was of interest due to early settlements that are likely to preserve evidence of destruction. A trench excavated here exposed a 2000 to 2278-year-old sand dyke cutting through a 3000-year-old occupation level. This was interpreted as potential evidence for a previous earthquake, from an unknown source (Rajendran et al., 2003). Excavations at other distant sites, such as Mothibaru, Lothal and Kathana, also suggested the possibility for an event between 2353 and 2730 cal years BP. From the seismically induced soft-sediment deformation features, Maurya et al. (1998) inferred earthquakes within the Cambay rift basin during age brackets ranging between 3369 and 3729 and 2840 and 3265 cal years BP. Rajendran et al. (2008) reported sand dykes from Bhavnagar, which could not be dated due to absence of any dateable material. However, it was seen as crosscutting the late Harappan occupation level dated as 3350- and 3850-years BP, as reported by Ratnakar (2001), which gives a possible age constraint. A moderate event has been reported from this region in 1705 CE (Iyengar et al., 1999), suggesting the possibility of a potential local source. Based on the available paleoseismic data, Rajendran et al. (2008) concluded that apart from the repeated activity associated with the 2001 source, other unknown sources exist in the Kachchh basin.

4.5 The Bhuj Earthquake as an Analog for Other Intra-Continental Earthquakes

The Bhuj earthquake provided an opportunity to further our understanding of the mechanism of the mid-continental, paleorift-related earthquakes. From the earthquake history of the region, it is evident that the 2001 earthquake was neither the first major one to occur there, nor was it going to be the last. As discussed in Chap. 3, an earthquake of M > 7.5 had occurred in 1819 in the northwestern part of the Rann and it was preceded by an earlier event whose location remains uncertain. Though the 2001 earthquake had an unprecedent impact and came as an unexpected shock, it offered many valuable lessons, both in terms of scientific understanding and developing a disaster management perspective.

Hosted by the reactivated faults associated with the failed rift tectonic architecture, both the 1819 and 2001 events were rift-related, although sourced at tectonically distinctive parts of the Kachchh basin. Despite the commonality in their tectonic geneses within a rift, these earthquakes show substantial differences in their deformational characteristics. As some models of deformation suggest, the 1819 earthquake sourced on the northern boundary fault of the rift is marked by an east–west running linear surface scarp. This scarp possibly represents the tip of a gently north-dipping fault that reached close to the ground surface but remains blind. The morphology of the epicentral and neighboring area has been interpreted as showing evidence of fault-propagation fold (Rajendran & Rajendran, 2003; Thakkar et al., 2012).

Unlike the 1819 earthquake, which produced a > 80-km-long surface scarp, with a maximum elevation of 4–6 m, the 2001 earthquake was noted for the absence of any primary surface deformation. It was hosted on a steep south-dipping blind fault in the middle of the rift basin. From the contrasting faulting/deformation characteristics of these two earthquakes, it appears that the geometrical disparities of the causative faults must have influenced their respective patterns of surface deformation. It is tempting to argue that the north-dipping buried low angle fault that hosted the 1819 earthquake could be a part of a fold-and-thrust structure in its nascent stage, in contrast to the 2001 event sourced on a steeply south-dipping fault within the basin.

The Bhuj earthquake attracted wide global attention of the seismologists owing primarily to the fact that it was potentially comparable to the 1811–12 New Madrid earthquake series (M_W 7.5–8.0) in the northeast Arkansas, USA (Ellis et al., 2001). Some of the shared properties of the Bhuj and New Madrid earthquakes begin with their association with the failed intra-continental rifts. The similarities in ground deformation include upwarping and subsidence of land, and absence of any primary surface rupture. Aftershocks recorded from the Bhuj earthquake indicate that deep rupture can propagate to the lower crust (Bodin & Horton, 2004). Johnston (1996) has proposed that the New Madrid earthquakes must have ruptured into the lower crust, despite the observation that current seismicity is limited to the top 15 km (Ellis et al., 2001). These observations would point to the different characteristics of the rift-related earthquakes, not just between rifts that, but also within the same rift.

Thus, despite some shared characteristics with other intraplate regions, the Kachchh earthquakes cannot be grouped along with typical continental events like those in New Madrid. One important distinction to be made is the distance from the nearest plate boundary. The Kachchh seismic zone is located within 400 km, whereas the New Madrid seismic zone is >1000 km away from the nearest active plate boundary. Thus, some workers point out that the Kachchh seismic zone is part of a diffuse plate boundary that is subjected to broad diffusion of deviatoric stresses originating from the intracontinental thrusting along the north-western corner of the Indian plate (Stein et al., 2002). It is argued that the internal rheological change within narrow zones can cause the rigid block to move differentially as a sliver even while continuing to be within a continental plate, eventually breaking up as a microplate. Numerical models also suggest that factors like the mechanical heterogeneity between the oceanic and continental parts of the Indian plate and the structural weakening of the rift basins (internal rheologic or thermal perturbations) may lead to the concentration of seismicity in the Kachchh basin (Li et al., 2002). Also, in terms of seismic productivity, the Kachchh seismic zone stands apart, in comparison to the New Madrid Seismic zone.

Yet another issue is to understand if the northern and southern parts of the rift basin differ in their seismic productivity and style of deformation. Perhaps the closer proximity to the plate boundary makes the northern part of the rift relatively more vulnerable to the collisional stresses. Thus, it may be treated as a diffused zone of higher seismic productivity showing relatively shorter recurrence interval between large earthquakes in comparison to the rest of the rift basin. While the orientation of maximum principal stress derived from the focal mechanism of the 2001 earthquake is consistent with the plate motion, its source characteristics mimic typical SCR earthquakes. These are some of the interesting issues that need to be addressed in the future studies.

References

Antolik, M., & Dreger, D. S. (2003). Rupture process of the 26 January 2001 Mw 7.6 Bhuj, India, earthquake from teleseismic broadband data. *Bulletin of the Seismological Society of America, 93*, 1235–1248. https://doi.org/10.1785/0120020142

Bendick, R., Bilham, R., Fielding, E., Gaur, V. K., Hough, S. E., Kier, G., Kukarni, M. N., Marin, S., Mueller, K., & Mukul, M. (2001). The 26 January 2001 "Republic Day" earthquake, India. *Seismological Research Letters, 72*, 328–335.

Biswas, S., & Grasemann, B. (2005). Quantitative morphotectonics of the southern Shillong Plateau (Bangladesh/India). *Australian Journal of Earth Sciences, 97*, 82–03.

Biswas, S. K. (1982). Rift basins in western margin of India and their hydrocarbon prospects with special reference to Kutch basin. *American Association of Petroleum Geologists Bulletin, 66*, 1497–1513.

Biswas, S. K. (1987). Regional tectonic framework, structure and evolution of the western marginal basins of India. *Tectonophysics, 135*, 307–327.

Biswas, S. K. (2016). Tectonic framework, structure and tectonic evolution of Kutch Basin, Western India. *Special Publication of the Geological Society of India, 6*, 129–150.

Biswas, S. K., & Khattri, K. N. (2002). A geological study of earthquakes in Kutch, Gujarat, India. *Journal of the Geological Society of India, 60*, 131–142.

Bodin, P., & Horton, S. (2004). Source parameters and tectonic implication of aftershocks of the Mw 7.6 Bhuj earthquake of 26 January 2001. *Bulletin of the Seismological Society of America, 94*, 818–827.

Chandrasekhar, D. V., & Mishra, D. C. (2002). Some geodynamic aspects of Kutch basin and seismicity: An insight from gravity studies. *Current Science, 83*, 492–498.

Chandrasekhar, D. V., Mishra, D. C., Singh, B., Vijayakumar, V., & Bürgmann, R. (2004). Source parameters of the Bhuj earthquake, India of January 26, 2001 from height and gravity changes. *Geophysical Research Letters, 31*, L19608. https://doi.org/10.1029/2004GL020768

Chopra, S., Kumar, D., Rastogi, B. K., Choudhury, P., & Yadav, R. B. S. (2012). Deterministic seismic scenario for Gujarat region, India. *Natural Hazard, 60*, 517–540.

Chung, W.-P., & Gao, H. (1995). Source parameters of the Anjar earthquake of July 21, 1956, India, and its seismotectonic implications for the Kutch rift basin. *Tectonophysics, 242*, 281–292.

Copley, A., Avouac, J. P., Hollingsworth, J., & Leprince, S. (2011). The 2001 Mw 7.6 Bhuj earthquake, low fault friction, and the crustal support of plate driving forces in India. *Journal of Geophysical Research, 116*, B08405. https://doi.org/10.1029/2010JB008137.

Das, A., Bhattacharya, F., Rastogi, B. K., Chauhan, G., Ngangom, M., & Thakkar, M. G. (2016). Response of a dryland fluvial system to climate-tectonic perturbations during the Late Quaternary: Evidence from Rukmawati River basin, Kachchh, Western India. *Journal of Earth System Science, 125*(6), 1119–1138.

Dumka, R. K., Chopra, S., & Prajapati, S. (2019). GPS derived crustal deformation analysis of Kachchh, zone of 2001(M7.7) earthquake, Western India. *Quaternary International, 507*, 295–301.

Ellis, M., Gomberg, J., & Schweig, E. (2001). Indian earthquake may serve as analog for New Madrid earthquakes. *Eos (Transactions, American Geophysical Union), 82*, 345–350. https://doi.org/10.1029/01EO00211.

Hough, S. E., Martin, S., Bilham, R., & Atkinson, G. M. (2002). The 26 January 2001 M 7.6 Bhuj, India, earthquake observed and predicted ground motion. *Bulletin of the Seismological Society of America, 92*(6), 2061–2079.

Iyengar, R. N., Sharma, D., & Siddiqui, J. M. (1999). Earthquake history of India in medieval times. *Indian Journal of History of Science, 34*, 181–237.

Johnston, A. C. (1996). Seismic moment assessment of earthquakes in stable continental regions, I. *Geophysical Journal International, 124*, 381–414.

Joshi, V. P., & Bisht, R. S. (1994). *India and the Indus civilization*. New Delhi: National Museum Institute.

Kandregula, R. S., Kothyari, G. C., Swamy, V., Taloor, A. K., Lakhote, A., Chauhan, G., Thakkar, M. G., Pathak, V., & Malik, K. (2021). Estimation of regional surface deformation post the 2001 Bhuj earthquake in the Kachchh region, Western India using RADAR interferometry. *Geocarto International*.

Kayal, J. R., Zhao, D., Mishra, O. P., De, R., & Singh, O. P. (2002). The Bhuj earthquake: Tomographic evidance for fluids at the hypocenter and its implications for rupture nucleation. *Geophysical Research Letters, 29*(24), 2152.

Kothyari, G. M., Kandregula, R. S., Chauhan, G., & Thakkar, M. G. (2020). Geomorphic and paleoseismological evidence of active Kachchh Mainland Fault, Kachchh, India. *Arabian Journal of Geosciences, 13*(12). https://doi.org/10.1007/s12517-020-05350-6

Kothyari, G. C., Rastogi, B. K., Morthekai, P., Dumka, R. K., & Kandregula, R. S. (2016). Active segmentation assessment of the tectonically active South Wagad Fault in Kachchh, Western Peninsular India. *Geomorphology, 253*, 491–507.

Kumar, G. P., Mahesh, P., Nagar, M., Mahender, E., Kumar, V., Mohan, K., & Kumar, R. M. (2017). Role of deep crustal fluids in the genesis of intraplate earthquakes in the Kachchh region, northwestern India.*Geophysical Research Letters, 44*, 4054–4063. https://doi.org/10.1002/2017GL072936

Li, Q., Liu, M., & Yang, Y. (2002). The 01/26/2001 Bhuj earthquake: Intraplate or interplate? In S. Stein & J. Freymueller (Eds.), *Plate boundary zones: Washington* (pp. 255–264). American Geophysical Union.

Malik, J. N., Morino, M., Mishra, P., Bhuiyan, C., & Kaneko, F. (2008). First active fault exposure identified along Kachchh Mainland Fault: Evidence from trench excavation near Lodai village, Gujarat, Western India. *Journal of the Geological Society of India, 71*, 201–208.

Mandal, P. (2013). Seismogenesis of the uninterrupted occurrence of the aftershock activity in the 2001 Bhuj earthquake. *Natural Hazards, 65*, 1063–1083. https://doi.org/10.1007/s11069-012-0115-7

Mandal, P., & Pujol, J. (2006). Seismic imaging of the aftershock zone of the 2001 Mw7.7 Bhuj earthquake. *Geophysical Research Letters, 33*(L05309), 1–4.

Mandal, P., & Horton, S. (2007). Relocation of aftershocks, focal mechanisms and stress inversion: Implications toward the seismo-tectonics of the causative fault zone of Mw7.6. (2001). Bhuj earthquake (India). *Tectonophysics, 429*(2007), 61–78.

Mandal, P., Rastogi, B. K., Satyanarayana, H. V. S., Kousalya, M., Vijayraghavan, R., Satyamurthy, C., Raju, I. P., Sarma, A. N. S., & Kumar, N. (2004). Characterization of the causative fault system for the 2001 Bhuj earthquake of Mw 7.7. *Tectonophysics, 378*, 105–121.

Maurya, D. M., Chowksey, V., Tiwari, P., & Chamyal, L. S. (2017a). Tectonic geomorphology and neotectonic setting of the seismically active South Wagad Fault (SWF). *Western India Using Field and GPR Data Acta Geophysics, 65*, 1167–1184.

Maurya, D. M., Chowksey, V., Patidar, A. K., & Chamyal, L. S. (2017b). A review and new data on neotectonic evolution of active faults in the Kachchh Basin. Western India: Legacy of post-deccan trap tectonic inversion. *Geological Society, London, Special Publications, 445*, 237–268.

Maurya, D. M., Rachana, R., & Chamyal, L. S. (1998). Seismically induced deformational structures (Seismites) from the Mild-Late Holocene Terraces, Lower Mahi Valley, Gujarat. *Journal of the Geological Society of India, 51*, 755–758.

Maurya, D. M., Thakkar, M. G., & Chamyal, L. S. (2003). Implications of transverse fault system on tectonic evolution of Mainland Kachchh, Western India. *Current Science, 85*, 661–667.

McCalpin, J. P., & Thakkar, M. G. (2003). 2001 Bhuj-Kachchh earthquake: Surface faulting and its relation with neotectonics and regional structures, Gujarat, Western India. *Annales Geophysicae, 46*, 937–956.

Mohan, K., Chaudhary, P., Patel, P., Chaudhary, B. S., & Chopra, S. (2018). Magnetotelluric study to characterize Kachchh Mainland Fault (KMF) and Katrol Hill Fault (KHF) in the western part of Kachchh region of Gujarat, India. *Tectonophysics, 726*.

Mohan, K. (2014). Seismic-hazard assessment in the Kachchh region of Gujarat (India) through deterministic modeling using a semi-empirical approach. *Seismological Research Letters, 85*(1), 1–9.

Morino, M., Malik, J. N., Gadhavi, M. S., Khalid, A., Bhuiyan, C., Mishra, P., & Kaneko, F. (2008). Active low-angle reverse fault and wide Quaternary deformation identified in Jhura trench across the Kachchh Mainland Fault, Kachchh, Gujarat, India. *Journal of Active Fault Research, 29*, 71–77.

Narula, P. L., Chaubey, S. K., & Sinha, S. (2002). Macroseismic surveys. *Earthquake Spectra, 18*, 45–50. https://doi.org/10.1193/1.2803905

Negishi, H., Mori, J., Sato, T., Singh, R., Kumar, S., & Hirata, N. (2002). Size and orientation of the fault plane for the 2001 Gujarat, India earthquake (Mw7.7) from aftershock observations: A high stress drop event. *Geophysical Research Letters, 29*(20), 1949.

Pollitz, F. F., & Kellogg, L., & Bürgmann, R. (2001). Sinking mafic body in a reactivated lower crust: A mechanism for stress concentration at the New Madrid Seismic Zone. *Bulletin of the Seismological Society of America, 91*(6), 1882–1897.

Prizomwala, S. P., Solanki, T., Chauhan, G., Das, A., Bhatt, N., Thakkar, M. G., & Rastogi, B. K. (2016). Spatial variations in tectonic activity along the Kachchh Mainland Fault, Kachchh, western India: Implications in seismic hazard assessment. *Natural Hazards (Dordrecht), 82*(2), 947–961.

Rajendran C. P., & Rajendran K. (2002). Historical constraints on previous seismic activity and morphologic changes near the source zone of the 1819 Rann of Kachchh earthquake: Further light on the penultimate event Seismol Res Lett. *73*, 470–479.

Rajendran, C. P., & Rajendran, K. (2003). The surface deformation and earthquake history associated with the 1819 Kachchh earthquake. *Memoirs of the Geological Society of India, 54*, 87–142.

Rajendran, C. P., Rajendran, K., Thakkar, M., & Goyal, B. (2008). Assessing the previous activity at the source zone of the 2001 Bhuj earthquake based on the near-source and distant paleoseismological indicators. *Journal of Geophysical Research, 113*, B0531.

Rajendran, C. P., Singh, T., Mukul, M., Thakkar, M., Kothyari, G., John, B., & Rajendran, K. (2020). Paleoseismological studies in India (2016–2020): Status and prospects. *Proceedings of the National Academy of Sciences, Indian, 86*(1), 585–607.

Rajendran, K., Rajendran, C. P., Thakkar, M. G., & Gartia, R. K. (2002). Sand blows from the 2001 Bhuj earthquake reveal clues on past seismicity. *Current Science, 83*, 603–610.

Rajendran, K., Rajendran, C. P., Thakkar, M., & Tuttle, M. P. (2001). The 2001 Kachchh (Bhuj) earthquake: Coseismic surface features and significance. *Current Science, 80*, 1397–1405.

Rao, N. P., Tsukuda, T., Koruga, M., Bhatia, S. C., & Suresh, G. (2002). Deep lower crustal earthquakes in central India: Inferences from analysis of regional broadband data of the 1997 May 21 Jabalpur earthquake. *Geophysical Journal International, 148*, 132–138.

Rastogi, B. K., Aggarwal, S. K, Rao, N., & Choudhury, P. (2011). Triggered/migrated seismicity due to the 2001 M w7.7 Bhuj earthquake, Western India. *Natural Hazards, 65*. https://doi.org/10.1007/s11069-011-0083-3.

Ratnakar, S. (2001). *Understanding Harappan civilization in the Indus valley* (p. 166). New Delhi: Tulika Books.

Schmidt, D. A., & Bürgmann, R. (2006). InSAR constraints on the source parameters of the 2001 Bhuj earthquake. *Geophysical Research Letters, 33*, L022315.

Scholz, C. H., Aviles, C. A., & Wesnousky, S. G. (1986). Scaling differences between large interplate and intraplate earthquakes. *Bulletin of the Seismological Society of America, 76*(1), 65–70.

Silpa, K., & Anil, E. (2021). Revisiting the seismogenic characteristics of stable continental interiors: The case of three Indian events. Quart. Int. *585*, 152–162. https://doi.org/10.1016/j.quaint.2020.12.035

Snelgrove, A. K. (1979). Migrations of the Indus river, Pakistan, in response to plate tectonic motion. *Journal of the Geological Society of India, 20*, 392–403.

Stein, S., Sella, G. F., & Okal, E. A. (2002). The January 26, 2001, Bhuj earthquake and the diffuse western boundary of the Indian plate. In S. Stein & J. Freymueller (Eds.), *Plate boundary zones: Washington* (pp. 243–254). American Geophysical Union.

Talwani, P., & Gangopadhyay, A. (2001). Tectonic framework of the Kachchh earthquake of 26 January 2001. *Seismological Research Letters, 72*, 336–345.

Thakkar, M. G., Ngangom, M., Thakker, P. S., & Juyal, N. (2012). Terrain response to the 1819 Allah Bund earthquake in western Great Rann of Kachchh, Gujarat, India. *Current Science, 103*(2), 208–212.

To, A., Bürgmann, R., & Pollitz, F. (2004). Postseismic deformation and stress changes following the 1819 Rann of Kachchh, India earthquake: Was the 2001 Bhuj earthquake a triggered event? *Journal of Geophysical Research, 31*, L13609.

Wesnounsky, S. G., Seeber, L., Rockwell, T. K., Thakur, V. C., Briggs, S., Kumar, S., & Ragona, D. (2001). Eight days in Bhuj: Field report bearing on surface rupture and genesis of the January 26, 2001 Republic Day earthquake of India. *Seismological Research Letters, 72*, 514–524.

Zoback, M. L., & Richardson, R. M. (1996). Stress perturbation associated with the Amazonas and other ancient continental rifts. *Journal of Geophysical Research, 101*, 5459–5475.

Chapter 5
Killari (Latur) 1993

5.1 Introduction

The Killari (Latur) earthquake of September 30, 1993 (M_w 6.3) that ruptured the surface of the Deccan Plateau was a surprising event in a continental interior region, quite far from the plate boundaries. In some papers, this earthquake is named after Latur, the town nearest to the affected area, but here it is named the Killari earthquake, after the village where its epicenter was located (Fig. 5.1). Sourced in the eastern margin of the Deccan plateau in the state of Maharashtra with an estimated death toll of 11,000 human lives and heavy damage to property, it is one of the most devastating earthquakes to have occurred in India (Gupta, 1993). It drew much attention because of its occurrence in the heartland of India, considered relatively free from damaging earthquakes. Not surprisingly, the 1993 earthquake created such renewed interest in revisiting the global sites of SCR earthquakes, that the American Geophysical Union chose it as a theme for the 1998 Chapman Conference held in India (Gupta & Johnston, 1998).

At the time of the earthquake, the nearest seismological observatory was at Hyderabad, which was about 200 km east of the epicenter, a station that has been operational since December 11, 1967. This observatory had not recorded any tremors from the Killari region till 1991, but it recorded 26 shocks (magnitude 2.0 to 4.0) from this region during 1992 (Baumbach et al., 1994). Following the earthquake, the seismic network in Peninsular India was expanded and modernized. The Killari earthquake also generated scientific interest that led to a flurry of site-specific investigations, including an initiative involving drilling of fault zones, perhaps the first site of any SCR earthquake activity to be studied in such detail (Gupta et al., 1999). Damaging earthquakes in plate interiors are much less frequent than those along plate margins, and thus, it is rare to have a documented record of a succeeding event at the same continental site for thousands of years. Thus, historical precedents are rarely known for most continental intracratonic earthquakes. As such, moderate-size events in

C. P. Rajendran and K. Rajendran, *Earthquakes of the Indian Subcontinent*,
GeoPlanet: Earth and Planetary Sciences,
https://doi.org/10.1007/978-981-16-4748-2_5

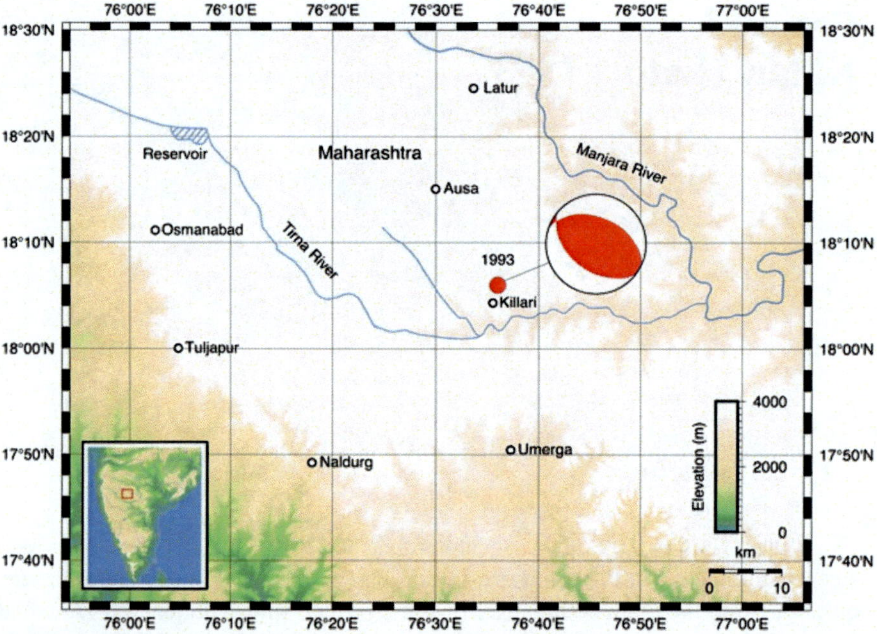

Fig. 5.1 Location of the 1993 Killari earthquake; focal mechanism from Seeber et al. (1996)

the continental interior regions happen so infrequently that every such earthquake becomes a subject of detailed study, compared to their "interplate" counterparts.

The Killari earthquake, gave an opportunity to study the mechanism of an intracratonic earthquake that was not associated with any rifted/passive tectonic regime. It was also an opportunity to evaluate the somewhat underestimated earthquake hazard in populated areas, with low level of preparedness. Whether in a failed rift or in the non-rifted crust, a dominant theory holds that the pre-existing faults, that are optimally oriented in the prevailing stress regime are suitable candidate structures to host SCR earthquakes. As discussed in Chap. 2, besides the tectonic stressing, fault strength changes due to external forcing could be a dominating factor. An increase in pore fluid pressure, for example, can trigger earthquakes in critically stressed regions, including those located within otherwise stable SCRs.

In the continental interior regions of Australia, earthquakes are generated on discrete faults, and their repeat intervals run into tens of thousands of years (Crone et al., 1992; Machette et al., 1993). These earthquakes are not related to any passive rifts, and it is not clear whether the pattern observed at some sites (e.g., Tennant Creek) can be considered as representative of a seismically active, non-rifted SCR region. The general understanding of the mechanism of SCR earthquakes requires the existence of a critically stressed pre-existing fault, but there are also other researchers who hold the view that cratonic earthquakes could originate on new faults. Thus, failure on a new fault, possibly influenced by the hydrological changes due to the proximity to the Tirna Reservoir, has been suggested as a potential mechanism for the Killari earthquake (Seeber et al., 1996). As the seismogenic fault at Killari is outside the direct influence of the reservoir load, the question is whether a stress change as low as 0.1 bars would be sufficient to drive the fault to failure. Such questions had stimulated a debate on 'new fault' versus 'pre-existing fault' in the case of the Killari earthquake (Rajendran & Rajendran, 1997; Seeber, 1997). The hypothesis of the new fault had also included the possibility of a preexisting causative structure, which could be buried in the Precambrian basement with little movement since the emplacement of the Deccan flows. This chapter provides a summary of the current understanding of the 1993 Killari earthquake and how it furthered our knowledge of the seismic sources in the continental interiors.

5.2 Tectonic Setting and Background Seismicity

The epicenter of the Killari earthquake falls in the Tirna River basin near the eastern margin of the Late Cretaceous–Eocene basalt flows (Deccan Traps). The Traps, which covers an area of > 600,000 km^2 in central India, consist of several episodic flows ranging in thickness from a few meters up to about 100 m, with the successive flows separated by the red bole or inter-trappean beds (Gupta & Dwivedy, 1996; Subbarao & Sukheswala, 1981). The flat lying stacks of basalt flows produce a staircase-like elevated topography in the Deccan region, and the Killari area typifies the much-subdued fringe of Trappean morphology, with no apparent topographic signatures of neotectonism.

The epicentral zone of the earthquake is at an elevation of about 0.5 km from the mean sea level. No topographically high areas exist more than 50 m above the broad valley formed by the Tirna River. The low level of tectonism, together with high erosional rates, is believed to have shaped the gently rolling and relatively featureless terrain and in such a landscape, the structural imprints generally remain subdued (Rajendran et al., 1996a, b). However, structurally controlled lineaments have been reported from the region. Based on the configuration of the drainage network, Babar et al. (2012) identified structurally controlled lineations along the NE-SW, NW–SE, E-W, and WNW-ESE directions, which are reported to have influenced the overall drainage network of the area and the tributaries of the Tirna River. A sudden high-velocity flow regime has also been identified in the central part of the basin near Killari, where the river locally attains a higher gradient (Lakshmi et al., 2020). This structural high is attributed to the tectonic uplift along the NW–SE direction.

Although the Trappean flows (560–620 m) appears to present a flat terrane, the Bouguer anomaly map of the area indicates an uneven basement, marked by bedrock ridges and basins with several local gravity highs and lows of 3–5 mGal (Kailasam, 1993). The epicentral zone of the 1993 earthquake is located on the flanks of one such 'high,' forming part of a NW–SE regional trend. Thus, it is reasonable to assume that the basement may preserve some expressions of earlier tectonic episodes, suppressed by the overlying basalt layers. The scientific drilling and geophysical studies in the epicentral area reveal a thinner crustal basement composed of amphibolite to granulite facies rocks that are characterized by high upper crustal temperatures derived from the mantle (Pandey, 2016).

Earthquake catalogs for Peninsular India (Chandra, 1977; Rao & Rao, 1984) indicate that no significant earthquakes (M > 6) have occurred in the source zone of the Killari earthquake during recent or historic times, except for a few events of intensity III–IV. Rajendran et al. (1996b) reported the existence of a 400-km-long, NW-oriented corridor of historic earthquakes in this region with at least five of intensity IV during the past 150 years (Fig. 5.2). However, only a small part of this zone was activated during the 1993 earthquake, as evident from the limited rupture area inferred from the distribution of aftershocks (Baumbach et al., 1994). The observation that only a small segment of the NW-trending structure had ruptured in 1993 raised the question about the earthquake history of the rest of the segments of this feature. Rajendran et al. (1996a) noted that no comparable-size earthquakes had occurred in the epicentral area of the 1993 event, at least for the past 1000 years. They based this assumption primarily on the longevity of historical monuments dating back to 1000–1200 CE in the vicinity of Killari that had remained intact until the 1993 earthquake.

However, the previous occurrences of similar earthquakes cannot be ruled out elsewhere along the NW-trending structure in the Deccan region of central India. One such source along this belt was uncovered in Ter, 40 km northwest of Killari (see Fig. 5.2 for location). Though presently a quiet village, during the early part of the first millennium it was a busy market town located on an ancient trade route. There are other sites from where historical earthquakes have been reported. Graham (1854), for example had reported the possibility of a damaging earthquake sometime between the thirteenth and fourteenth centuries in the vicinity of the town of Kolhapur in western Maharashtra (Fig. 5.2), which had damaged some of the ancient temples near this town. These observations are indicative that the Killari-type earthquakes are not exceptional, but joins a continuum from the distant past, either as a part of the NW-trending belt of seismicity or as sporadic events observed elsewhere in the Deccan Province.

Geological evidence for an earthquake, that occurred about 1500 years ago was reported from the village of Ter (Rajendran, 1997). In a later study, Sukhija et al. (2006) reported a Killari-type event that may have occurred here between 190 BCE and 410 CE—an observation based on the tectonic signatures, like fault offsets and liquefaction preserved in the alluvial deposits of Tirna and Manjira river valleys. Babar et al. (2012) have explored different locations in the Tirna Valley and recorded stratigraphic offsets and morphologic evidence for multiple earthquakes bracketed between 353 and 1183 CE. Their magnitudes are unclear but the current understanding of the range of magnitudes in peninsular India suggests that they might not exceed M 6.5.

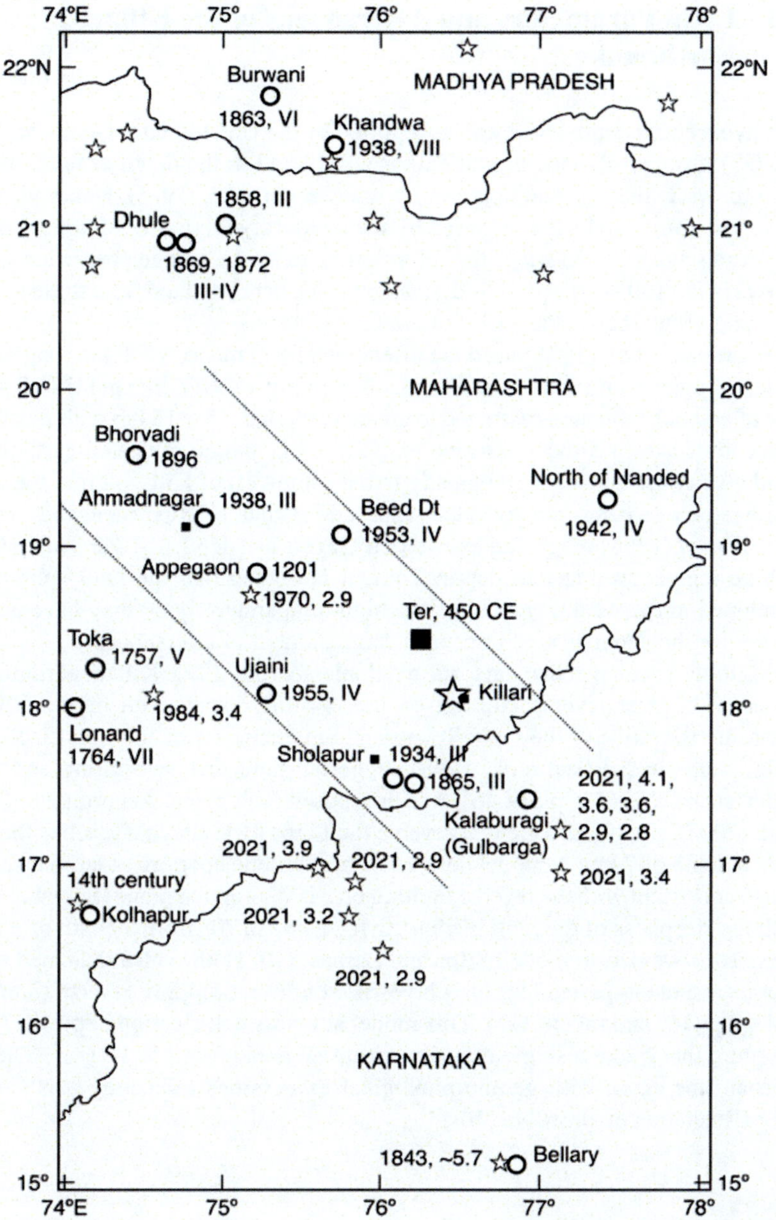

Fig. 5.2 Historical seismicity of the Killari region. Filled circles are historically documented events and stars are instrumentally located events. The location of Ter is shown by a filled square (after Rajendran et al., 1996b)

5.3 Focal Parameters and Aftershocks of the Killari Earthquake

The hypocentral depth of 6.8 km, as reported by the United States Geological Survey (USGS), from worldwide digital data is considered as the depth of the initiation of rupture, while the centroid depth is 2.6 km (Seeber et al., 1996). Hundreds of aftershocks that followed were recorded by the temporary recording stations established by various agencies. Much of the information about the source properties and aftershocks of the 1993 earthquake is derived from the data obtained from these temporary networks (Baumbach et al., 1994; Kayal et al., 1996).

Baumbach et al. (1994) used the aftershock focal plane, ~ 5.5 km long and 6 km wide, to suggest a northwest-striking fault, dipping 45° SW, like the USGS solution. The aftershock zone was restricted to an area less than 15 × 15 km with most activity concentrated toward the southwest of Killari. Sixty-nine relocated earthquakes of good quality are also confined mostly to the southwest of Killari. From the analyses of the remote sensing data, local tectonics, and available geophysical and earthquake data, Kumar (1998) suggested low velocity layers at a depth of 12 ± 3 km, attributed to high temperature at that depth. Thus, it is argued that the fault splits (splays) developed in the weak zone above the high-temperature body may have acted as a trigger for the generation of unusually large number of aftershocks.

Multiple focal mechanisms are available for the 1993 Killari earthquake, all of which suggest reverse faulting on the NW–SE plane, with minor differences regarding the strike of the fault. Teleseismic moment tensor solution (Seeber et al., 1996; Ramesh & Eastabrook, 1998) suggested pure reverse faulting and the NW plane (strike N 126°; dip SW 46°) was the chosen fault plane. According to Gahalaut et al. (2003), the N112° oriented reverse fault dips 42° NW, and reaches the surface from a depth of 7 km. A model derived from synthetic aperture radar interferometry is also consistent with the reverse faulting on a NW-oriented plane (Satyabala, 2006), with the steepness of the fault increasing from 45° to 70° from a depth of 6 km. In a teleseismic waveform model, Silpa and Earnest (2021) have also indicated the same geometry and suggested slip on a NW–SE southwest-dipping reverse fault (strike: 134°, dip: 44° and rake: 112°). This model suggests a nucleation depth of 7 km and a rupture that broke a single asperity within an area of ~13 × 13 km^2. These fault mechanisms agree with geomorphological expressions indicating a NW-oriented fault (Rajendran et al., 1996a, b).

5.4 Investigations of the Surface Rupture

The 1993 Killari event is one of the rare examples of intracratonic earthquakes to have produced a surface rupture. Mapped initially by Seeber et al. (1996), the rupture zone was about 500 m long and 100 m wide, located about 4 km northwest of Killari, with its subdued continuation traceable for about 0.5 km to the east and west (Fig. 5.3a, b). Prominent among the deformation features mapped are the sets of

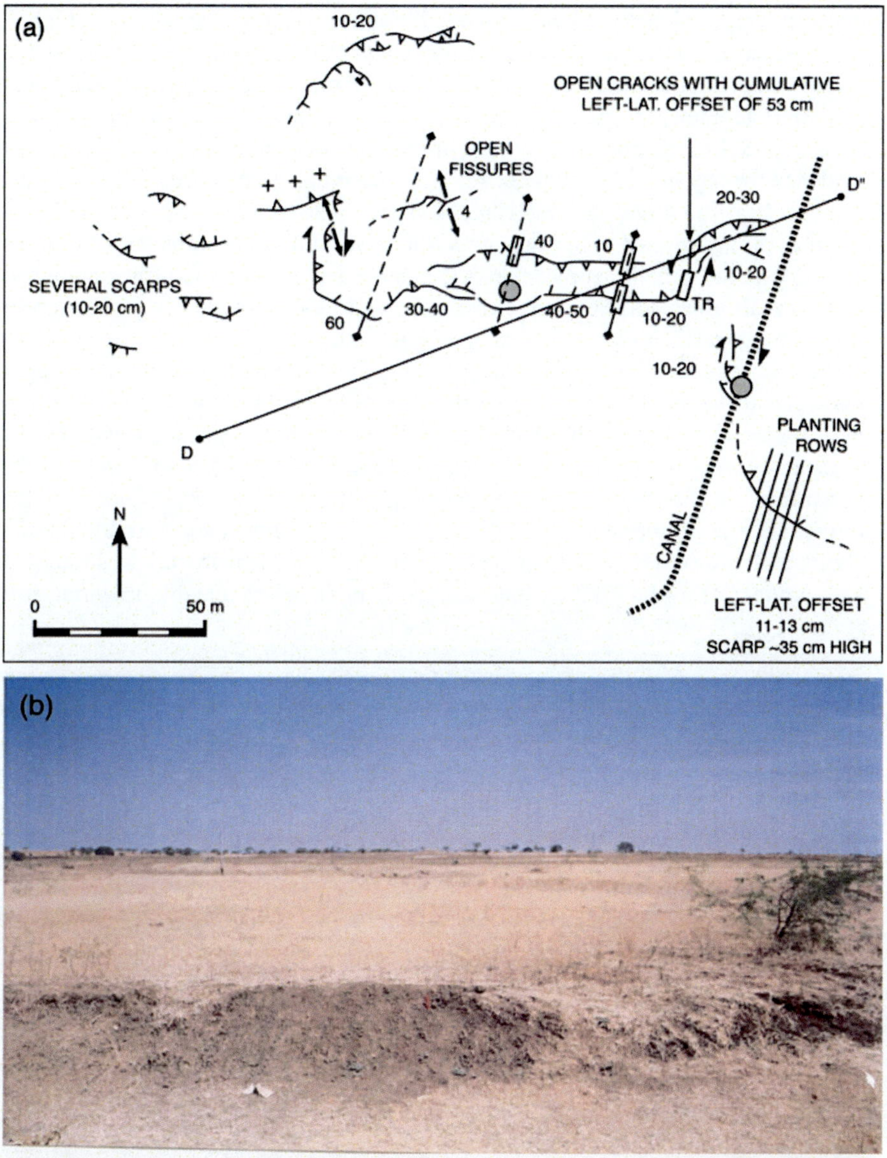

Fig. 5.3 **a** Map of the surface rupture showing the scarps (a view is given in 5.3b) and profiles (dashed lines). Locations of trenches excavated by Seeber et al. (1996) are shown as unfilled rectangles along the profiles. Circles filled in grey indicate localized uplift. TR shows the location of the trench excavated by Rajendran et al. (1996b) (view of the trench wall is given in Fig. 5.4). DD" is the profile showing the morphology of the scarp in Fig. 5.5 (modified after Seeber et al., 1996). **b** View of the surface rupture observed near the central part of the rupture zone. The maximum height of the scrap is 80 cm

opposite-verging scarps and the most notable among them was about 70 cm high. The deformation expressed on the country rocks was quite complex, probably due to the layered nature of the weathered basalt flows and it may not be a direct reflection of the structural geometry of the basement fault. Three trenches excavated in the rupture zone exposed small offsets of 50 cm in the soil-rock interface, but no convincing evidence for any previous earthquakes was observed. In the absence of any breccia or mineralization along the reactivated foliation planes within the basalt layers, the rupture was interpreted to have occurred on a new fault, an argument that apparently agreed with the lack of geomorphic evidence for any previous faulting and/or tilting in the basalt layers. Although the reservoir could have potentially played a role in the seismogenesis, it was left as an open question.

In a subsequent work, Rajendran et al. (1996b) estimated the maximum height of the scarp as 80 cm and reported that the best-developed part of the surface rupture, with fissures and cracks that followed a dominant east-northeast trend. Trenches excavated across the scarp generally revealed a low angle (~15°), south-west dipping thrust fault on the southern wall, suggesting that the southwestern block had been thrust onto the northeastern block (Fig. 5.4). A wide impact zone around the thrust sheets with crushed basalt embedded in yellowish and whitish clay was interpreted as a product of a pre-1993 earthquake. In absence of any datable material, it was

Fig. 5.4 The section exposed in the trench (TR in Fig. 5.3a) showing the displacement of rock–soil interface. Hanging wall is on the southern side; the footwall is on the northern side. Arrow indicates movement along the contact between the two blocks (after Rajendran et al., 1996b)

difficult to estimate the timing of the event, but from the nature of the soil profile, appearance of the gouge, and the general morphological characteristics, the penultimate earthquake is inferred to have occurred atleast tens of thousands of years ago. Additional information on previous faulting comes from the boreholes drilled on either side of the fault that suggest a vertical displacement of 4–6 m in the 350-m-thick Deccan Traps (Gupta et al., 1999). This cummulative displacement must have resulted from several Killari-type earthquakes in the past, as a single event accounts only for a maximum vertical slip of about 1.0 m (Ramesh and Estabrook (1998).

Although the geological signatures argued for the thrusting of the southwest block, there was a geomorphological paradox because the hanging-wall block was at a lower elevation, despite its being overthrust to the northeast (Fig. 5.5). Rajendran et al. (1996b) interpreted this feature as an obsequent fault-line scarp, which has resulted from repeated thrusting from the southwest but was also subject to higher rate of erosion. The massive and compact footwall, being less vulnerable to erosion, remains at a slightly higher elevation, creating a geomorphologically inverted feature. Taking cues from the subtle scarp formed in 1993, it is evident that the morphological evidence from past earthquakes is poorly preserved, also considering the extent of anthropogenic activity. Digital imagery (Landsat TM) enhanced using multispectral

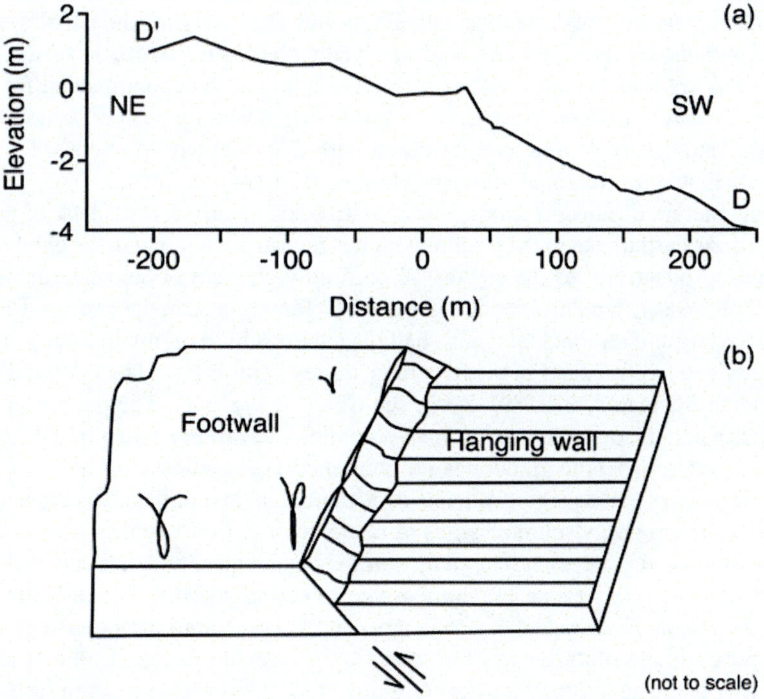

Fig. 5.5 **a** Topographic profile along DD'as shown in Fig. 5.3a; **b** schematic sketch of the hanging wall and footwall blcks (Rajendran et al., 1996b)

ratio techniques for bands 7 and 4 was used to recognize the northeast and southwest blocks based on their tonal difference. With its sparse vegetation and thin soil cover, the southwest block appeared in a lighter tone, compared to the darker tone of the northeast block, which had a thicker soil profile and denser vegetation. The geomorphologic inversion of the hanging and footwall blocks observed along the Killari fault scarp is a clear demonstration of how erosional processes outstrip the rate of tectonism, as observed in other intracratonic settings (e.g., Crone et al., 1992; Machette et al., 1993), leaving the perception of low-level neotectonic activity.

5.5 Importance of the 1993 Killari Earthquake

The 1993 earthquake is one of the rare continental interior events that caused extensive damage and loss of life. The damage was confined to ~15 km around the epicenter but its impact was too severe considering the intermediate range of magnitude. Damage surveys that focused on engineered structures reported that the collapse of traditional stone-and-mud buildings in the meizoseismal area was nearly total. It was reported that the single-story dwellings with wood-plank roofs, typically topped with a layer of clay (30–60 cm thick), performed poorly, and their collapse was responsible for the high number of casualties. Dwellings with thatched roof and panel, some plastered with mud, performed well, and survived with only minor cracks on the walls. A few brick-masonry houses with concrete lintel bands, as recommended for seismic zones IV and V, suffered no damage. Most elevated water tanks performed well, but for the one that collapsed straight down with displacement of about 0.5 m, which was attributed to rotational vibration (Jain et al., 1994).

The picture of massive damage and destruction clearly reflects lack of preparedness for an earthquake in the region, which is not surprising, given the extremely low seismic productivity of the region. Accordingly, the earlier seismic zonation map (IS: 1893–984), which classified areas of different earthquake potential based on past activity and regional tectonics, had placed the Killari region in Zone I. This map was revised in 2002, and in the new map, zones I and II have been merged (Bureau of Indian Standards, BIS1893-2002; see Chap. 1, Fig. 1.4). The pattern of damage intensity underscores the need to map potentially hazardous faults in the region and provide region-specific guidelines for construction activities.

The Killari earthquake occurred in a region of insignificant background seismicity, showing only subdued signs of neotectonic activity, typical of areas marked by slow tectonic processes. The shallow focal depth and abundant aftershock activity associated with this event, are comparable to earthquakes in the Australian shield, like the one at Tennant Creek (Bowman, 1992). The Killari earthquake generated a complex surface rupture comparable to that at Tennant Creek (Crone et al., 1992) and Ungava (Canada) earthquakes (Adams et al., 1991). These examples show that the Killari-type events, which show apparent randomness, may occur unexpectedly on discrete faults in regions that may remain inactive for a long time. It must be recognized that with the slow slip rates and low background seismicity, which are characteristic in SCR regions, the causative faults may not be easily detectable. An earthquake in the early part of the first millennium near the village of Ter and a historical reference to a fourteenth century earthquake near Kolhapur, are reminders that similar events could occur in other parts of the Deccan Trap region. Studies in Killari and its neighborhood provide rare insights into an enigmatic SCR earthquake source.

The lesson learned from Killari is that conventional hazard assessments relying on statistical rates of seismicity may not apply in continental interiors. In such areas where the background information on earthquakes is sparse and surface expressions are not suggestive of any tectonism, hazard assessments need to adopt new approaches. A comprehensive evaluation of geologic data would help in identifying potentially hazardous faults and their patterns of long-term activity. As the borehole data from the vicinity of the 1993 rupture indicates, the cumulative fault slip is relatively small, suggesting a low rate of strain accumulation. The studies, however, also suggest that the region overall is nominally active, as evidenced by an earthquake dating back to the beginning of the first millennium and the uplifted the riverbed of Tirna. These observations may suggest that there could be multiple seismic sources and earthquakes may occur episodically with long inter-event intervals. Geological methodologies might help to detect potential hazardous faults in such regions characterized by. subdued geomorphologic and tectonic features. The lessons learned from the Killari experience provide a template for gaining insights into such 'out-of-the-blue' events in the continental interiors marked by low rate of strain accumulation.

References

Adams, J., Wetmiller, R. J., Hasegava, H. S., & Drysdale, J. (1991). The first surface faulting from historical intraplate earthquakes in North America. *Nature, 352*, 617–619.

Babar, Md., Chunchekar, R.V., Yadava, M. G., & Ghute, B. B. (2012). Quaternary geology and geomorphology of Terna River Basin in west central India. *E&G Quaternary Science Journal, 61*(2), 159–168. https://doi.org/10.3285/eg.61.2.04

Baumbach, M., Grosser, H., & Schmidt, H. G., et al. (1994). Study of the foreshocks and aftershocks of the intraplate Latur earthquake of September 30, 1993. *India. Mem. Geol. Soc. India, 35*, 33–63.

Bowman, J. R. (1992). The 1988 Tennant Creek, northern territory, earthquakes: A synthesis. *Australian Journal of Earth Sciences, 39*, 651–669. https://doi.org/10.1080/08120099208728056

Chandra, U. (1977). Earthquakes of peninsular India—a seismotectonic study. *Bulletin of the Seismological Society of America, 67*, 1387–1413.

Crone, A. J., Machette, M. N., & Bowman, J. R. (1992). Geologic investigations of the 1988 Tennant Creek, Australia earthquakes—implications for paleoseismicity in stable continental regions. *U.S. Geological Survey Bulletin*, 2032-A, pp. 51.

Gahalaut, V., & Raju, P. et al. (2003). Rupture mechanism of the 1993 killari earthquake, India: Constraints from aftershocks and static stress change. *Tectonophysics, 369*, 71–78.

Graham, D. C. (1854). *Statistical report on the principality of Kohlapur, Bombay Education Sociey's Press* p. 335

Gupta, H. K. (1993). The deadly latur earthquake. *Science, 262*(5140), 1666–1667.

Gupta, H. K., & Dwivedy, K. K. (1996). Drilling at Latur earthquake region exposes a peninsular gneiss basement. *Journal of Geological Society of India, 47*, 129–131.

Gupta, H. K., & Johnston, A. (1998). Chapman conference on stable continental region (SCR) earthquakes. *Journal of Geological Society of India, 52*, 115–117.

Gupta, H. K., Rao, R. U. M., & Srinivasan, R. et al. (1999). Anatomy of surface rupture zones of two stable continental region earthquakes, 1967 Koyna and 1993 Latur, India. *Geophys Res Lett, 26*, 1985–1988. https://doi.org/10.1029/1999GL900399

Jain, S. K., Murthy, C. V. R., Chandak, K., et al. (1994). The September 29, 1993, M6.4 Killari, Maharashtra, Earthquake in Central India. *EERI Special Earthquake Report,* 1–8.

Kailasam, L. N. (1993). Geophysical and geodynamical aspects of the Maharashtra earthquake of September 30, 1993. *Curr Sci, 65*, 736–739

Kayal J. R., De, R., Das, B., & Chowdhury, S. N. (1996). After shock monitoring and focal mechanism studies, Killari earthquake, 30 September 1993. In: P. L. Narula, S. K. Sharma, & B. S. R. Murthy (Eds.) *Geol. Surv. India Sp. Publ.* 37, 165–185.

Kumar, S. (1998). Intraplate seismicity and geotectonics near the focal area of the Latur earthquake (Maharashtra). *India, Journal of Geodynamics, 25*, 109–128.

Lakshmi, B. V., Deenadayalan, K., Gawali, P. B. et al. (2020). Effects of Killari earthquake on the paleo-channel of Tirna River basin from central India using anisotropy of magnetic susceptibility. Sci Rep 10. 20587. https://doi.org/10.1038/s41598-020-77542-9

Machette, M. N., Crone, A. J., & Bowman, J. R., (1993). Geologic investigations of the 1986 Marryat Creek, Australia earthquake—Implications for paleoseismicity in stable continental regions: U.S. Geological Survey Bulletin 2032-B, p. 29.

Pandey, O. P. (2016). Deep scientific drilling results from Koyna and Killari earthquake regions reveal why Indian shield lithosphere is unusual, thin and warm. *Geoscience Frontiers, 7*, 851–858.

Rajendran, C. P., Rajendran, K., Unnikrishnan, K. R., & John, B. (1996a). Paleoseismic indicators in the rupture zone of the 1993 Killari (Latur) earthquake. *Cursos e Congresos Da Universidade De Santiago De Compostela, 70*, 385–390.

Rajendran, C. P., Rajendran, K., & John, B. (1996b). The 1993 Killari (Latur), Central India earthquake: An example of fault reactivation in the Precambrian crust. *Geology, 24*, 651–654.

Rajendran, C. P. (1997). Deformational features in river bluffs at Ter, Osmanabad district, Maharashtra, evidence for an ancient earthquake. *Current science, 82*, 750–755.

Rajendran, C. P., & Rajendran, K., (1997). Comments on "The 1993 Killari earthquake in Central India: A new fault in the Mesozoic basalt flows?" by Seeber et al., *Journal of Geophysical Research, 102*(B11), 561–564.

Ramesh, D., & Estabrook, C. (1998). Rupture histories of two stable continental region earthquakes of India. *Journal of Earth System Science, 107*, L225–L233.

Rao, B. R., & Rao, S. (1984). Historical seismicity of peninsular India. *Bulletin of the Seismological Society of America, 74*, 2519–2533.

Satyabala, S. P. (2006). Coseismic ground deformation due to an intraplate earthquake using synthetic aperture radar interferometry: The M_w6.1 Killari, India, earthquake of 29 September 1993, *Journal of Geophysical Research: Solid Earth,* 111. https://doi.org/10.1029/2004JB003434

Seeber, L. (1997). The 1993 Killari earthquake in central India: A new fault in Mesozoic basalt flows? Reply *Journal of Geophysical Research,* 102, 24565–24570.

Seeber, L., Ekstrom, G., & Jain, S. K., et al. (1996). The 1993 Killari earthquake in central India: A new fault in Mesozoic basalt flows? *Journal of Geophysical Research, 101*, 8543–8560.

Silpa, K., & Earnest, A. (2021). Revisiting the seismogenic characteristics of stable continental interiors: The case of three Indian events. *Quaternary International, 585*, 152–162. https://doi.org/10.1016/j.quaint.2020.12.035

Subbarao, K. V., & Sukheswala, R. N. (Eds.). (1981). Deccan volcanism and related basalt provinces in other parts of the world: Geological Society of India Memoir 3, p. 474.

Sukhija, B. S., Bandaru, V. L., Rao, M., et al. (2006). Widespread geologic evidence of a large Paleoseismic event near the Meizoseismal area of the 1993 Latur earthquake, Deccan Shield, India. *Journal Indian Geophysical Union, 10*, 1–14

Chapter 6
Jabalpur 1997

6.1 Introduction

The May 22, 1997 (M_w 5.8) earthquake occurred ~8 km southeast of the city of Jabalpur in the state of Madhya Pradesh in the central India (Fig. 6.1). Occurrence of this event four years after the devastating 1993 Killari earthquake reinforced the hitherto underestimated seismic vulnerability in Peninsular India. As a sequel to the Killari earthquake, this event had a multiplying effect insofar as widening the distribution of broadband stations in peninsular India by the Department of Science and Technology (Government of India). The expanded broadband network established after the Killari earthquake also included a station at Jabalpur, which provided high-quality broadband data. Thus, the Jabalpur earthquake had the distinction of being the first moderate earthquake in the Indian peninsular shield to have been studied using near and far-field digital records. Its focal depth, estimated as 36 km (Bhattacharya et al., 1997; Singh et al., 1999), was considered unusual for a SCR setting. The local network established after the main event recorded a few aftershocks, which were also quite deep and in the range of 25–45 km (Acharya et al., 1998). The proximity of its epicenter to a populous city also gave an additional opportunity to study the response of modern and traditional built environments to a continental earthquake (Jain et al., 1997). With its lower crustal hypocentral depth and spatial association with the well-mapped Narmada Rift, the Jabalpur earthquake attracted much scientific attention.

Field investigations indicated that much of the damage was confined to an area of about 15 × 35 km^2, to the north of the epicenter. Significant variations in the MSK intensity estimates were documented, but the maximum value of VIII was reported from the city of Jabalpur. Damage was reported also from other cities such as Mandla, located to the southeast of Jabalpur (see Fig. 6.1 for locations). According to the reports, 8546 houses collapsed completely and 52,690 were badly damaged. A total of 887 villages were affected, leaving 39 people dead and about 350 injured (Jain et al., 1997).

C. P. Rajendran and K. Rajendran, *Earthquakes of the Indian Subcontinent*,
GeoPlanet: Earth and Planetary Sciences,
https://doi.org/10.1007/978-981-16-4748-2_6

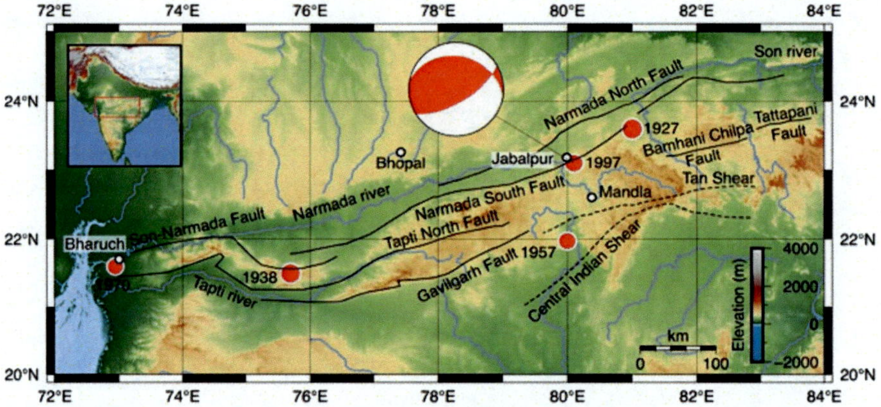

Fig. 6.1 Major tectonic features of the Narmada Son lineament zone and previous significant earthquakes in the region from published sources discussed in the text. The focal mechanism of the 1997 earthquake is from Ramesh and Estabrook (1998)

The Jabalpur earthquake was spatially associated with the Narmada-Son Lineament/Fault (NSL or NSF, as abbreviated variously in the literature), a prominent structure that cuts across the central India in an NNE-SSW direction (Fig. 6.1). While most well-known SCR earthquakes have occurred on unmapped faults, the unambiguous spatial association of the Jabalpur earthquake with the morphologically conspicuous NSL is unique. The fact that the NSL has been seismically active, hosting similar moderate earthquakes during the documented history, is a rarity in the SCRs. Another point of interest is the deep crustal source (~40 km) observed in the NSL zone, which is quite unusual for intra-continental earthquakes. This chapter reviews the tectonic setting of the Jabalpur earthquake and provides an evaluation of its source parameters and faulting mechanism.

6.2 Tectonic Setting of the Narmada-Son-Fault

The Indian shield is considered as an agglomeration of three landmasses—the Dharwar, Aravalli and Singhbhum protocontinents. The accretion dates back to ~3.5 billion years when various terranes joined together to form a mega-shear zone (Naqvi et al., 1974). This region was subjected to series of compressional and extensional tectonic processes that included three generations of rifting and magmatic activity from the Middle to Late Archaean to Meso-Proterozoic (Das & Patel, 1984; West, 1962). A conspicuous ENE-WSW linear structural trend cuts through the Indian subcontinent, dividing the central India as two distinct geological blocks. This linear structure is referred to as the Son-Narmada-Tapti lineament zone, a name derived after the Son, Narmada, and Tapti Rivers that occupy these structurally controlled valleys. Together, the Son-Narmada-Tapti Valleys represent a composite lineament zone, generally referred to as the NSL or the Son–Narmada–Tapti (SONATA) rift zone. Geological and geophysical studies have unequivocally described this lineament as a deep fault zone, cutting across almost the entire breadth of central India (Verma & Banerjee, 1992; West, 1962).

Deep-seated faulting along NSL was identified initially by West (1962), who pointed out that the Meso–Neoproterozoic Vindhyan rocks exist only to its north, while the Gondwana Supergroup (6- to 7-km thick succession of fluvial and lacustrine deposits from the Permian to Cretaceous Periods ranging from 298 to 66 Ma) are found to its south. It has been suggested that repeated movements have been occurring on the bounding faults at different periods since the Early Proterozoic. The elevated Satpura Range, considered as the horst, is flanked by two grabens – the Narmada graben in the north and the Tapti–Purna graben to the south, both of which host thick piles of Quaternary sediments (Shanker, 1988). The horst–graben structure of this rifted zone is evident also from deep seismic sounding (Kaila et al., 1989).

Repeated reactivation along the Son rift zone, coupled with the step-faulting associated with the northward movement of the Indian plate since the Miocene (~23 to 5 million years ago) is believed to have shaped the structural setting of the Son-Narmada-Tapti lineament zone. The last event of active rifting, also linked with alkaline magmatism, is believed to have occurred during the Paleocene (66–56 million years ago), as evident from the vents of eruption at various locations along the valley (Srinivasan et al., 1998). The post-Himalayan collisional processes (~40 Ma) caused a reversal of stress in the Indian shield, marking an end to regime of extensional tectonics. The combined geomorphologic, stratigraphic, and structural evidence also suggests tectonic inversion in the Holocene due to the incremental in the compressional stress caused by the northward drift of the Indian plate (e.g., Chamyal et al., 2002; Copley et al., 2014).

Heat flow data values from various locations in Indian peninsula suggest large variations in surface heat flow (Roy & Rao, 2000). High heat flow anomalies and numerous hot springs have been observed within the Narmada Valley, which are linked to an anomalous thermal structure in the region (Shanker, 1988). Rogers and Callahan (1987) link the heat production in the Narmada Valley with longevity of the rifting process, and argue that the sustained orogenic processes have led to the thickening and stacking of crustal slabs. The episodic heating and rifting in the Narmada rift were also overprinted by a variety of tectonic processes, including the transition from extensional to compressional regimes following the India-Eurasia collision. Perhaps, such sequences of tectonic events have made the Narmada rift different from other failed rifts such as the Reelfoot Rift, which hosts the seismically active New Madrid seismic zone (Missouri, USA). Although both are failed rifts, as discussed in the earlier chapters, there are differences too, including the presence of an active collision boundary that binds the Indian plate. Such differences need to be taken into account, while interpreting their relative seismic productivities and the bearing on the causative structures.

Significant gravity anomalies have been observed in many parts of central India, including a chain of gravity highs, 25–30 km wide, running parallel to NSL for almost 1000 km (Mishra, 1992). The residual Bouguer anomaly map shows a major gravity high in the Jabalpur area, which forms part of the long chain of gravity highs observed in the regional gravity map. Verma and Banerjee (1992) had inferred the presence of a massive high-density intrusive body that could explain this feature. The results from the gravity surveys along three N–S profiles to the west of Jabalpur helped Singh and Meissner (1995) to infer the presence of high-density bodies at a depth range of 24–32 km between the Narmada and Tapti Rivers. Deep seismic sounding surveys have added further confirmations on the local structure of the Moho and intra-crustal layers, including a high velocity (7.2 km/s) in the Narmada zone (Kaila et al., 1989; Tewari & Kumar, 2003). The crustal velocity model by Murty et al., (2008) using travel time inversion of wide-angle reflections reconfirmed an uplifted crustal block and Moho up-warp beneath the NSL, possibly indicating that the NSF and NNF are deeply penetrating faults.

Receiver function models suggest a distinct 15-km-thickening of the crust and a complex nature of the Moho within the rifted zone (Kumar et al., 2015), along with substantial evidence for magmatic underplating in the lower crust (Fig. 6.2). Teleseismic receiver functions computed for a 250-km-long profile also suggest down-warping of Moho at ~52 km across the width of the rifted zone, in contrast to an average of 40 km depth elsewhere (Rai et al., 2005). Further, it is also noted that the crust beneath the NSL has a higher V_p/V_s of 1.84 compared to ~1.73 in the surrounding regions. This anomaly has been interpreted as being due to the presence of a high-density mafic body at depth, which may lead to gravitationally

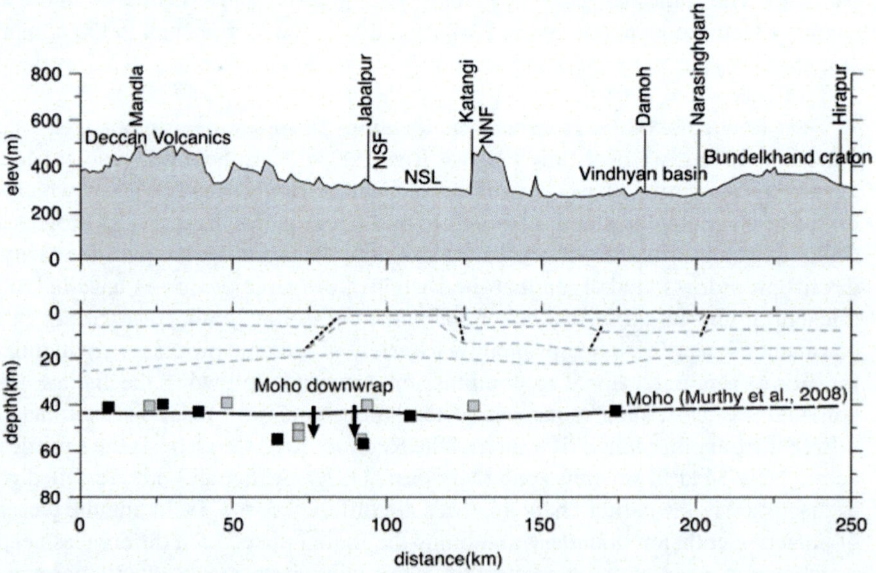

Fig. 6.2 Summary of results from seismic imaging across a profile across the Narmada rift. The part shows the topography and the bottom image shows various interpretations. Solid and grey squares represent the Moho mapped by Kumar et al. (2015) and Singh et al. (2015), respectively (adapted from Kumar et al., 2015)

induced stresses in the lower crust and failure along the pre-existing faults. For such a model, Kumar et al. (2015) refer to other continental flood basaltic provinces where magmatic underplating is recognized as a dominant mechanism of crustal growth and evolution.

Finite element models that used a range of material properties for the Narmada Rift suggested that density heterogeneities alone are not sufficient to enhance the stress concentration at the hypocentral region (Manglik et al., 2008). However, it is argued that a mechanically strong lower crust overlying a relatively weak sub-Moho layer could possess the ability to enhance the stress concentration, thus implying a weaker mantle, in comparison to the lower crust. Other geophysical studies also suggest the presence of an anomalous zone of a hot upper mantle with temperatures higher than the solidus of basalt (1200 °C) within the lower crust (Anand & Rajaram, 2004). More recently, Koulakov et al. (2018) have put forth a fresh argument that the Narmada-Son lineament, marked by step-shaped structures comprising thick and thinner lithosphere, may define a zone of potential subduction initiation which will evolve in a few million years. In summary, all of these models suggest an anomalous crustal architecture in the Narmada rift region, which possible accounts for its deep crustal seismogenic processes.

6.3 Background Seismicity

Seismically, the Narmada rift is relatively active, with more than 30 earthquakes, falling in the magnitude range of 3.0 and 6.5, during the last 70 years (Chandra, 1977; Rao & Rao, 1984). Some of the known earthquakes associated with this structure are the Son Valley (1927, M 6.5), Satpura (1938, M 6.3), Balaghat (1957, M 5.5), and Broach (1970, M5.4) (Fig. 6.1). Chattopadhyay et al. (2020) provide an updated review of earthquakes in central India, including the SONATA. Among the historical events, the 1938 Satpura earthquake, with its epicenter located within the Satpura Hills, north of the Tapti River Valley is noted for its hypocentral depth of 40 km (Mukherjee, 1942). However, the Broach earthquake on the western terminus of the SONATA was shallower at 11 km (Chung, 1993).

Rao and Rao (2006) have reviewed the historical earthquakes compiled from various sources, and they relate the earthquakes in the Narmada rift zone to exhumed active faults. The overall picture of background seismicity associated with the NSL/SONATA shows occasional moderate and small earthquakes at shallow to lower crustal depths and a distinct spatial association with the mid-continental structure. Focal depth estimates of earthquakes in the NSL zone cannot be considered as reliable, except for the 1970 Broach (Chung, 1993) and the 1997 Jabalpur earthquakes. Despite these limitations, a study by Gahalaut & Bürgmann, (2004), based on focal depths reported in the ISC catalogue, indicate existence of earthquake sources within the lower crust under the entire NSL zone (Fig. 6.3).

While there is no documented history of any large earthquakes in central India, the morphological cues on fault scarps indicate the presence of active faults in the

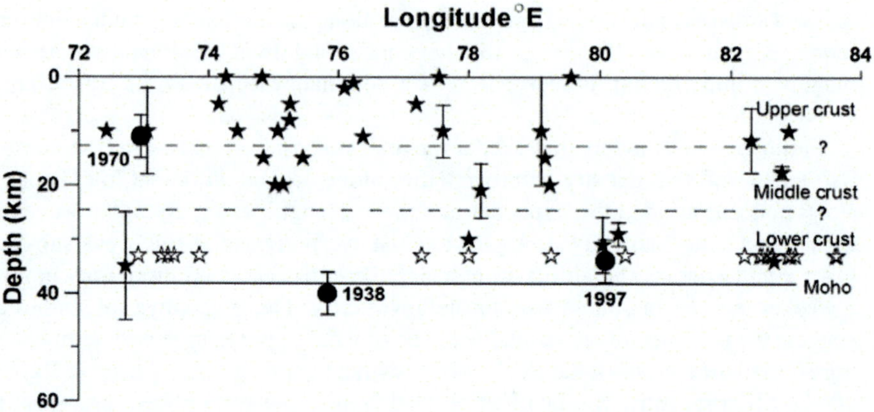

Fig. 6.3 Depth distribution of earthquakes ((ISC catalogue: 1967–1999) in the NSL zone. Unfilled stars represent the restricted focal depths at 33 km. Earthquakes with error bars indicate focal depth derived from three or more consistent depth phases (adapted from Gahalaut & Bürgmann, 2004)

region. It is reported that thrust earthquakes in the magnitude range of 7.6–8.4 have occurred in the western Tapti region of the rift (Copley et al., 2014; Fig. 6.1). The geological evidence for active tectonics along these faults, which, according to this study, implies significant seismic hazard in central India. Thus, from the point of understanding of the mechanism of reactivation of ancient rift-related fault systems, as well as for the seismic hazard assessment, the Narmada-Tapti rift zone calls for more detailed studies.

Source Parameters and Proposed Mechanism

The network of ten new broadband stations installed after the 1993 Killari earthquake was already in place when the Jabalpur earthquake occurred. The then existing stations of the India Meteorological Department (IMD) also recorded the event, the nearest being the stations at Nagpur (231 km) and Bilaspur (237 km). The plane oriented N80°E with a dip of 66° SE, supported by its spatial association with the NSL, was preferred solution (Bhattacharya et al., 1997). Earthquake-induced ground cracks and fissures are also oriented in the ENE–WSW direction, on the downdip extension of the NSL (Acharya et al., 1998). Results of modelling obtained by various workers using waveform inversion are also comparable with the structural trend of the NSL. These solutions generally suggest an ENE-WSW fault plane, with a steep dip (Fig. 6.1). The solutions provided by various workers are quite similar (Ramesh & Estabrook, 1998: 62°N; 64°; Singh et al., 1999: 61°N, dip 64°; Rao et al., 2002: 62°N, dip 68°; Saikia, 2006: 65°N, 68°; Silpa & Earnest, 2021: 59°N, 67°). Among these, the latest by Silpa and Earnest suggest a smaller rupture and breakage of a single asperity, differing from the two subevent model proposed by Singh et al. (1999). Twenty-eight aftershocks of magnitude ≥ 3.0 were recorded by the local network of stations, with their hypocentres in the depth range of 25–45 km. A composite fault plane solution based on these events suggests a mechanism similar to the mainshock (Acharya et al., 1998).

A rift pillow-related mechanism for the Jabalpur earthquake was initially proposed by Rajendran and Rajendran (1998). This suggestion was mostly inferred from the velocity anomalies and the presence of high-density bodies inferred from Bouguer gravity images (Reddy et al., 1997). Later models propose alternate mechanisms that do not necessarily require a rift pillow. For example, Rao et al. (2002) adopted a more quantitative approach and suggested stress accumulation from a horizontally elongated or elliptical body, which possibly represents a serpentinized mafic intrusive in the lower crust. Gahalaut & Bürgmann, (2004) suggested the presence of high pore pressure from the dehydration of serpentinite. The combined effect of lower frictional coefficient of fractured rocks, high strain rate and favourable N-S oriented stress from the plate motion are important in driving the failure. That the brittle/ductile transition may occur at deeper levels within in the rift, could explain the lower crustal depth.

Seismic imaging using receiver functions has provided more clarity on the nature of subsurface structures within the Narmada rift. Using teleseismic receiver functions from nine short-period stations installed along a 250-km-long profile across the NSL, Rai et al. (2005) presented the first image of this region, which suggests density inhomogeneities. Down-warping of the Moho and the presence of high-density mafic mass in the deep crust are capable of generating gravitationally induced stresses in the lower crust, leading to failure along the pre-existing faults within the Narmada rift. The model by Kumar et al. (2015) also supports the proposal of an anomalous structure in the source zone of the 1997 earthquake. Whether it is described as the down-warping of the Moho (Rai et al., 2005), or as underplating (Kumar et al., 2015), both the models suggest that the structural complexity inherited from rifting and the ensuing magmatism provides the fundamental causative structural setting for the origin of lower crustal earthquakes.

6.4 Outstanding Questions

The mid-continental Narmada rift has a history of moderate earthquakes, which itself makes it a special case of SCR seismicity. The modern-day deep crustal earthquake and the historically documented events, with their hypocentral depths closer to 40 km, make the Narmada rift unique among its counterparts elsewhere. The Kachchh rift has also generated mid-crustal earthquakes, the 2001 Bhuj event being the latest example. Seismic sources in both these rifts seem to be associated with high-density mantle material in the hypocentral depths. The stress concentrations arising from such anomalies, together with the hydro-mechanical weakening induced by pore pressure changes could be driving the earthquakes in both these rifts. However, the 1997 and 2001 earthquakes also exhibit some differences in the seismogenesis within these two rifts. The Narmada rift with its deep crustal sources is known for moderate earthquakes with relatively much shorter regional recurrence whereas the Kachchh rift regime is known to generate larger earthquakes. Whether these differences are linked to their respective histories of evolution and their structural architectures or their relative placement with respect to the nearest plate boundary is a point to be explored, as more examples become available.

As we understand today, the Narmada rift is the outcome of both extensional and compressional tectonics of longer duration, from the middle to late Archaean to Meso-Proterozoic. In contrast, the Kachchh rift is a relatively younger structure within the Precambrian craton. Whether the age of a paleo rift has a control on the depth-wise positioning of the frozen rift pillows, leading to their seismogenic depths is an idea that requires validation. The Holocene morphological scarps are known to exist in the Narmada rift and it has been argued that they might hold potential for large earthquakes. The possibility for greater magnitude earthquakes along the border faults cannot be ignored and their potential impact need more focussed studies. The Jabalpur earthquake is a reminder of the level of destruction and damage that even a moderate earthquake can cause. And the future holds more such challenges.

References

Acharya, S. K., Kayal, J. R., Roy, A., & Chaturvedi, R. K. (1998). Jabalpur earthquake of May 22, 1997: Constraint from aftershock study. *Journal of the Geological Society of India, 51*, 295–304.

Anand, S. P., & Rajaram, M. (2004). Crustal structure of Narmada-Son lineament: An aeromagnetic perspective. *Earth Planets and Space, 56*, e9–e12.

Bhattacharya, S. N., Ghosh, A. K., & Suresh, G., et al. (1997). Source parameters of Jabalpur earthquake of 22 May 1997. *Current Science, 73*, 855–863.

Chamyal, L. S., Maurya, D. M., Bhandari, S., & Raj, R. (2002). Late Quaternary geomorphic evolution of the lower Narmada valley, Western India: Implications for neotectonic activity along the Narmada-Son fault. *Geomorphology, 46*, 177–202.

Chandra, U. (1977). Earthquakes of Peninsular India—a seismotectonic study. *Bulletin of the Seismological Society of America, 67*, 1387–1413.

Chattopadhyay, A., Bhattacharjee, D., & Srivastava, S. (2020). Neotectonic fault movement and intraplate seismicity in the central Indian shield: A review and reappraisal. *Journal of Mineralogical and Petrological Sciences, 115*, 138–151.

Chung, W.-Y. (1993). Source parameters of two rift-associated intraplate earthquakes in peninsular India: The Bhadrachalam earthquake of April 13, 1969 and the Broach earthquake of March 23, 1970. *Tectonophysics, 225*, 219–230.

Copley, A., Mitra, S., Alastair Sloan, R., Gaonkar, S., & Reynolds, K. (2014). Active faulting in apparently stable peninsular India: Rift inversion and a Holocene-age great earthquake on the Tapti Fault. *Journal of Geophysical Research: Solid Earth, 119*. https://doi.org/10.1002/2014JB011294.

Das, B., & Patel, N. P. (1984). Nature of the Narmada-Son lineament. *Journal of Geological Society of India, 25*, 267–276.

Gahalaut, V. K., & Bürgmann, R. (2004). Constraints on the source parameters of the 26 January 2001 Bhuj, India, earthquake from satellite images. *Bulletin of the Seismological Society of America, 94*, 2407–2413. https://doi.org/10.1785/0120040021.

Jain, S. K., Murty, C. V. R., & Arlekar, J., et al. (1997) Some observations on engineering aspects of the Jabalpur earthquake of 22 May 1997. *EERI Special Earthquake Report, EERI Newsletter, 31*(8), 8.

Kaila, K. L., Murthy, P. R. K., & Mall, D. M. (1989). The evolution of the Vindhyan basis vis-a'-vis the Narmada-Son lineament, central India, from deep seismic soundings. *Tectonophysics, 162*, 277–289.

Koulakov, I., Gerya, T., & Rastogi, B. K., et al. (2018). Growth of mountain belts in central Asia triggers a new collision zone in central India. *Science and Reports, 8*, 10710. https://doi.org/10.1038/s41598-018-29105-2.

Kumar, R. M., Singh, A., Kumar, N., & Sarkar, D. (2015). Passive seismological imaging of the Narmada paleo-rift, central India. *Precambrian Research, 270*(2015), 155–164.

Manglik, A., Thiagarajan, S., Mikhailova, A., & Rebetsky, Y. (2008). Finite element modelling of elastic intraplate stresses due to heterogeneities in crustal density and mechanical properties for the jabalpur earthquake region, central india. *Journal of Earth System Science, 117*(2), 103–111.

Mishra, D. C. (1992). Mid-continent gravity 'high' of central India and the Gondwana tectonics. *Tectonophysics, 212*, 153–161.

Mukherjee, S. M. (1942). Seismological features of the Satpura earthquakes of the 14th March 1938. In *Proceedings of the Indian Academy of Sciences-Section A* (Vol. 16, pp. 167–175).

Murty, A. S. N., Sain, K., Tewari, H. C., & Prasad, B. R. (2008). Crustal velocity inho-mogeneities along the Hirapur-Mandla profile, central India and its tectonic implications. *Journal of Asian Earth Sciences, 31*, 533–545.

Naqvi, S. M., Rao, D., & Narain, H. (1974). The protocontinental growth of the Indian shield and the antiquity of its rift valleys. *Precambrian Research, 1*, 345–398.

Rai, S. S., Vijay Kumar, T., & Jagadeesh, S. (2005). Seismic evidence for significant crustal thickening beneath Jabalpur earthquake, 21 May 1997, source region in Narmada—Son lineament, central India. *Geophysical Research Letters, 32*, L22306.

Rajendran, K., & Rajendran, C. P. (1998). Characteristics of the 1997 Jabalpur earthquake and their bearing on its mechanism. *Current Science, 74*, 168–174.

Ramesh, D., & Estabrook, C. (1998). Rupture histories of two stable continental region earthquakes of India. *Journal of Earth System Science, 107*, L225–L233.

Rao, B. R., & Rao, S. (1984). Historical seismicity of peninsular India. *Bulletin of the Seismological Society of America, 74*, 2519–2533.

Rao, B. R., & Rao, V. K. (2006). Influence of fluids on deep crustal Jabalpur earthquake of 21, May 1997: Geophysical evidences. *Journal of Seismology, 10*, 301–314. https://doi.org/10.1007/s10950-006-9018-y.

Rao, N. P., Tsukuda, T., & Koruga, M., et al. (2002). Deep lower crustal earthquakes in central India: Inferences from analysis of regional broadband data of the 1997 May 21 Jabalpur earthquake. *Geophysical Journal International, 148*, 132–138.

Reddy, P. R., Sain, K., & Murthy, A. S. N. (1997). On the seismic vulnerability of Jabalpur region: Evidence from deep seismic sounding. *Current Science, 72*, 796–800.

Rogers, J. W., & Callahan, E. J. (1987). Radioactivity, heat flow and rifting of the Indian continental crust. *The Journal of Geology, 95*, 829–836.

Roy, S., & Rao, R. U. M. (2000). Heat flow in the Indian shield. *Journal of Geophysical Research: Solid Earth, 105*(25), 25587–25604.

Saikia, C. (2006). Modeling of the 21 May 1997 Jabalpur earthquake in central India: Source parameters and regional path calibration. *Bulletin of the Seismological Society of America, 96*, 1396–1421. https://doi.org/10.1785/0120050120.

Shanker, R. (1988). Heat flow map of India and discussions on its geological and economic significance. *Indian Minerals, 42*, 89–110.

Silpa, K., & Earnest, A. (2021) Revisiting the seismogenic characteristics of stable continental interiors: The case of three Indian events. *Quaternary International, 585*, 152–162. https://doi.org/10.1016/j.quaint.2020.12.035.

Singh, S. K., Dattatrayam, R. S., & Shapiro, N. M., et al. (1999). Crustal and upper mantle structure of peninsular India and source parameters of the 21 May 1997 Jabalpur earthquake (Mw = 5.8): Results from a new regional broadband network. *Bulletin of the Seismological Society of America, 89*, 1631–1641.

Singh, A. P., & Meissner, R. (1995). Crustal configuration of the Narmada- Tapti region (India) from gravity studies. *Journal of Geodynamics, 20*, 111–127.

Singh, A., Singh, C., & Kennett, B. L. N. (2015). A review of crust and upper mantle structure beneath the Indian subcontinent. *Tectonophysics, 644–645*, 1–21. https://doi.org/10.1016/j.tecto.2015.01.007.

Srinivasan, R., Jaffri, S. H., Rao, G. V., & Reddy, G. K. (1998). Phreatomagmatic eruptive centre from the Deccan Trap Province, Jabalpur, central India. *Current Science, 74*, 787–790.

Tewari, H. C., & Kumar, P. (2003). Deep seismic sounding studies in India and its tectonic Implications. *Journal of the Virtual Explorer, 12*, 30–54.

Verma, R. K., & Banerjee, P. (1992). Nature of continental crust along the Narmada-Son lineament inferred from gravity and deep seismic sounding data. *Tectonophysics, 202*, 375–397.

West, W. D. (1962). The line of Narmada-Son valley. *Cursos e Congresos Da Universidade De Santiago De Compostela, 31*, 143–144.

Part II
Plate Boundary Earthquakes: Northwest and Central Himalaya

Chapter 7
Seismotectonics of the Himalayan Fold and Thrust Belt

7.1 Introduction

The India-Eurasia collision during the Cenozoic has led to the evolution of the Himalaya, one of the most intensely deforming continental collision zones in the world. The eastern and western terminations of the Himalayan arc are marked respectively by the eastern and western syntaxes, which strongly bend around a vertical axis and form distinctive collisional belts. In contrast to the dominant arc-normal compression and thrust faulting along the Himalaya arc, strike-slip tectonics prevail along the east and west of the two Himalayan syntaxes (Ding et al., 2001). West of the western syntaxial bend (around Nanga Parbat), the Chaman fault system has accommodated the northward motion of India with respect to Afghanistan and Iran via left-slip motion (e.g., Lawrence et al., 1981). East of the eastern syntaxial bend (around Namche Barwa), right-slip faults dominate along the Indo-Burman Ranges (see Figs. 7.1 and 7.5).

The Himalayan syntaxes not only define the boundaries between diverse tectonic regimes, but the physiography also hosts two opposite flowing rivers, the Indus and the Tsangpo-Brahmaputra, the transcontinental rivers, which transport huge loads of sediment from the Tibetan plateau. The high denudation rates associated with these rivers and climatic processes are believed to have provided a positive feedback mechanism for localizing significant crustal deformation (e.g., Burg et al., 1988). In particular, the NW Himalaya is considered a classic example of the interaction of climate and tectonics and its role in mountain building (Clift, 2017). Hirschmiller et al. (2014) suggest that the morphology of the Himalayan foreland fold-and-thrust belt is primarily controlled by climatically induced erosional process, thus reducing the orogenic load on the subducting plate. But beyond the proposal of climate-tectonics interaction, uncertainty persists on the relative contribution of each of these processes in shaping mountain landscape (Mandal et al., 2021). The building of high topography following the early Cenozoic collision of India with Eurasia has played

C. P. Rajendran and K. Rajendran, *Earthquakes of the Indian Subcontinent*,
GeoPlanet: Earth and Planetary Sciences,
https://doi.org/10.1007/978-981-16-4748-2_7

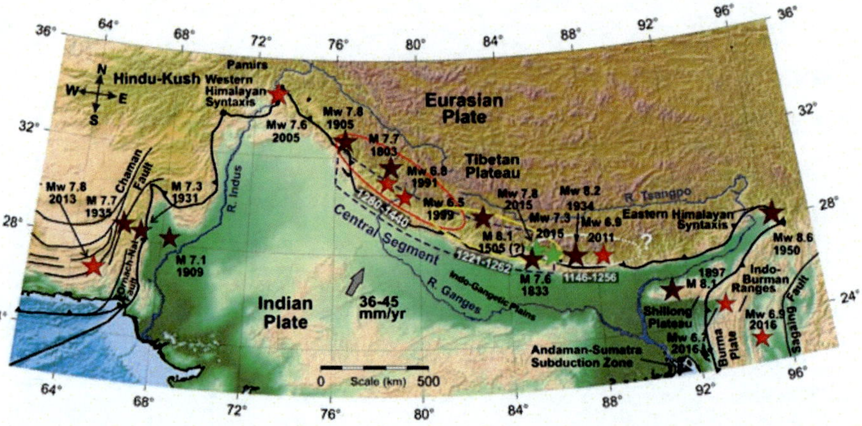

Fig. 7.1 The Himalaya plate boundary showing large/great earthquakes since the historically documented times. Black line shows India-Eurasia plate boundary. Red stars are historical events of M ≥ 8; brown stars are historical events of M ≥ 7 and green stars are modern events (2015, M_W 7.8 and M_W 7.3). Grey arrow shows the GPS convergence rates in mm/yr (Stevens and Avouac, 2015). The ellipses show the inferred rupture zones of the 1100 (white), 1255 (yellow) and 1344 (red) CE events (Wesnousky et al., 2017a, 2017b; Rajendran et al., 2019)

a fundamental role in strengthening the Asian monsoon in South and East Asia, an evident by-product of climate-tectonics coupling.

The ~2500-km-long Himalayan arc and the contiguous emergent structures associated with the collision is the seat of some of the most damaging earthquakes in the recent global history. Among the large/great plate boundary earthquakes in the last century are the 1905 Kangra (M_W 7.8), 1934 Bihar-Nepal (M_W 8.2), and 1950 Upper Assam (M_W 8.6), the last one being the largest to have occurred on a continental plate boundary after seismic instrumentation came into vogue (Fig. 7.1). The 1897 Shillong earthquake is not a true plate boundary event, although it has often been grouped among the great Himalaya earthquakes (e.g., Seeber & Armbruster, 1981).

The Himalaya is an orogenic wedge formed by a stack of thrust sheets scraped off the Indian crust (Le Fort, 1975), which was being underthrust beneath Asia after the closure of the Tethys Ocean (e.g., Nabelek et al., 2009; Searle et al., 1987). From the north to south, the Main Central Thrust (MCT) and the younger Main Boundary Thrust (MBT) are presently passively displaced above the basal Main Himalayan Thrust (MHT). The youngest is the Main Frontal Thrust (MFT), also known as the Himalayan Frontal Thrust (HFT), emerges along the Siwalik Hills (the Outer or the lesser Himalaya) (Mugnier et al., 2013). As discussed by several authors, the imbricated thrust stack, comprising the MCT, MBT and MFT along with the MHT define the overall structural architecture of the Himalayan fold-and-thrust belt. All the thrust systems within the wedge sole into a main basal décollement, a mid-crustal reflector at a depth of about 40 km beneath southern Tethyan Himalaya near Tibet (e.g., Hodges, 2016; Zhao et al., 1993; Fig. 7.2). In this geometry and crustal structure,

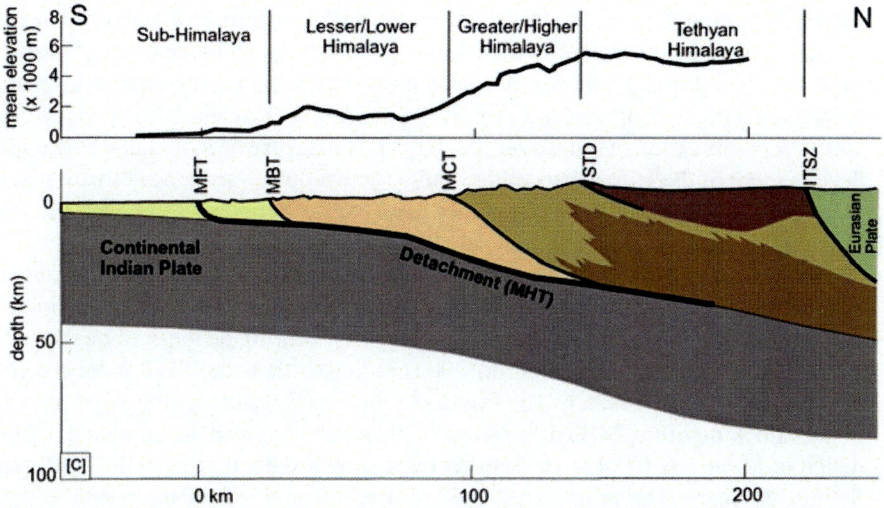

Fig. 7.2 Generalized cross section of the central Himalaya, showing the Main Central Thrust (MCT), Main Boundary Thrust (MBT), Main Frontal Thrust (MFT), South Tibetan Detachment (STD), and Indo-Tsangpo Suture Zone (ITSZ). Detachment plane coincides with the Main Himalayan Thrust (MHT) (after Lavé & Avouac, 2001)

large earthquakes originate on the MHT, at the depth of the brittle-ductile transition (Fig. 7.3).

The MHT absorbs about 20 mm/yr of convergence in Nepal, which is nearly half of the present convergence between India and Eurasia (Bettinelli et al., 2006). Locally imaged by geophysical data (e.g., Zhao et al., 1993), the MHT is known to display a crustal ramp, inferred also from balanced cross-sections. This zone, marked

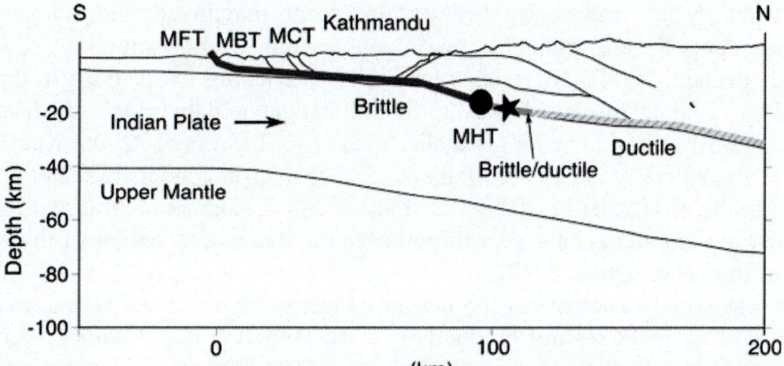

Fig. 7.3 Schematic cross-section through central Nepal. Dark band shows the inferred rupture zone, and the grey band is the ductile regime. The position of the black dot indicates the tip of dislocation, and the star shows the source of great earthquakes (after Mugnier et al., 2013)

by the increased dip of the MHT, coincides with clustering of micro-seismicity, higher topographic gradients, and fluvial incision. The steep range of the Himalayan front developed about 100 km north of its southern periphery is also intimately associated with an evolving mid-crustal ramp (Lavé & Avouac, 2001; Pandey et al., 1995). A combination of ramp overthrusting and underplating associated with duplex development of the Himalayan wedge and/or active out-of-sequence thrusting at the front of the high Himalaya are the other mechanisms proposed to explain the steep topographic front (e.g., Bollinger et al., 2004; Wobus et al., 2005).

In the aftermath of the 2015 earthquake, the Nepal Himalaya has been subjected to extensive investigations and these studies have provided new seismotectonic insights on the structural complexities of the MHT and their bearing on the earthquake generation. The model by Jouanne et al. (2004) indicates that the MHT is locked from the surface to approximately 100 km down dip, corresponding to a depth of 15 to 20 km. Locking of the MHT is believed to have resulted in an accumulated moment deficit of $M_0 = 6{:}6 \pm 0{:}4 \times 10^{19}$ Nm/yr (assuming a shear modulus of 30 GPa) over the last 20 years. The moment released by the seismicity over the past 500 years, amounts to only 0.9×10^{19} Nm/yr, indicating a large shortfall of seismic slip. A fraction of the accumulated moment is assumed to be released elastically by transient aseismic slip event (slow slip events) or after-slip, following large earthquakes. However, no large slow slip event has been observed over the 20 years covered by geodetic measurements in the Nepal Himalaya (Ader et al., 2012).

7.2 An Overview of Seismicity

The large and great earthquakes in many parts of the Himalaya had occurred prior to the widespread seismic instrumentation, and in the absence of recorded data, information on these events has been based mostly on archival records. Details gathered by meticulously sifting through a variety of documents remain the major source of information about these earthquakes (e.g., Iyengar et al., 1999; Ambraseys & Jackson, 2003; Bilham, 2004). There have been many trenching excavations in the recent decades using paleoseismological tools, and the gap in data prior to the historically documented period is being gradually filled (e.g., Lave et al., 2005; Kumar et al., 2006; Rajendran et al., 2015; Malik et al., 2016; Jayangondaperumal et al., 2017; Wesnousky et al., 2017a, 2017b). Studies of archaeological records and the proxy evidence using stalagmite growth perturbations have complemented these efforts (Rajendran et al., 2013, 2016).

It was widely known that the rate of convergence along the Himalaya arc is a critical factor in the seismic productivity of the Himalaya that straddles a very active plate boundary (Bilham et al., 2001, Bilham, 2019). With more information coming in about the past earthquakes, there is further clarity on their spatial and temporal association with the causative faults. The instrumentally recorded data and their interpretations formed the basis for some of the early models on the mechanism of earthquakes in the Himalaya and their relation to the major structures (e.g., Seeber &

Armbruster, 1981). With the advent of GPS geodesy, it is now possible to obtain more quantitative estimates of convergence and slip-based average earthquake recurrence rates. Thus, earlier models inferred a slip rate of 15 mm/yr and a coseismic slip of > 6 m per event, and the proposed recurrence interval for great earthquakes (M $\geq$ 8.0) would be a few hundred years (e.g., Jackson & Bilham, 1994). As more GPS data became available, the earlier estimates are being revised, also considering the various segments of the Himalaya. It must be mentioned here that the paleoseismological data from the Himalaya, which project a much longer elapsed time for the great earthquakes often do not conform to the first order approximations derived from the GPS data.

With an average rate of convergence of ~17 mm/yr, the 2000 km $\times$ 100 km, the Main Himalayan Thrust zone accumulates seismic moment sufficient for a M_w = 7.3 earthquake every year (Bilham, 2019). The revised estimation also suggests that 10 years of stress accumulation is enough to create an earthquake of magnitude $M_w = 8$, and in 100 years the stress would buildup to generate a $M_w = 8.6$ earthquake. Going by these projections, the Himalayan plate boundary is overdue for multiple great future earthquakes along some of its segments, which have undergone a prolonged period of quiescence. The past 70 years since the 1950 earthquake have been exceptionally quiet, but for the occurrence of two large earthquakes-the 2005 Kashmir (M_w 7.6) and 2012 Gorkha (Nepal) earthquakes (M_w 7.8). The seismic quiescence along some parts of the Himalaya arc has been discussed widely. Prominent among the segments is the central seismic gap in the Garhwal-Kumaun Himalaya - the segment between the ruptures of the 1905 and 1934 earthquakes (Khattri & Tyagi, 1983; Khattri, 1987). The Kashmir gap, west of the 1905 Kangra earthquake and the Assam gap between the ruptures of the 1897 and 1950 earthquakes are the other potential gaps lacking in ruptures, although some recent studies report about gap-fillers, to be discussed in the respective chapters.

The early geological reports on the 1905 and 1934 earthquakes in the northwest and the central-eastern Himalaya do not carry any mention of coseismic surface ruptures, and thus the relation of these earthquakes with their causative structures remained ambiguous (Dunn et al., 1939; Middlemiss, 1910). The apparent absence of surface ruptures from these great earthquakes was central to one of the earliest influential models on the mechanism of the great Himalayan earthquakes, as proposed by Seeber and Armbruster (1981). In this model, the 1934-type great earthquakes are assumed to be associated with blind thrusts, rooted on the MHT and their ruptures terminated near the Main Frontal Thrust (MFT) without breaking the surface. Later investigations that applied modern tools of mapping and analyses have claimed evidence for surface ruptures for some of the medieval earthquakes that are attributed to the 1505, 1803 and 1905 events (Malik et al., 2015, 2016) and the 1934 Bihar earthquake (Sapkota et al., 2013). Despite such reported discoveries of surface ruptures of the historical earthquakes, those conclusions have been critiqued later and the topic remains debated requiring more resolution of the field data (e.g., Wesnousky et al., 2018; Rajendran et al., 2018a).

Modern-day earthquakes in different structural settings of the various parts of the Himalaya demonstrate the complexity of their structural relations. For example, the

2005 Kashmir earthquake, sourced on an out-of-sequence thrust that produced a 70-km-long surface rupture (Avouac et al., 2006), makes a significant deviation from the blind-thrust and the conventional décollment models. The recently proposed model for the 1950 Assam earthquake argues for a complex rupture, involving multiple faults (Coudurier-Curveur et al., 2020). The modern central Himalayan segment earthquakes, the 1991 Uttarkashi (M_W 6.8), 1999 Chamoli (M_W 6.6), and the 2015 Gorkha (Nepal) (M_W 7.8) are believed to have occurred on the ramp-flat on the down-dip part of the Main Himalayan Thrust (see Rajendran et al., 2017 and references therein).

The segment of the Himalaya from eastern Nepal to Bhutan is noted for its strike-slip earthquakes sourced at depth ranges of 40 to 120 km, in response to the ongoing intra-slab deformation (Sunilkumar et al., 2019). The 18 September 2011 (M_W 6.9) Sikkim event ~130 km north of the main Himalayan frontal collision zone, sourced at a hypocentral depth of 45 km, is one of the recent examples of intra-slab deformation. Two earlier strike-slip earthquakes also indicate intraslab deformation. One is the November 19 (mb 6.0) earthquake near the city of Gangtok that caused minor structural damage, and was widely felt in eastern and north-eastern India, Bangladesh, Bhutan, and Nepal (Drukpa et al., 2006). The other is the August 20, 1988, Udaipur earthquake (M_W 6.8), that claimed more than 700 lives, and caused heavy damage in both Nepal and India, and was sourced at upper mantle depth, beneath the MFT (Bilham et al., 2003) (Fig. 7.4).

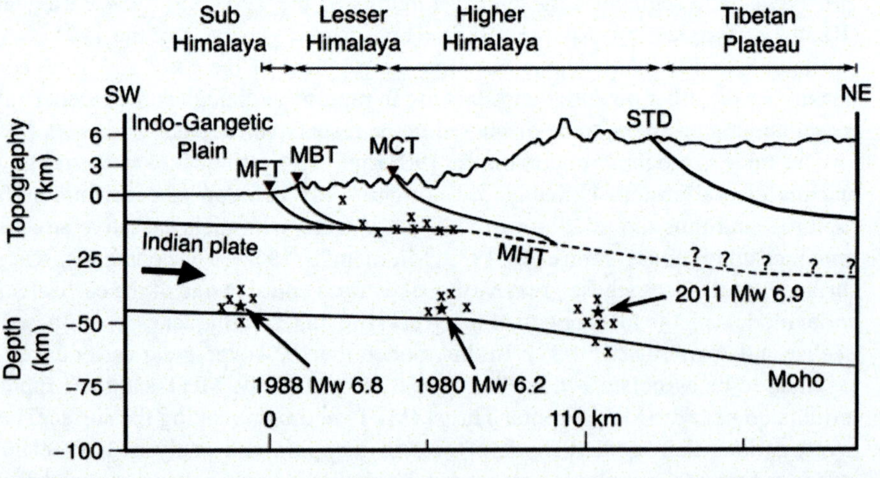

Fig. 7.4 Schematic section of the underthrusting Indian lithosphere beneath the Sikkim Himalaya (85–90° E parallel). The arrow represents the ~36 mm/year Indian plate motion relative to Eurasia. The crosses indicate notable clusters of earthquakes on the mid-crustal ramp. Hypocentral depths of the 1980 M_W 6.2, 1988 M_W 6.8, and 2011 M_W 6.9 earthquakes showing strike-slip faulting mechanisms are suggestive of intra-slab deformation (figure from Sunilkumar et al., 2019)

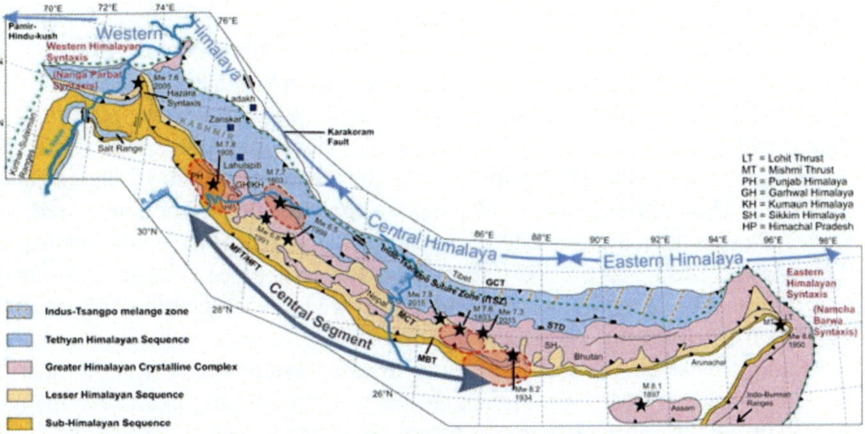

Fig. 7.5 The Himalaya plate boundary showing the lithology, structue and the western, central, and eastern segments. Major earthquakes and structural units such as the MCT, MBT, MFT/HFT, STD, and the GCT are marked. The central segment occupies the region between the 1905 Kangra and 1934 Bihar-Nepal ruptures. Geological sequence is adapted from Yin (2006)

7.3 Structural Setting

The structure of the Himalaya has been visualized as a stack of northward dipping thrust sheets, which have resulted from the progressive under-thrusting of the Indian lithosphere along the dècollement (detachment plane/MHT) (Le Fort, 1975; Zhao et al., 1993) (Figs. 7.2 and 7.5). The continuing southward movement and the sequential abandonment of the laterally continuous and spatially displaced thrust faults are considered as a defining character of the Himalayan orogeny (Gansser, 1981; Molnar, 1988; Valdiya, 1991, 1992). Structurally, the Himalayan orogen is bordered by the Indus-Tsanpo Suture Zone (ITSZ), to its north, the sinistral Chaman fault (CF) to its west, the dextral Sagaing fault (SF), to its east, and the Main Frontal Thrust (MFT), to its south. The Great Counter Thrust (GCT) is a regional north directed thrust system that has been mapped across the entire length of the Himalaya from northwest India to Eastern Tibet (e.g., Gansser, 1964).

A discussion on the structural setting of the Himalaya must include not just the convergent boundary, but also its contiguous areas. For example, the Indo-Gangetic basin that accommodates 4—5-km-thick basin-fill, and the cratonic basement that dips 2–3° toward the Himalaya is an important consequence of the orogeny (Lyon-Caen & Molnar, 1985). The northward translation of India and its anticlockwise rotation as it welded with Asia have resulted in complex structures along the collision boundaries (e.g., Molnar & Tapponnier, 1975). Prominent among them are the arc-parallel thrust systems and the arc-perpendicular transpressional fault systems along its western and eastern margins, both of which are seismically active. The western transpressional boundary defined by the Chaman transform fault accommodates 18 ± 1 mm/yr of convergence (Mohadjer et al., 2010; Tapponnier et al., 1981). It has generated four M > 7 earthquakes in the last 120 years - the 1909 (M 7.1) Kachhi (Central Balochistan) and the Quetta 1935 (M 7.7; Ambraseys and Bilham, 2003), the 1931 (M 7.3) Mach (Bolan District in Balochistan Province; Szeliga et al., 2009), and the 2013 M_w 7.7 Balochistan (southwestern Pakistan; Avouac et al., 2014). A major tectonic structure in the Northwest Himalaya, the Salt Range thrust (SRT) is an equivalent of the MFT and it marks the southern margin of the Hazara arc, reportedly absorbing at least 9–14 mm/yr of north–south shortening (Baker et al., 1988). Studies have demonstrated that at the millennial timescale, intervening thrust between the MBT and the MFT must have accommodated significant amounts of this deformation (Gavillot et al., 2016). It is within this structurally complex region that the 2005, Kashmir earthquake had occurred.

Along the eastern boundary, the Eastern Himalaya Syntaxis (EHS), regarded as the source of the great 1950 Assam earthquake, the Indian Plate abuts against the Eurasian and the Burma Plates (see Chap. 13). Approximately 50% of the resulting motion (~18 mm/year) is accommodated by strike-slip motion on the Sagaing Fault (Vigny et al., 2003). The plate boundary extends southward along the ~500-km-long Indo-Burman arc where the Indian lithosphere plunges eastwards forming a subduction zone and a fold-and-thrust belt, with west-verging thrusts (Mukhopadhyay & Dasgupta, 1988; Ni et al., 1989; Gahalaut & Gahalaut, 2007).

To continue from an earlier discussion, the present-day structure of the Himalaya is dictated by the under-thrusting of the Indian Plate beneath the high mountains along the plane of detachment (MHT; Hodges, 2016; Zhao et al., 1993). The MHT absorbs ~2 cm/yr of the shortening, which is nearly half the convergence between India and Eurasia (Bilham et al., 1997; DeMets et al., 1994; Jouanne et al., 2004). Following the critical taper model by Dahlen (1990), the progressive southward thrusting of the Eurasian plate has resulted in the formation of the three major thrust systems (MCT, MBT, and MFT). As discussed earlier, these thrusts are rooted in the MHT (Main Himalayan Thrust), the detachment plane, along which the convergence between India and Asia is taking place (Célérier et al., 2009; Yin, 2006 and references therein). The ongoing convergence has given rise to a series of thrust faults and associated structures along the Himalaya. The most visible and significant shortening structures are a series of east–west striking thrust systems that separate the great-, lesser-, and sub- Himalayan systems from one another. The highest and the oldest of these is the Main Central Thrust (MCT), which marks the contact between the greater and the lesser Himalaya (Gansser, 1964) (see Figs. 7.2 and 7.5).

Most studies describe the MCT as a broad shear zone, ranging from several hundred meters to several kilometers in thickness, with a complex deformational history. This oldest thrust structure seems to have developed during 20–23 Ma, and it has transported crystalline rocks from the Higher Himalayan zone to the top of the Paleozoic sediments. The resulting large displacements are evident from the series of half-klippen, with a cumulative slip of about 150–250 km. The MCT defines a sharp physiographic and tectonic boundary between the Himalayan foothills and the southern alluvial plain, and it is traceable from the Salt Range in Pakistan to the eastern syntaxis in Arunachal (see Fig. 7.5 for locations). The MCT joins the transpressional systems in the east and the Kirthar-Sulaiman ranges in the west (Thakur, 2013, Schelling, 1999, Hodges, 2016).

The MBT marks the contact between the lesser- and the sub- Himalaya and lies to the south of the MCT, flattening with depth and emplacing the metasedimentary rocks of the Lesser Himalaya. Traceable for larger distances than the MCT, it is believed that the MBT might have developed as early as 11–9 Ma. The total amount of thrust on the MBT is not known because of the lack of exposures of matching rocks on the hanging- and foot-wall blocks. The constitutive pre-Tertiary clastic formations, however, show signs of brittle faulting in the bedrock and occasionally some evidence of transport over the Quaternary deposits (Nakata, 1989; Valdiya, 1992). Reconstructions based on balanced cross-sections suggest the throw on the MBT to be at least several tens of kilometers. It is suggested that the lesser Himalaya are made up of a far-traveled thrust sheet, the basal thrust serving as a buried duplex (Srivastava & Mitra, 1994).

The low foothills of the sub-Himalaya, consisting of Tertiary and Quaternary clastic formations (0.5–18 Ma), are separated from the Gangetic Plains by a series of thrusts known as the Himalayan Frontal Thrust (HFT), as mentioned earlier. Also referred to as the Main Frontal Thrust (MFT), this is the southern-most and the youngest thrust that separates the sub-Himalaya from the lesser Himalaya (Lyon-Caen & Molnar, 1985; Powers et al., 1998; Valdiya, 1992; Lavé & Avouac, 2000). Actual exposures of this gentle north dipping (~30°) thrust are rare, but the few existing exposures show well-defined scarps that cut river terraces and alluvial fans (Nakata, 1989).

A set of anticlinal ridges and synclinal valleys are found to be associated with the MFT in some places, especially along the central Himalayan front. These structures owe their origin to the décollement, forcing transport of sedimentary wedges over an unyielding basement surface, an observation supported by well data. Some researchers argue that the low dip of the faults formed within the thrust belt may lead to the evolution of the surface folds and not discrete scarps. This mechanism makes the surface slips on the MFT rather elusive, as in the case of the 1905 earthquake, and it complicates the interpretation of coseismic deformation features (Yeats & Lillie, 1991).

In the overall tectonic framework, the Himalayan Fold-Thrust-Belt (FTB) expresses itself as southward verging thrust sheets (Avouac, 2003; Schelling & Arita, 1991), dominated by a large antiformal duplex located directly south of the MCT. The relatively well-constrained initiation ages for the MCT (Early Miocene) and the MBT (Late Miocene-Pliocene), as well as the initiation age of the MFT (Pliocene-Holocene) are consistent with the traditional models of fold-and-thrust-belts with progressive propagation toward the foreland (e.g., Srivastava & Mitra, 1994). The evolutionary model (e.g., Le Fort, 1975) is based on this kind of mechanism, wherein the older thrusts become inactive, as the new ones evolve. The steady-state model on the other hand, treats the MCT and the MBT to be contemporaneous and merging at depths with a common detachment surface (Seeber & Armbruster, 1981).

Well-documented historical and recent earthquakes show significant structural variability among their ruptures, as summarized by Mugnier et al. (2013), and reproduced in Fig. 7.6. Great earthquakes could nucleate in the deeper part of the brittle creeping zone and affect the whole brittle MHT and propagate all the way to the MFT as in the case of the 1934 earthquake. However, all the great earthquakes do not rupture towards the MFT as believed to be in the case of the earthquake of 1833 CE or the recent example of 2015 in Nepal. The extent of rupture seems to be controlled by the location of the brittle creeping zone with respect to the crustal ramp. Thus, the ramp might act like a large-scale asperity and the ruptures affecting the lower flat may not necessarily propagate southward along the ramp as in the case of the moderate magnitude Kumaon–Garhwal earthquake of 1991 event. For the 1999 Chamoli earthquake a different mechanism involving reactivation of faults that branch off from the brittle/ductile transition zone has been suggested. This event is considered as an example of reactivation of a blind roof thrust fault in the Lesser Himalaya duplex (Satyabala & Bilham, 2006, Rajendran et al. 2018b). Marking a difference from the examples discussed above, the 2005 Muzaffarabad (Kashmir) earthquake was slip on a steep fault splayed from the MHT, typifying an 'out-of-sequence' event.

It is important to note that while the millennial-scale-deformation dominated by folding and faulting has been modeled using geologic and geodetic data, the internal deformation of the wedge is poorly understood. Passive translation along the ramp and out-of-sequence thrusting within the wedge are suggested as potential mechanisms for internal deformation (Avouac, 2003; Wobus et al., 2005; Morell et al., 2015). The role of the mid-crustal ramp in promoting stress accumulation has also been highlighted in a study of earthquakes in Nepal Himalaya (Pandey et al.,

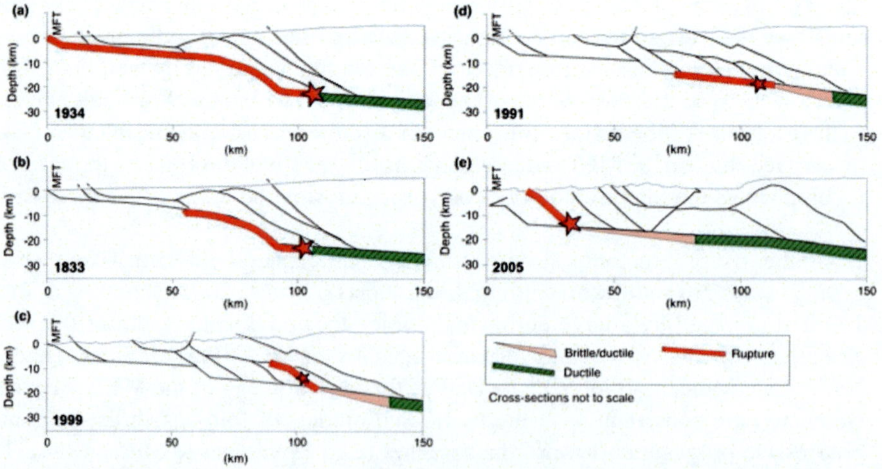

Fig. 7.6 Schematic cross-sections of the central and Kashmir Himalaya showing rupture propagations of historical and some recent earthquakes. **a** and **b** Nepal, 1934 M_W 8.2; 1833 M 7.6; **c** 1999 M_W 6.5 Chamoli and **d** 1991 M_W 6.8, Uttarkashi **e** Kashmir Himalaya, 2005 M_W 7.6 (adapted from Mugnier et al., 2013)

1995). However, the largest ramp associated with the basal detachment with the Himalayas is located north of the MCT under the Kangmar Dome and revealed in the INDEPTH project, it is assumed to be deforming aseismically (Nelson et al., 1996).

With the advances in GPS geodesy, the estimates of convergence are being revised. Approximations of India–Eurasia convergence rates and directions have been predicted since the earliest days of GPS geodesy (Paul et al., 2001; Sella et al., 2002). A more recent review suggests that the convergence rates between N10°E and N19°E vary from 336 to 43.4 mm yr^{-1} (De Mets et al., 2020 and references therein). Earlier models had predicted that the deformation within Asia reduces India's arc-normal convergence with Tibet to ~18 mm/yr (Wang et al., 2001). As also mentioned earlier, it is argued that the 18 mm/yr convergence would add up to the potential slip available to drive large earthquakes at an interval of about three centuries (Bilham, 2004). It appears that there are many segments of the Himalaya that are undergoing predicted slip deficit much more than three centuries, and are primed for future earthquakes (Bilham, 2019 and references therein). The geological studies, however, indicate that persistent seismic quiescence in terms of earthquakes could be real in some parts of the Himalaya, like the central Himalaya (Rajendran et al., 2015). Here, the temporal clustering of great earthquakes in the medieval times may have accommodated millennial convergence. Considering the long-elapsed time (600–700 years) since the last event, the repeat of this cycle is likely to be potentially imminent in the central Himalaya. The subsequent chapters discuss the major Himalayan earthquakes to provide a comprehensive review of their source mechanisms and the previous seismic history of their source regions.

B1. Imbricate Fans and Duplex structures

The architecture of thrust systems in a collisional orogeny is shaped by the interconnections and relationships between the individual faults. The fundamental element of the thrust system is the **thrust sheet**, which is a volume of rock bound below by a thrust fault. If the tip of the fault does not reach the surface, the thrust is **blind**, which occurs near the frontal margin of the thrust belt. Several individual thrusts may stack up, along frontal margins and the thrust sheets may overlap like roof tiles. They all dip in the same direction and that is what creates an **imbricate thrust system**. Faults with large displacement generally die out in a set of smaller, and usually sub-parallel **splay faults**. They branch off in the same direction out of the main, deeper décollement and form what is known as an **imbricate fan**, which spreads displacement over a large volume of rock. A **leading fan** is one in which the youngest is the front thrust and in a **trailing fan** the youngest will be at the rear (Fig. 7.7a, b).

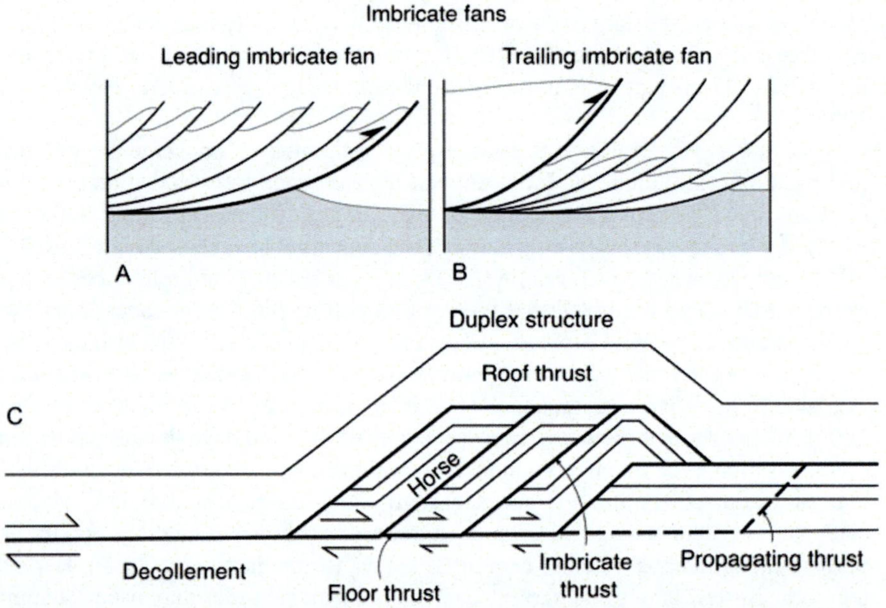

Fig. 7.7 Types of imbricate fans. **a** leading, and **b** Trailing (Boyer and Elliot, 1982). **c** Kinematic model showing the successive stages in the growth of a fault duplex (Mitra and Boyer, 2020)

A thrust duplex structure consists of a series of sub-parallel ramps that branch off a relatively flat, lower **floor thrust or the sole thrust** and merge upward into **the upper roof thrust.** The whole structure encloses a package of S-shaped, detached slices of rocks surrounded by faults on all sides called **horses,** stacked in a systematic manner (Boyer and Elliot, 1982). In the Himalayan fold-thrust belt, a significant amount of shortening is accomplished through the deformation within the Lesser Himalayan Duplex (e.g., Mitra and Boyer, 2020) (Fig. 7.7c). The presence the duplex zones with abrupt and complex changes in dip angles along the MHT is considered as a characteristic feature of crustal shortening, resulting in the physiographic transition from Lesser Himalaya into Greater Himalaya (Cattin and Avouac, 2000)

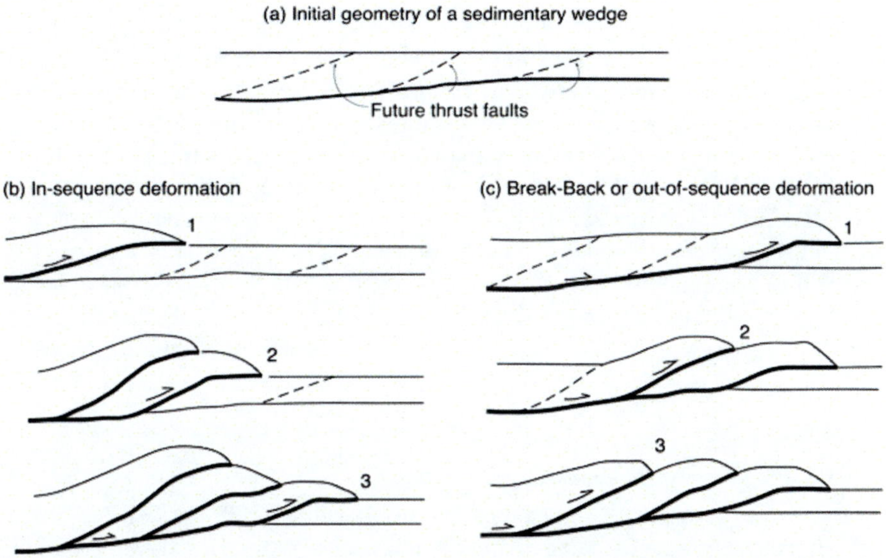

Fig. 7.8 **a** initial geometry; **b** in-sequence deformation where deformation propagates toward the foreland with time, and **c** break-back or out-of-sequence thrusts where deformation propagates toward the hinterland with time (from Moreley, 1988)

B2. Out-of-sequence thrusts

In a collisional orogen, the hinterland to foreland propagation of deformation occurs in a sequence. Thrusts commonly propagate and climb up-section in the direction of slip and in the process, bringing older rocks above the younger rocks. Younger footwall imbricates keep forming progressively and the slip along the older imbricate is transferred to the younger one as the older thrusts get deactivated and the younger, normal sequence thrusts progress toward the foreland. The thrust system evolves in this manner and such foreland directed progression sequence is accepted as the normal deformation sequence in thrust belts and are considered as **in-sequence**. Any new thrust that is part of the same deformation event, and does not belong to foreland progressing sequence, but forms towards the hinterland is said to be **out-of-sequence or** break-back thrust (Fig. 7.8).

The term 'out-of-sequence' implies that in-sequence-thrusting is the normal sequence and it applies to one random thrust, or to a series of break-back thrusts, and incorporates a wide variety of structures including folding (Moreley, 1988). In the Himalaya, deformation progressed southward from the MCT through MBT to the MFT in sequence, with all the three merging on the gently-dipping, Main Himalayan Thrust (MHT) (Yin, 2006). However, several out-of-sequence thrusts have been reported and these are seismically active (as in the case of the 2005 Kashmir earthquake); they accommodate part of the shortening produced by India–Eurasia collision (Mukul et al., 2007).

B3. Physiographic Transitions

There is a striking contrast between the surface uplift rates in the high Himalayan ranges as compared to the lower rates in the Himalayan foothills, which has given rise to a very distinct physiographic transition. Three physiographic transitions (PT1, PT2, and PT3) are mapped along the southern margin of the Tibetan Plateau. The northernmost of these, PT1 separates the high peaks of the main Himalayan ranges from the subdued topography of the Tibetan Plateau. The second is PT2, which marks the transition from the main topographic front of the central Himalaya from the foothills that occur at a lower mean elevation. A third one, the less significant front PT3 separates the foothills from the Gangetic Plain. Of these the transition, PT2 is best explained by a mid-crustal ramp in the MHT, which is most prominent in the central Nepal Himalaya and has attracted much attention as it is believed to indicate a tectonic change (Hodges et al., 2001; Fig. 7.9). While the ramp hypothesis is consistent

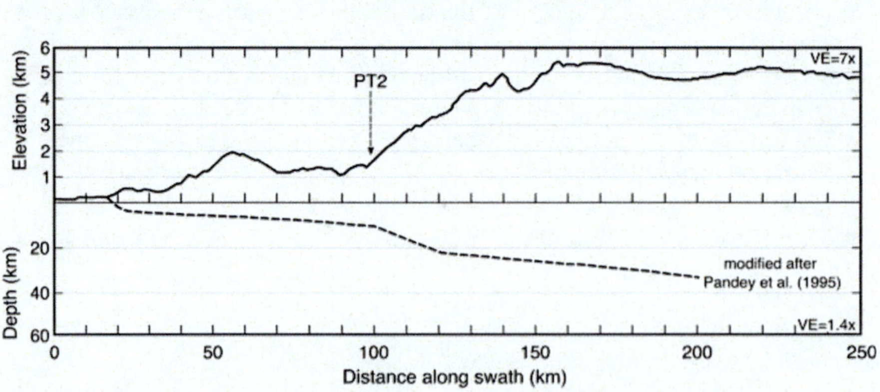

Fig. 7.9 Topographic swath profiles oriented across central Nepal showing mean elevations along the swaths. The physiographic break is marked as PT2 (adapted from Harvey et al., 2015)

with most geological and geophysical data, the sharpness, and the relatively abrupt change in surface uplift rate across the physiographic transition cannot be reconciled with a broader transition zone due to a ramp (e.g., Cattin and Avouac, 2000).

Geodetic, geomorphic, and thermo-chronologic studies confirm that a zone of rapid uplift and exhumation lies immediately north of PT2, in contrast to the relatively low rates within the Lesser Himalaya (e.g., Lavé & Avouac, 2001). The lower boundary of this transition has been delineated based on morphology, hillslope gradients, channel gradients and the extent of thick alluvial fill deposits. With its location 20 and 30 km south of the surface trace of the MCT, some authors consider it as an unmapped, out-of-sequence thrust fault (Wobus et al., 2003, 2005).

References

Ader, T., Avouac, J.-P., & Zeng, J. L., et al. (2012). Convergence rate across the Nepal Himalaya and interseismic coupling on the main Himalayan thrust: Implications for seismic hazard. *Journal of Geophysical Research, 117*, B044403. https://doi.org/10.1029/2011JB009071.

Ambraseys, N., & Jackson, D. (2003). A note on early earthquakes in northern India and southern Tibet. *Current Science, 84*, 570–582.

Avouac, J.-P. (2003). Mountain building, erosion and the seismic cycle in the Nepal Himalaya. *Advances in Geophysics, 46*, 1–80.

Avouac, J. P., Ayoub, F., & Leprince, S., et al. (2006). The 2005, Mw 7.6 Kashmir earthquake: sub-pixel correlation of ASTER images and seismic waveforms analysis. *Earth and Planetary Science Letters, 249*(3), 514–528.

Avouac, J.-P., Ayoub, F., & Wei, S., et al. (2014). The 2013, mw 7.7 Balochistan earthquake, energetic strike-slip reactivation of a thrust fault. *Earth and Planetary Science Letters, 391*, 128–134. https://doi.org/10.1016/j.epsl.2014.01.036

Baker, D. M., Lillie, R. J., & Yeats, R. S., et al. (1988). Development of the Himalayan frontal thrust zone: Salt Range. *Pakistan Geology, 16*(1), 3–7.

Bettinelli, P., Avouac, J. P., & Flouzat, M., et al. (2006). Plate motion of India and interseismic strain in the Nepal Himalaya from GPS and DORIS measurements. *Journal of Geodesy, 80*, 567–589.

Bilham, R. (2019). Himalayan earthquakes: a review of historical seismicity and early 21st century slip potential. In: P. J. Treloar, & M. P. Searle (Eds.), *Himalayan tectonics: A modern synthesis. geological society* (Vol. 483, pp. 423–482). London: Special Publications.

Bilham, R. (2004). Earthquakes in India and the Himalaya: Tectonics, geodesy and history. *Annals of Geophysics, 47*, 839–858.

Bilham, R., Bendick, R., & Wallace, K. (2003). Flexure of the Indian plate and intraplate earthquakes. *Proceedings of Indian Academy of Sciences (Earth and Planetary Science), 112*, 315–329.

Bilham, R., Larson, K., , Freymuller, & , J., Project IDYLIM members. (1997). Indo-Asian collision rates in Nepal. *Nature, 386*, 61–64.

Bilham, R., Gaur, V. K., & Molnar, P. (2001). Himalayan seismic hazard. *Science 293*, 1442–1444.

Bollinger, L., Avouac, J. P., & Beyssac, O., et al. (2004). Thermal structure and exhumation history of the Lesser Himalaya in central Nepal. *Tectonics, 23*, TC5015. https://doi.org/10.1029/2003TC001564.

Boyer, S. E., & Elliot, D. (1982). Thust systems. *American Association of Petroleum Geologists, 86*, 1196–1230.

Burg, J. P., Nievergelt, P., Oberli, F. (1988). The Namche Barwa syntaxis: Evidence for exhumation related to compressional crustal folding. *Journal of Asian Earth Sciences, 16*, 239–252.

Célérier, J., Harrison, M. T., & Webb, A. A. G (2009). The Kumaun and Garhwal lesser Himalaya, India: part 1. Structure and stratigraphy. *Geological Society of America Bulletin, 121*(9–10), 1262–1280. https://doi.org/10.1130/B26344.1.

Cattin, R., & Avouac, J.-P. (2000). Modeling mountain building and the seismic cycle in the Himalaya of Nepal. *Journal of Geophysical Research, 105*, 13389–13407. 10.1186/s40645-017-0151-8.

Clift, P. D. (2017). Cenozoic sedimentary records of climate-tectonic coupling in the Western Himalaya. *Progress in Earth and Planetary Science, 4*, 39. https://doi.org/10.1186/s40645-017-0151-8

Coudurier-Curveur, A., Tapponnier, P., & Okal, E., et al. (2020).A composite rupture model for the great 1950 Assam earthquake across the cusp of the East Himalaya syntaxis *Earth and Planet. Sci. Lett, 531*, 115928. https://doi.org/10.1016/j.epsl.2019.115928.

Dahlen, F.A (1990) Critical taper model of fold-and-thrust belts and accretionary wedges. *Annual Review of Earth and Planetary Sciences, 18*, 55–99.

Ding, L., Zhong, D. L., & Yin, A., et al. (2001). Cenozoic structural and metamorphic evolution of the eastern Himalayan syntaxis (Namche Barwa). *Earth and Planetary Science Letters, 192*, 423–438. https://doi.org/10.1016/S0012-82X(01)00463-0.

DeMets, C., Gordon, R. G., Argus, D. F., & Stein, S. (1994). Effect of recent revisions to the geomagnetic reversal time scale on estimates of current plate motions. *Geophysical Research Letters, 21*(20), 2191–2194.

DeMets, C., Merkouriev, S., & Jade, S. (2020). High-resolution reconstructions and GPS estimates of India-Eurasia and India-Somalia plate motions: 20 Ma to the present. *Geophysical Journal International, 220*(2), 1149–1171. https://doi.org/10.1093/gji/ggz508

Drukpa, D., Velasco, A. A., & Doser, D. I. (2006). Seismicity in the Kingdom of Bhutan (1937–2003): Evidence for crustal transcurrent deformation. *J Geophys Res 111*, B06301. https://doi.org/10.1029/2004JB00308

Dunn, J. A., Auden, J. B., & Ghosh, A. M. N., et al. (1939) The Bihar–Nepal Earthquake of 1934. Memoirs of the Geological Survey of India, 73. Survey of India, Calcuta (p. 391)

Gansser, A (1981) The geodynamic history of the Himalaya. Zagros Hindu Kush Himalaya Geodynamic Evolution (pp. 111–121).

Gansser, A. (1964). *Geology of the Himalayas* (p. 289p). Interscience.

Gahalaut, V. K., & Gahalaut, K. (2007). Burma plate motion. *Journal of Geophysical Research, 112*, B10402. https://doi.org/10.1029/2007JB004928

Gavillot, Y., Meigs, A., & Yule, D., et al. (2016). Shortening rate and Holocene surface rupture on the Riasi fault system in the Kashmir Himalaya: Active thrusting within the Northwest Himalayan orogenic wedge. *Bulletin, 128*(7–8), 1070–1094. https://doi.org/10.1130/B31281.1.

Harvey, J. E., Burbank, D. W., Bodo Bookhagen, B. (2015). Along-strike changes in Himalayan thrust geometry: Topographic and tectonic discontinuities in western Nepal. *Lithosphere 7*, 511–518. https://doi.org/10.1130/L444.1

Hirschmiller, J., Grujic, D., & Bookhagen, B., et al. (2014). What controls the growth of the Himalayan foreland fold-and-thrust belt? *Geology, 42*(3), 247–250.

Hodges, K. V., Hurtado, J. M., & Whipple, K. X. (2001). Southward extrusion of Tibetan crust and its effect on Himalayan tectonics. *Tectonics, 20*, 799–809. https://doi.org/10.1029/2001TC001281

Hodges, K. V. (2016). Crustal decoupling in the collisional orogenessis: Examples from the East Greenland and Caledonides and Himalaya. *Annual Review of Earth and Planetary Sciences, 44*, 685–708.

Iyengar, R. N., Sharma, D., & Siddiqui, J. M. (1999). Earthquake history of India in medieval times. *Indian Journal of History of Science, 34*, 181–237.

Jayangondaperumal, R., Daniels, R. L., & Niemi, T. M. (2017). A paleoseismic age model for large-magnitude earthquakes on fault segments of the Himalayan Frontal Thrust in the Central Seismic Gap of northern India. *Quaternary International*. https://doi.org/10.1016/j.quaint.2017.04.008.

Jackson, M., & Bilham, R. (1994). Constraints on Himalaya deformation inferred from vertical velocity fields in Nepal and Tibet. *Journal of Geophysical Research, 99*, 13897–13912.

Jouanne, F., Mugnier, J. L., Gamond, J. F., et al. (2004). Current shortening across the Himalayas of Nepal. *Geophysical Journal International, 157*(1), 1–14.

Khattri, K. N. (1987). Great earthquakes, seismicity gaps and potential for earthquakes along the Himalayan plate boundary. *Tectonophysics, 38*, 79–92.

Khattri, K. N., & Tyagi, A. K. (1983). Seismicity patterns in the Himalayan plate boundary and identification of areas of high seismic potential. *Tectonophysics, 96*, 281–297.

Kumar, S., Wesnousky, S. G., & Rockwell, T. K. et al. (2006). Paleoseismic evidence of great surface rupture earthquakes along the Indian Himalaya. *Journal of Geophysical Research, 111*. https://doi.org/10.1029/2004JB003309

Lawrence, R. D., Khan, S. H., & Dejong, K. A., et al. (1981). Thrust and strike slip fault interaction along the Chaman transform zone, Pakistan. In K. R. McClay, & N. J. Price (Eds.), *Thrust and nappe tectonics* (pp. 363–370). London: Geological Society London Special Publications 9.

Lave, J., Yule, D., & Sapkota, S. N., et al. (2005). Evidence for a Great Medieval Earthquake (~1100 A.D.) in the Central Himalayas. *Nepal Science, 307*, 1302–1305. https://doi.org/10.1126/science.1104804.

Lavé, J., & Avouac, J. P. (2000). Active folding of fluvial terraces across the Siwaliks Hills, Himalayas of central Nepal. *Journal of Geophysical Research, 105*, 5735. https://doi.org/10.1029/1999JB900292

Lavé, J., & Avouac, J. P. (2001). Fluvial incision and tectonic uplift across the Himalayas of central Nepal. *Journal of Geophysical Research, 106*(B11), 26561–26591. https://doi.org/10.1029/2001JB000359

Le Fort, P. (1975). Himalayas: The collided range. Present knowledge of the continental arc. *American Journal of Science, 275-A*, 1–44.

Lyon-Caen, H., & Molnar, P. (1985). Gravity anomalies, flexure of the Indian plate, and the structure, support and evolution of the Himalaya and the Ganga basin. *Tectonics, 4*, 513–538.

Malik, J. N., Naik, S. P., Santiswarup, S., & Okumora, K. (2016). Paleoseismic evidence of the 1505 CE (?) and 1803 CE earthquakes from the foothill zone of the Kumaun Himalaya along the Himalayan frontal thrust (HFT). *India, Tectonophysics*,. https://doi.org/10.1016/j.tecto.2016.07.026

Malik, J. N., Sahoo, S., Satuluri, S., & Okumura, K. (2015). Active Fault and Paleoseismic Studies in Kangra Valley: Evidence of Surface Rupture of a Great Himalayan 1905 Kangra Earthquake (M_w 7.8), Northwest Himalaya, India. *Bulletin of the Seismological Society of America, 105*(5), 2325–2342. https://doi.org/10.1785/0120140304.

Mandal, S. K., Scherier, D., & Wittmann, H. (2021). Tectonic accretion controls erosional cyclicity in the Himalaya. *AGU Advances*, 2e2021AV000487. https://doi.org/10.1029/2021AV000487

Middlemiss, C. S. (1910). The Kangra earthquake of 4th April 1905. *Memoirs of the Geological Survey of India, 38*, 409.

Mitra, G., Boyer, S. E. (2020). Duplexes as slip transfer zones: Application to the Himalayan fold-thrust belt. *Journal of Asian Earth Sciences, 190*, 104214. https://doi.org/10.1016/j.jseaes.2019.104214.

Molnar, P. (1988). A review of geophysical constraints on the deep structure of the Tibetan plateau, the Himalaya and the Karakoram, and their implications. *Philosophical Transactions of the Royal Society of London Series A, 326*, 33–88.

Molnar, P., & Tapponnier, P. (1975). Cenozoic tectonics of Asia: Effects of a continental collision. *Science, 189*, 419–426. https://doi.org/10.1126/science.189.4201.419

Mohadjer, S., Bendick, R., & Ischuk, A., et al. (2010). Partitioning of India-Eurasia convergence in the Pamir-Hindu Kush from GPS measurements. *Geophysical Research Letters, 37*(4). https://doi.org/10.1029/2009GL041737.

Morell, K. D., Sandiford, M., & Rajendran, C. P., et al. (2015). Geomorphology reveals active décollement geometry in the central Himalayan seismic gap. *Lithosphere, 7*(3), 247–256.

Moreley, C. K. (1988). Out-of-sequence thrusts. *Tectonics, 7*, 539–561. https://doi.org/10.1029/TC007i003p00539

Mugnier, J. L., Gajurel, A., & Huyghe, P., et al. (2013). Structural interpretation of the great earthquakes of the last millennium in the central Himalaya. *Earth-Science Reviews, 127*, 30–47.

Mukul, M., Jaiswal, M., & Sighvi, A. K. (2007). Timing of recent out-of-sequence active deformation in the frontal Himalayan wedge: insights from the Darjiling sub-Himalaya, India. *Geology, 35*, 999–1002.

Mukhopadhyay, M., & Dasgupta, S. (1988). Deep structure and tectonics of the Burmese arc: Constraints from earthquake and gravity data. *Tectonophysics, 149*, 299–322.

Nabelek, J. G., Hetényi, J., & Vergne, S., et al. (2009). Underplating in the Himalaya-Tibet collision zone revealed by the Hi-CLIMB experiment. *Science, 325*(5946), 1371–1374. https://doi.org/10.1126/science.1167719.

Nakata, T. (1989). Active faults of the Himalaya of India and Nepal. In L. L. Malinconica Jr., & R. L. Lillie (Eds.), *Tectonics of the Western Himalayas: Geological Society of America Special Paper* (Vol, 232, pp. 243–264).

Nelson, K. D., Zhao, W., & Brown, L. D., et al. (1996). Partially molten crust beneath southern Tibet: Synthesis of project INDEPTH results. *Science, 274*, 1684–1688.

Ni, J. F., Guzman-Speziale, M., & Bevis, M., et al. (1989). Accretionary tectonics of Burma and the three-dimensional geometry of the Burma subduction zone. *Geology, 17*(1), 68–71.

Pandey, M. R., Tandukar, R. P., Avouac, J. P. (1995). Interseimic strain accumulation on the Himalayan crustal ramp (Nepal). *Geophysical Research Letters, 22*, 751–754.

Paul, J., Burgmann, R., & Gaur, V. K., et al. (2001). The motion and active deformation of India. *Geophysical Research Letters, 28*, 647–650.

Powers, P. M., Lillie, R. J., & Yeats, R. S. (1998). Structure and shortening of the Kangra and Dehra Dun Reentrants, Sub-Himalaya. *India Geological Society of America Bulletin, 110*, 1010–1027.

Rajendran, C. P., Sanwal, J., & John, B., et al. (2019). Footprints of an elusive mid-14th century earthquake in the central Himalaya: Consilience of evidence from Nepal and India. *Geological Journal, 54*, 2829–2846. https://doi.org/10.1002/gj.3385.

Rajendran, C. P. John, B., & Anandasabari, K., et al. (2018a). On the paleoseismic evidence of the 1803 earthquake rupture (or lack of it) along the frontal thrust of the Kumaun*Himalaya, Tectonophysics, 722*, 227−234. ISSN 0040-1951.

Rajendran, K., Parameswaran, R. M., & Rajendran, C. P. (2018b). Revisiting the 1991 Uttarkashi and the 1999 Chamoli, India, earthquakes: Implications of rupture mechanisms in the central Himalaya. *Journal of Asian Earth Sciences, 162*, 107−120.

Rajendran, K., Parameswaran, R., & Rajendran, C. P. (2017). Seismotectonic perspectives on the Himalayan arc and contiguous areas: Inferences from past and recent earthquakes. *Earth-Science Reviews, 173*, 1–30. https://doi.org/10.1016/j.earscirev.2017.08.003.

Rajendran, C. P., John, B., & Rajendran, K. (2015). Medieval pulse of great earthquakes in the central Himalaya: viewing past activities on the frontal thrust. *Journal of Geophysical Research: Solid Earth* 1623–1641. https://doi.org/10.1002/2014JB011015.

Rajendran, C. P., Rajendran, K., Sanwal, J., & Sandiford, M. (2013). Archaeological and historical database on the medieval earthquakes of the central Himalaya: Ambiguities and inferences. *Seismological Research Letters, 87*, 1098–1108.

Rajendran, C. P., John, B., Rajendran, K., Sanwal, J. (2016). Liquefaction record of the great 1934 earthquake predecessors from the north Bihar alluvial plains of India. *Journal of Seismology, 20*, 733- 745. https://doi.org/10.1007/s10950-016-9554-z.

Sapkota, S., Bollinger, L., & Klinger, Y., et al. (2013). Primary surface ruptures of the great Himalayan earthquakes in 1934 and 1255. *Nature Geoscience, 6*, 71–76.

Satyabala, S. P. & Bilham, R. (2006). Surface deformation and subsurface slip of the 28 March 1999 $Mw = 6.4$ west Himalayan Chamoli earthquake from InSAR analysis. *Geophysical Research Letters, 33*, L23305. https://doi.org/10.1029/2006GL027422.

Schelling, D., & Arita, K. (1991). Thrust tectonics, crustal shortening and the structure of the Far-Eastern Nepal Himalaya. *Tectonics, 10*, 851–862.

Schelling, D. (1999). Frontal structural geometries and detachment tectonic of the northeastern Karachi arc, southern Kirthar Range, Pakistan. In A. Macfarlane, R. B. Sorkhabi, & J. Quade (Eds.), *Himalaya and Tibet: Mountain Roots to Mountain Tops* (Vol. 328, pp. 287–302). Geol Soc Am Special Papers.

Searle, M. P., Windley, B. F., & Coward, M. P., et al. (1987). The closing of Tethys and the tectonics of the Himalaya. *Geological Society of America Bulletin, 98*, 678–701.

Seeber, L. & Armbruster, J. (1981). Great detachment earthquakes along the Himalayan arc and long-term forecasting. In D. W. Simpson & P. G. Richards (Eds.), *Earthquake prediction-an international review. Maurice Ewing Series* (Vol. 4, pp. 259−277). Am. Geophys. Union.

Sella, G. F., Dixon, T. H., & Mao, A. (2002). REVEL: A model for Recent plate velocities from space geodesy. *Journal of Geophysical Research, 107*(B4), 2081. https://doi.org/10.1029/2000JB000033

Srivastava, P., & Mitra, G. (1994). Thrust geometries and deep structure of the outer and lesser Himalaya, Kumaon and Garhwal (India): Implications for evolution of the Himalayan fold-and-thrust belt. *Tectonics, 13*, 89–109.

Stevens, V. L., Avouac, J.-P. (2015). Interseismic coupling on the main Himalayan thrust. *Geophysical Research Letters, 42*, 5828–5837. https://doi.org/10.1002/2015GL06484

Sunilkumar, T. C., Earnest, A., Silpa, K., & Andrews, R. (2019). Rupture of the Indian slab in the 2011 Mw 6.9 Sikkim Himalaya earthquake and its tectonic implications. *Journal of Geophysical Research: Solid Earth, 124*, 2623–2637. https://doi.org/10.1111/1755-6724.13935

Szeliga, W., Bilham, R., & Schelling, D., et al. (2009). Fold and thrust partitioning in a contracting fold belt: insights from the 1931 Mach earthquake in Baluchistan. *Tectonics, 28*(5). https://doi.org/10.1029/2008TC002265.

Tapponnier, P., Mattauer, M., Proust, F., & Cassaigneau, C. (1981). Mesozoic ophiolites, sutures, and arge-scale tectonic movements in Afghanistan. *Earth and Planetary Science Letters, 52*(2), 355–371.

Thakur, V. C. (2013) Active tectonics of Himalayan frontal fault system. *Internationl Journal of Earth Sciences, 102*, 1791–1810.

Valdiya, K. S. (1991). Quaternary tectonic history of northwest Himalaya. *Current Science, 61*, 664–668.

Valdiya, K. S. (1992). The main boundary thrust zone of the Himalaya. *India Ann Tecton, 6*, 54–84.

Vigny, C., Socquet, A., & Rangin, C., et al. (2003). Present-day crustal deformation around Sagaing fault, Myanmar. *Journal of Geophysical Research, 108*(B11), 2533.

Wang, Q., Zhang, P. Z., & Freymueller, J. T., et al. (2001). Present day crustal deformation in China constrained by global positioning measurements. *Science, 294*, 574–577.

Wesnousky, S. G., Kumahara, Y., & Chamlagain, D., et al. (2017a). Geological observations on large earthquakes along the Himalayan frontal fault near Kathmandu, Nepal. *Earth and Planetary Science Letters, 457*, 366–375.

Wesnousky, S. G., Kumahara, Y., & Chamlagain, D., et al. (2017b). Large paleoearthquake timing and displacement near Damak in eastern Nepal on the Himalayan frontal thrust. *Geophysical Research Letters* https://doi.org/10.1002/2017GL074270.

Wesnousky, S. G., Kumahara, Y., & Nakata, T., et al. (2018). New observations disagree with previous interpretations of surface rupture along the Himalayan Frontal Thrust during the great 1934 Bihar-Nepal earthquake. *Geophysical Research Letters, 45*, 2652–2658.

Wobus, C., Heimsath, A., Whipple, K., & Hodges, K. (2005). Active out-of-sequence thrust faulting in the central Nepalese Himalaya. *Nature, 434*(7036), 1008–1011. https://doi.org/10.1038/nature03499.

Wobus, C. W., Hodges, K. V., & Whipple, K. X. (2003). Has focused denudation sustained active thrusting at the Himalayan topographic front? *Geology, 31*, 861–864. https://doi.org/10.1130/G19730.1.

Yeats, R. S., & Lillie, R. J. (1991). Contemporary tectonics of the Himalayan frontal fault system. Folds, blind thrusts and the 1905 Kangra earthquake. *Journal of Structural Geology, 13*, 227–233.

Yin, A. (2006). Cenozoic tectonic evolution of the Himalayan orogen as constrained by along-strike variation of structural geometry, exhumation history, and foreland sedimentation. *Earth-Science Reviews, 76*, 1–131.

Zhao, W., Nelson, K.D., et al. (1993). Deep seismic reflection evidence for continental underthrusting beneath southern Tibet. *Nature, 366*, 557–559.

Chapter 8
Uttarkashi 1803

8.1 Introduction

The devastating earthquake that struck the temple town of Uttarkashi at the midnight of September 1, 1803, is the largest historically documented from the central Himalaya. It caused massive damage and loss of life in parts of central Himalaya as well as some locations in the Indo-Gangetic (Ganga) Plains. The IMD's (India Meteorological Department) database on Indian earthquakes places this event in the central Himalaya, with some uncertainty. Some earlier catalogues located it near Mathura, in the Gangetic Plains, about 100 km south of Delhi (Oldham, 1883). This event is reported to have strongly affected Garhwal and Kumaun, two previous principalities, currently part of the Indian state of Uttarakhand. Ballore's (1904) memoir on the Indian earthquakes is the earliest publication that unequivocally attributes it to a central Himalayan source, near Srinagar (Sirmur). Most accounts mention of severe damage in the old central Himalayan towns of Uttarkashi (Barahat) and Srinagar (Fig. 8.1). This event impacted not only the central Himalaya but also some of the distant and populated historic towns like Delhi, Mathura, and Aligarh, in the Gangetic Plains. Thus, judging by the extent of the affected area, it is considered as the most damaging Central Himalayan earthquake in the last > 200 years, although lower in damage potential as compared to the great 1934 Nepal-Bihar earthquake. The event, therefore, can only be viewed as a 'nominal' template for larger events likely to occur in the central Himalaya, a zone designated as a 'seismic gap', as elaborated in Chap. 7.

The secondary effects like soil liquefaction generated in the alluvial plains (as in the case of 1905, and 1934 earthquakes) suggest a focused relay of energy that is consistent with the model of southward propagation of rupture. The 1803 earthquake was, therefore, initially considered as a blind, great décollement event, whose rupture did not reach the surface (Seeber & Armbruster, 1981). However, the newly gained understanding of the décollement geometry consisting of the ramp-flat systems supports alternate models of source mechanisms, styles of faulting and

C. P. Rajendran and K. Rajendran, *Earthquakes of the Indian Subcontinent*,
GeoPlanet: Earth and Planetary Sciences,
https://doi.org/10.1007/978-981-16-4748-2_8

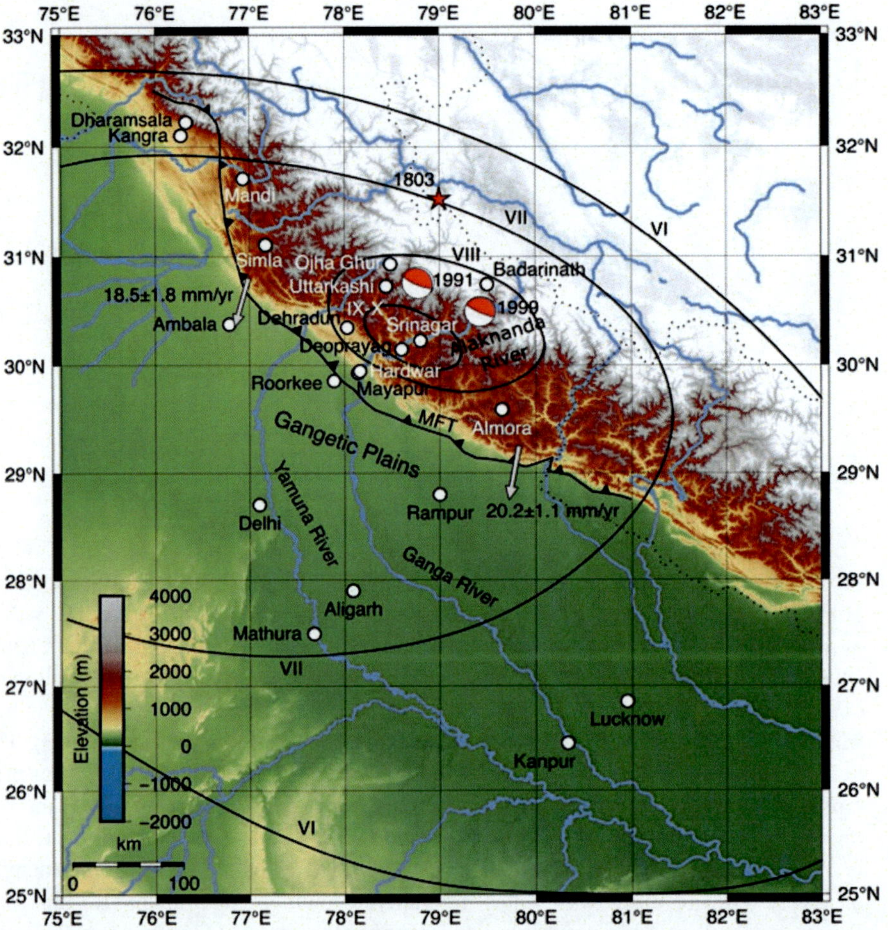

Fig. 8.1 The Garhwal-Kumaun region; inferred rupture extent of the 1803 earthquake and the epicentral location (Ambraseys & Douglas, 2004). Focal mechanisms (GCMT) of the 1991 Uttarkashi and the 1999 Chamoli earthquakes are shown above their epicenters. Convergence between India and South Tibet at a rate of 20.2 ± 1.1 mm/yr (Stevens and Avouac, 2015)

surface deformation, which might also apply to the 1803 earthquake (Avouac et al., 2006). It has been suggested that all the moderate to large earthquakes occupying the lower flat of the detachment plane (décollement) under the Himalaya do not necessarily propagate southward along the ramp to reach the Main Frontal Thrust (MFT) (Mugnier et al., 2013). To which tectonic category the 1803 earthquake might belong, and what could be its actual size, are issues that have been widely discussed. It is not easy to find answers for these questions as the event had occurred much before seismic instrumentation was in place and modern seismological principles were in vogue. Therefore, the only course of action for modern researchers to learn more about this earthquake is to rely on the historical accounts of its impact and to use geological and archeoseismological evidence to arrive at some reasonable conclusions. Observing the state of the temples in the Garhwal-Kumaun Himalaya, Rajendran et al. (2013) noted that their damage and reconstruction histories can be of great use to develop the history of large earthquakes in the region.

Various estimates of intensity, magnitude, and epicenter of the 1803 event are cited by different authors (e.g., Ambraseys & Douglas, 2004; Rajendran & Rajendran, 2005; Szeliga et al., 2010). The details derived from these sources, and how such data has been used to understand the characteristics of this earthquake will be discussed here. This chapter reviews a multitude of hypotheses on the 1803 earthquake and their usefulness in understanding its mechanism and the prognosis about earthquake hazard in the central Himalaya.

8.2 Felt Reports, Intensity, Location and Magnitude

The 1803 earthquake occurred in the pre-instrumental era, and thus, its magnitude and source characteristics have been inferred from the intensity distribution. Detailed information on its felt effects is available in the writings of Hodgson (1822), Baird-Smith (1843), and Atkinson (1882). Transcripts from the original documents have been reproduced by Dasgupta and Mukhopadhyay (2014). The accounts by F. V. Raper of the British military, who visited Garhwal during 1807–1808, is the most informative, in terms its spatial coverage. In his survey, Raper travelled through most of the villages situated along the Bhagirathi and Alaknanda Valleys, excluding the northern part of the Yamuna basin. The accounts cover the populated areas that were severely affected and provide details on the landslides and rock falls that followed the earthquake (Raper, 1810). The report by Raper carries a detailed narration of the severe destruction observed in the town of Srinagar (30.22°N: 78.78°E), situated on the left bank of Alaknanda River. Srinagar had remained the capital of the princely state of Garhwal, starting from 1519 until 1804.

Nearly one thousand houses, including the palace of the local ruler, were either badly damaged or destroyed during the earthquake. The palace, probably one among the best-made structures, suffered such serious damage that it had to be partly dismantled (Raper, 1810). The capital of the principality had to be relocated later because a flood caused by the breach of a landslide-induced dam (in 1893 or 1894) had destroyed what had remained intact after the earthquake. The Himalayan gazetteers (Atkinson, 1882) and the Imperial Gazetteer of India (Hunter et al., 1909) mention that a long sequence of aftershocks was felt in Srinagar. Hydrological responses like drying up of streams and emergence of new springs were also reported from Srinagar, possibly in response to the passage of seismic waves. The earthquake also affected the nearby town of Deoprayag, in somewhat equal measure. The town of Badarinath sustained damage conforming to VIII to IX in intensity scale. Further north, in Gangotri, the main temple was destroyed (Hodgson, 1822). The western part of the Uttarakhand Himalaya (Garhwal) suffered the severest damage. Barahat (currently known as Uttarkashi) was the most affected, killing about 300 people. The report of Baird-Smith (1843) also concurs with the intensity of damage that affected the temples and houses in Uttarkashi (see Fig. 8.1 for locations). Between Srinagar and Uttarkashi, it is likely that the former had more expanded built-environment (that too located within the river valleys) and higher density of population during the time of the earthquake.

The reports suggest that compared to the intensity of damage in the western Uttarakhand Himalaya (the Garhwal Province), the eastern part (Kumaun) was relatively free from severe impact. There were no reports of any serious damage even from its most populated town of Almora. Significantly, the temples of 10–12 century located in this region had survived the earthquake (Rajendran et al., 2013). Hodgson (1822) makes a specific mention of earthquake-related damage at Ojha Ghur, located ~20 km north of Uttarkashi, where settlements and an old fort were destroyed. Barahat and Ojha Ghur are the farthest points in the west where severe damage had occurred. The damage pattern shows many local high-intensity peaks, as observed in and around Mathura, Aligarh, and various other sites, which are in the Gangetic Plains. Some of these sites are quite far from the epicenter and the observed shaking effects could possibly be attributed to the amplification of seismic energy (see Fig. 8.1 for locations).

Those who have reviewed the felt reports on large, early 19th-century earthquakes have noted that their intensity estimates are often subject to errors, primarily because of the 'difficult-to-detect' biases in the felt reports (Ambraseys & Douglas, 2004). One common bias is because the densely populated areas tend to be over-represented. Further serious issue is that damage effects are often used without accounting for attenuation. Thus, discrepancies due to the lack of proper scrutiny are observed between various estimates as evident in the case of the 1803 earthquake. Rajendran and Rajendran (2005) used the documented felt reports to suggest that the description of the damage correspond to VIII to IX on the MSK scale and based of the descriptions the epicenter was located close to Srinagar. Ambraseys and Douglas (2004) assigned a maximum value of VIII to Srinagar, based on intensity estimates from 33 observation points and applying new scaling relations to account for attenuation, which were not incorporated in the previous models. It should, however, be considered that Srinagar is situated on the alluvial deposits of the Alaknanda River that could amplify seismic waves.

Uncertainties about the epicenter notwithstanding, most reports mention of the severe damages in the towns of Uttarkashi and Srinagar (Figs. 8.1 and 8.2) and relying on the intensity reports, the initial estimate of its epicenter was located close to Srinagar (e.g., Rajendran & Rajendran, 2005). Szeliga et al. (2010) moved the epicenter to Uttarkashi, closer to the epicenter of the 1991 earthquake. Based on reports of damage to ancient temples, Rajendran et al. (2013) also suggest that the source must be closer to Uttarkashi. This conclusion was based on the damage/reconstruction histories of the eighth-twelfth century temples in the central Himalaya, especially those which were reconstructed after the 1803 earthquake. If indeed the 1803 and 1991 earthquakes have a common source, the argument that the region has a history of recurring events by Szeliga et al. (2010) has important consequences. As with most pre-instrumentation era earthquakes, there are

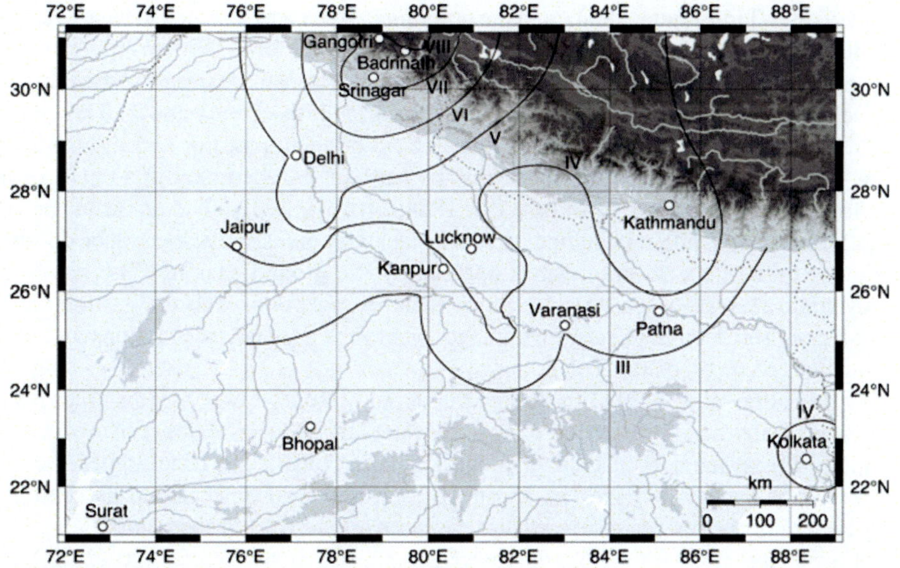

Fig. 8.2 Isoseismal map of the 1803 earthquake based on thirty-three intensity points (after Ambraseys & Douglas, 2004)

different magnitude estimates for the 1803 earthquake. Ambraseys and Jackson (2003) assigned a tentative magnitude of Ms 7.5, but subsequently, Ambraseys and Douglas (2004) and Rajendran and Rajendran (2005) assigned magnitudes close to M_w 7.5, whereas Szeliga et al. (2010) suggested a magnitude M_w 7.3.

8.3 Damage Pattern in the Gangetic Plains

The 1803 earthquake impacted the Gangetic (Ganga River) Plains rather severely in some parts, as reported by Baird-Smith (1843). For instance, it destroyed an ancient temple near Mayapur, near Hardwar. The town of Mathura experienced incidents of earthquake-induced ground features including fissuring and bubbling of water and many houses, including a principal mosque of the place was damaged (Piddington, 1804). These reports of intense ground failure might have led Oldham (1883) to conclude that the earthquake occurred near Mathura.

A similar pattern of damage was reported in Aligarh, a town, north of Mathura, where impact and the overall intensity is described to be 'violent' (ING, 1833). Minor form of damage occurred in some cities in the eastern part of the Gangetic Plains. In Varanasi (Kashi), some people lost their lives reportedly, as tiles fell over them and people might have been trapped indoors as the earthquake occurred in midnight (Athavale, 1995). Minor impact of the earthquake was felt also in the city of Calcutta

(Kolkata) (Piddington, 1804), and the outlying isoseist was reported to have extended all the way to Chittagong in the east (Oldham, 1899).

The 1803 earthquake caused loss of life and extensive damage to buildings in Delhi, the most notable being the Qutb Minar, a 13th-century tower. The cupola, "a plain square top on four stone pillars", built on the top of the 72.5 m-high tower is reported to have come down during the earthquake, according to a report of the Archaeological Survey of India (ASI, 1864; Sharma, 2001). The report of the ASI informs that the pillar, including its balustrades, balconies, and the entrance doorway at the base of the column were impacted by the ground shaking. The non-linear dynamic analysis of the structure shows that the upper two levels of the structure are the most susceptible to seismic damage, where the highest accelerations and drifts are noticed (Peña et al., 2010).

Rajendran et al. (2018) suggested that a tall masonry tower like the Qutb Minar could ideally be considered as a long-standing far-field regional recorder of ground motion from the central Himalaya earthquakes. Thus, it has been argued that any record of damage (or lack of it) on this structure can be regarded as central to developing the history of great/large earthquakes sourced in the central Himalaya. It is noted that prior to 1368, the tower was shorter by two stories (20–25 m), and its seismic response might have been considerably more robust. The topmost part of this structure consisting of a plain square top on four stone pillars toppled during the 1803 earthquake (Munshi, 1911; Rajendran et al., 2013) (Fig. 8.3).

According to Rajendran et al. (2019), this example can be considered as a template of the response of this tower to ground shaking from previous large/great earthquakes in the central Himalaya, adding a caveat that its construction history should be considered carefully. The top half of the fourth story appeared to have been reconstructed, also in marble, in 1368 CE, which suggests that the tower may have been damaged sometime between 1230 and 1368 CE. Accepting all the above caveats, there is a possibility that a great earthquake in the mid-fourteenth century (the historical candidate of 1344 CE) could have impacted this tower. This information is important from the perspective of the status of the central seismic gap. It implies that a long temporal gap of damaging earthquakes had existed in the region until the occurrence of the1803 earthquake.

8.4 Structural Setting

The geology and tectonics of the Garhwal-Kumaun Himalaya have been well studied. The Main Central Thrust (MCT) in this part of the orogeny, as defined by Heim and Gansser (1939), has been correlated across the orogen (e.g., Yin, 2006). Although widely recognized as the most important structure responsible for the development of the Himalaya, its exact placement remains debated. Multiple strands of the MCT have been traced as part of a crustal-scale south-vergent thrust fault system and as a zone of high ductile strain where the classic stratigraphic sequences of the Himalaya are exposed (Fig. 8.4). At its root zone, in the Garhwal Himalaya, the MCT is known

Fig. 8.3 View of the Qutb Minar (Delhi), showing dimensions of the different stories and their respective years of construction. The top half of the fourth storey, built in marble is believed to have been constructed in 1368CE

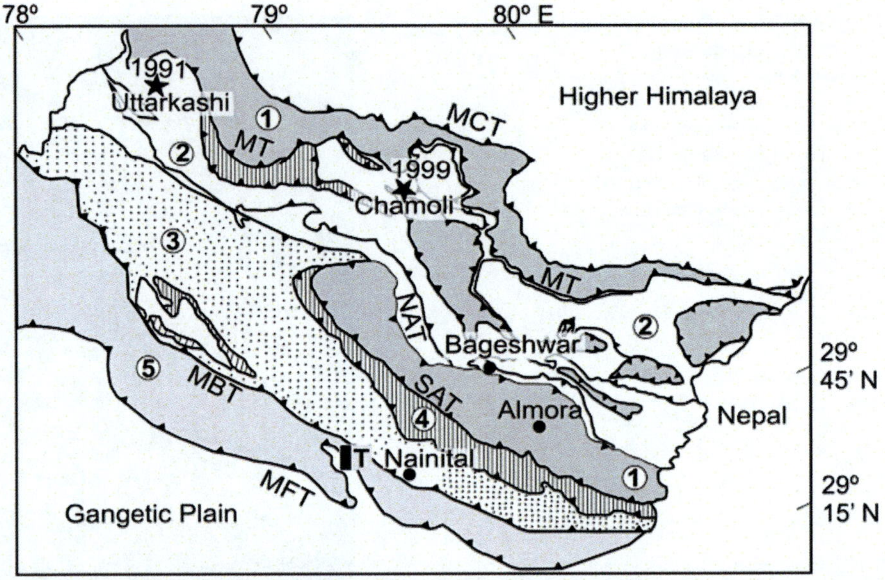

Fig. 8.4 Major structures and rock types in the Garhwal and Kumaun region. 1—outer Himalayan klippe; 2—limestone and dolomitic rocks; 3—crystallines and granites; 4—outer crystallines; 5—Siwaliks; MT: Munsiari Thrust, NAT: North Almora Thrust, SAT: South Almora Thrust. Locations of the 1992 Uttarkashi and 1999 Chamoli earthquakes and the trench (T) at Padampur are shown (modified after Valdiya, 1980)

as the Munsiari Thrust (MT), and at a higher structural position, it is named the Vaikrita thrust (Valdiya, 1980), and these are also known as MCT-I and MCT-II, respectively (Caldwell et al., 2013 and references therein). The thrust system also includes other structures, such as the Ramgarh Thrust (RT), North Almora Thrust (NAT), and the South Almora Thrust (SAT).

Structural reconstructions of the lower Garhwal Himalaya mostly interpret the Main Himalayan Thrust (MHT) as having a flat-ramp-flat geometry, which plays a significant role in the deformation of the Himalaya (e. g., Célérier et al., 2009; Srivastava & Mitra, 1994). Two competing tectonic models have been proposed to explain the present-day kinematics of the Central Himalaya. The deformation is centered on the MHT, and the rapid exhumation is a product of underplating along the MHT ramp, as proposed by Avouac (2003). This view is contested by those who favor an out-of-sequence thrusting in the MCT zone Hodges et al., 2004; Wobus et al., 2005). A physiographic transition, called NPT-2, identified along the lower–high Himalaya in the central Nepal, 10–30 km south of the MCT is suggested to be an outcome of out-of-sequence thrusting, The discontinuities in the thermo-chronologic ages across the MCT, variations in exhumation rates and differential uplift rates are regarded as the indications of an evolving out-of-sequence thrusting. This distinctive physiographic transition at the base of the high Himalaya is also traced further west in the state of Uttarakhand, India, which is characterized by abrupt strike-normal increases in channel steepness and a tenfold increase in erosion rates (Morell et al., 2015). However, these discontinuities in topography can also be attributed to overthrusting and progressive incremental growth of duplex over the crustal ramp (e.g., Robert et al., 2009).

Earlier researchers have described the mechanism as underplating at the ramp, wherein the Indian crust on the footwall of the MHT forms duplexes that are accreted onto the Himalayan wedge (Schelling & Arita, 1991). In continuation to the earlier work in the Nepal Himalaya on the physiographic transition zone, Morell et al. (2015) have identified similar feature in the Garhwal Himalaya that coincides with the Main Himalayan Thrust ramp. This zone is again characterized by distinct morphology, such as channel steepness, rock uplift, and shows spatial association with earthquakes. Although it remains unclear whether the ramp kinematics involves an emergent fault or duplex, Morell et al. (2015) aver that it could provide a coherent fault segment that may be capable of hosting a great earthquake. The question whether the 1803 earthquake was associated with the ramp kinematics becomes moot in this background. It is important to explore the role of out-of-sequence thrusts in the earthquake process because it has implications for the rupture propagation and generation of surface rupture. Based on the current understanding of the structural setting of the Garhwal Himalaya, it is also likely that the 1803 earthquake originated on the MHT, and its onward propagation was possibly arrested somewhere near the mid-part of the detachment plane (see Fig. 7.6 for the conceptual model).

The 1803 earthquake has recently been the subject of studies in the context of exploration for surface ruptures and evidence of past earthquakes. It was initially considered as a full-length de-collement earthquake (with a magnitude of > M 8.0), whose rupture did not reach the surface (Seeber & Armbruster, 1981). Later workers recalibrated its magnitude as $M_W \sim 7.5$ and suggested that it does not qualify to be a great plate boundary earthquake. However, it is inferred to have ruptured ~200 km, based on damage reports (Ambraseys & Douglas, 2004; Ambraseys & Jackson, 2003). Based on the similarity in the extent of strong ground shaking between the 1803 and 1905 Kangra earthquakes, it has been suggested that the 1803 event probably was devoid of any ground rupture (Ambraseys & Bilham, 2000).

The debate on surface faulting got a fresh start when a study by Malik et al. (2016) reported stratigraphic evidence for surface faulting on the frontal thrust during the time interval of 1750 and 1932 CE. Their study was carried out near Padampur (Dholia Village), ~10 km east of Ramnagar Town in the foothill zone of the Kumaun Himalaya on the forward part of the Main Frontal Thrust (Fig. 8.5). Two earlier faulting events were also found in the same section, dated at 467–570 and 1294–1587 CE, the latter suspected to be the 1505 earthquake (Fig. 8.4). Re-examination of the trench stratigraphy at the nearby site by Rajendran et al. (2018), however, posits a single episode of a low-angle displacement at this site. This could possibly be an event that occurred between 1266 and 1636 CE, which conforms, to the observations from neighboring sites located to the east and west of the Padampur trench. The trenches have presented evidence for a medieval great earthquake, which could either be the 1344 or 1505 CE event, both of which are historically known. While reiterating a great earthquake has indeed occurred in the medieval period, these findings also suggest

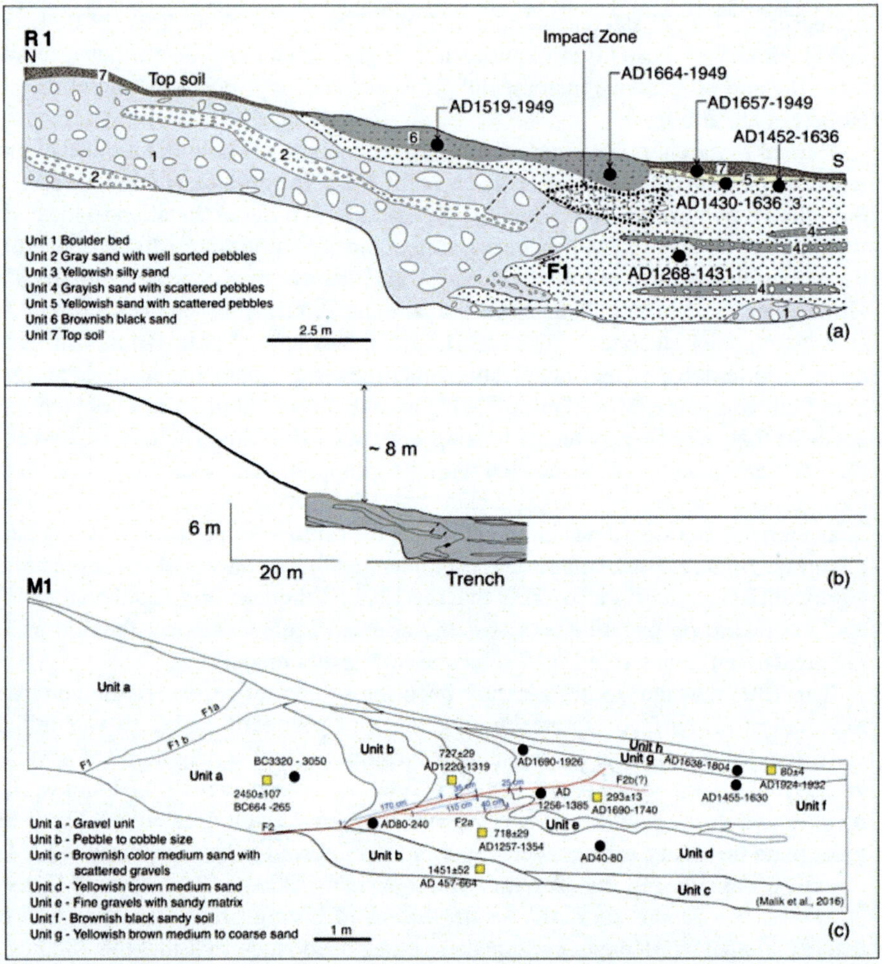

Fig. 8.5. **a** Log of the west wall of the trench R1 mapped by Rajendran et al. (2018). **b** Profile across the scarp at Padampur shown in Fig. 8.4 and the location of trench relative to the profile **c** a sketch of the section from the trench M1 by Malik et al. (2016)

that the 1803 earthquake most likely did not rupture the MFT. Further, it is inferred that like some of the other well-known décollement earthquakes (e.g., 1833 and 2015 Nepal earthquakes), the 1803 rupture might have been accommodated midway on some of the hinterland structures, falling short of reaching the Main Frontal Thrust (MFT).

8.5 Background Seismicity

Local seismic activity in the Garhwal region during recent times includes two moderate-sized earthquakes, the 1991 October 20, M_w 6.8 Uttarkashi and the 1999 March 28, M_w 6.6 Chamoli events (Fig. 8.1). This region had also not experienced any moderate or large events during the post-instrumentation period, and using the data from temporary networks, during 1984–1986, Khattri et al. (1989) had suggested the existence of a band of low/moderate earthquakes (ML < 5) sourced at about 13 km depth, spatially correlated with the general trend of the MCT. Source models of the 1991 Uttarkashi earthquake suggested thrust faulting at 15 km depth on a plane dipping ~14° oriented in the NW–SE (317°) direction. Cotton et al. (1996) had placed its source south of MCT-I, on the MHT flat. The gentle dip of their fault plane is consistent with the idea that the 1991-type earthquakes originate on the lower flat of the MHT. The effect of this earthquake was felt in New Delhi, although not as intensely as the 1803 earthquake, possibly due to its lower magnitude, small rupture area and its southwest-directed slip vector. However, the comparable directivity towards the Gangetic Plains, suggests that both these earthquakes followed similar slip geometry (Rajendran et al., 2017).

The March 28, 1999, Chamoli earthquake (M_w 6.6), located at ~50 km east of the 1991 and ~100 km southeast of the 1803 sources is one of the few modern-day moderate events in the Central Himalaya. This earthquake was studied using seismic networks established a few days after the earthquake and the aftershocks provided well-constrained depth estimates (Kayal et al., 2003). The hypocenters were located above the MHT and, from the pattern of the well-located aftershocks, Kayal et al. (2003) suggested re-activation of fault strands that branch off the MHT. The local network that operated during 2005–2008 recorded earthquakes that were clustered in the southern part of MCT, and depth-wise, above the MHT (Mahesh et al., 2012). The InSAR data (Satyabala & Bilham, 2006) point to a 15° north-dipping fault striking 300°, and a centroid depth of 13 km. A similar solution was suggested also by Xu et al. (2016). Utilizing the InSAR data, Xu et al. (2016) located the hypocenter at 15 ± 2 km at ~120 km north of the Main Frontal Thrust, on the northernmost detachment near the mid-crustal ramp of the MHT. The locking line of the MHT in the Garhwal Himalaya at about 17 km depth is nearly 120 km north of the surface trace of the MFT (Caldwell et al., 2013; Xu et al., 2016). These geometrical constraints indicate that the ruptures of both the 1991 and 1999 earthquakes originated at 15 km depth and located >100 km north of the trace of the MFT, and both seem to be associated with the flat-ramp geometry. The Coulomb stress modeling of the 1991 and 1999 earthquakes suggests lateral and up dip increase in stresses around the region and caused stressing of the MHT (Parija et al., 2021). The implication is that the coseismic Coulomb stress change along with the the regional stress field could trigger up-dip earthquakes in the central Himalaya and possibly the whole MHT during a great detachment earthquake.

8.6 Hazard Perspective on the 1803 Earthquake

The damage pattern directed onto the Gangetic Plains indicates the possibility that the 1803 earthquake might have been a detachment (décollement) event (Fig. 8.6). As suggested by some previous workers (e.g., Mugnier et al., 2013), all major earthquakes need not necessarily end up with a surface scarp, the 1833 (M_w ~ 7.7) Nepal earthquake being an example. It is argued that the extent of rupture is also controlled by the location of the crustal ramp, which acts like a structural asperity that inhibits the propagation of rupture. The April 2015 Gorkha (Nepal) (Mw 7.8) provides a well-authenticated example of a large earthquake whose rupture was arrested midway on the detachment plane, without advancing onto the Himalayan frontal thrust (Avouac et al., 2015).

Based on its postulated rupture pattern, it is possible that the 1803 earthquake may be grouped along with the 1833, and 2015 events, a suite of blind earthquakes (M_w 7.8) whose ruptures were arrested mid-way on the MHT, in the down-dip part of the seismogenic zone (Zilio et al., 2019). However, a nuanced difference between the 1833/2015 and the 1803 events could be related to their rupture propagation kinematics. In the case of the 2015 earthquake, it has been suggested that the rupture propagation was controlled by the subsurface fault geometry (Hubbard et al., 2016). There is insufficient data to develop models for the earlier events, but the possible role of structural control is somewhat evident from their felt effects. While the 1803 earthquake may have propagated southward, as implied by the isoseismals, the 2015 (Gorkha) earthquake rupture advanced in a southeastward direction and was the least damaging in the Gangetic Plains, as will be discussed in Chap. 12. The observed effects suggest that not every great or large earthquake necessarily leads to surface ruptures along the MFT.

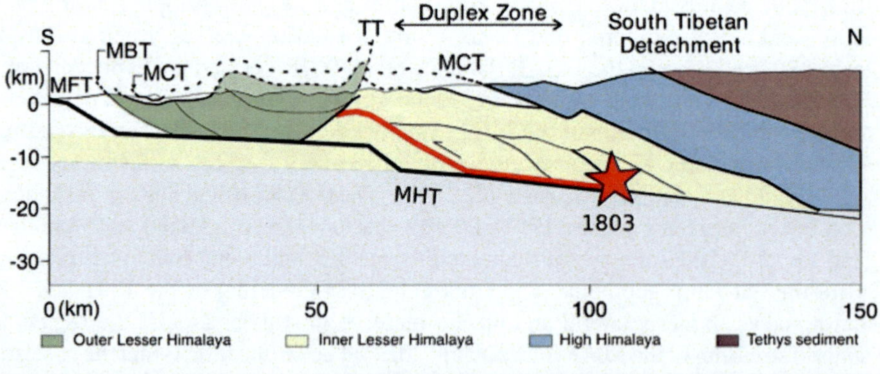

Fig. 8.6 A generalized cross section of the Garhwal-Kumaun Himalaya showing the possible southward rupture extent of the 1803 earthquake. The rupture propagates southward through the dècollement and is accommodated on the hinterland structures. The nucleation zone of the earthquake is tentative, assuming the width of the locked detachment in the Garhwal-Kumaun region not exceeding 100 km (adapted from Rajendran et al., 2018).

Analysis of the damage history of the 1803 earthquake provides many pointers for assessing the potential effects of future great/large events in the central Himalaya. An important question is whether the effects of this earthquake can be used as a template for assessing the damage from past events. The ancient capital town of Delhi with its rich heritage of monuments presents itself as an ideal candidate for testing such hypotheses. The 500-year-long history of Delhi suggests two types of events that this city has experienced; those located in its vicinity and those sourced along the plate boundary. Among the Himalayan earthquakes, it is the 1803 event that has caused the maximum damage in Delhi in the last >200 years. A much older Himalayan earthquake of greater magnitude may have impacted about 600 years ago, but there is not much physical evidence for its validation, primarily because of the sparse distribution of built structures. The ancient temples made of hard crystalline 'granitic' rocks in the Garhwal-Kumaun Himalaya can be considered as long-lived 'reflectors' of damage than the less durable brick structures of the Gangetic Plains that are amenable to complete destruction in an earthquake. Together with the examples of lesser magnitude, modern-day earthquakes sourced in the Garhwal Himalaya (e.g., 1991 Uttarkashi), the 1803 event provides some cues on the possible effects of such earthquakes in Delhi and the neighboring areas—the National Capital Region of India. In summary, the template provided by the 1803 earthquake can applied in developing seismic hazard mitigation strategies for the cities located on the Ganga alluvial plains.

References

Ambraseys, N., & Douglas, J. J. (2004). Magnitude calibration of north Indian earthquakes. *Geophysical Journal International, 159*, 165–206.

Ambraseys, N., & Jackson, D. (2003). A note on early earthquakes in northern India and southern Tibet. *Current Science, 84*, 570–582.

Ambraseys, N., & Bilham, R. (2000). A note on the Kangra Ms = 7.8 earthquake of 4 April 1905. *Current Science, 79*, 45–50.

Archaeological Survey. (1864). Report No. 111, *J. Asiatic Soc. Bengal, 33*, 59–60.

Athavale, R. N. (1995). Two undocumented historic earthquakes and additional information on the major earthquake of 1 September 1803 from north India. *Current Science, 69*, 279–280.

Atkinson, E. T. (1882). The Himalayan Gazetteer, II, Pt. 1, 2; reprint, Cosmo, Delhi.

Avouac, J.-P. (2003). Mountain building, erosion and the seismic cycle in the Nepal Himalaya. *Advances in Geophysics, 46*, 1–80.

Avouac, J. P., Ayoub, F., & Leprince, S., et al. (2006). The 2005, Mw 7.6 Kashmir earthquake: sub-pixel correlation of ASTER images and seismic waveforms analysis. *Earth and Planetary Science Letters, 249*(3), 514–528.

Avouac, J.-P., Meng, L., & Wei, S., et al. (2015). Lower edge of locked main Himalayan thrust unzipped by the 2015 Gorkha earthquake. *Nature Geoscience, 8*, 708–711. https://doi.org/10.1038/ngeo2518.

Baird-Smith, R. (1843). Memoir on Indian earthquakes. *The Journal of the Asiatic Society of Bengal, 2*(12), 1029–1056.

Ballore, M. (1904). The seismic phenomena in British India and their connection with its geology. *Memoirs of the Geological Survey of India, 3*(35), 153–194 (Calcutta).

Caldwell, W., Klemperer, S., & Lawrence, J., et al. (2013). Characterizing the main himalayan thrust in the Garhwal Himalaya, India, with receiver function CCP stacking: Earth planet. *Science Letters, 367*, 15–27.

Célérier, J., Harrison, M. T., Webb, A. A. G. (2009). The Kumaun and Garhwal lesser Himalaya, India: part 1. Structure and stratigraphy. *Bulletin of the Geological Society of America, 121*(9–10), 1262–1280. https://doi.org/10.1130/B26344.1.

Cotton, F., Campillo, M., Deschamps, A., & Rastogi, B. K. (1996). Rupture history and seismotectonics of the 1991 Uttarkashi, Himalaya earthquake. *Tectonophysics, 258*, 35–51.

Dasgupta, S., & Mukhopadhyay B. (2014). Historiography and commentary on the 16 June 1819 Kutch Earthquake, Gujarat, India, Indian. *Journal of Geosciences, 68*(1), 57–126

Heim, A. A., & Gansser, A. (1939). *Central himalaya: Geological observations of the swiss expedition, 1936: Delhi* (p. 26). Hindustan Publishing.

Hodges, K. V., Wobus, C., & Ruhl, K., et al. (2004). Quaternary deformation, river steepening, and heavy precipitation at the front of the Higher Himalayan ranges. *Earth and Planetary Science Letters, 220*, 379–389. https://doi.org/10.1016/S0012-821X(04)00063-9

Hodgson, J. A. (1822). Journal of a survey to the heads of the rivers, Ganges and Jumna. *Asiat Research, 14*, 60–152.

Hubbard, J., Almeida, R., & Foster, A., et al. (2016). Structural segmentation controlled the 2015 Mw 7.8 Gorkha earthquake rupture in Nepal. *Geology, 44*(8), 639–642.

Hunter, W. W., Cotton, J. S., Burn, R., & Meyer, W. S. (1909). *The imperial gazetteer of India* (Vol. XI, p. 518). Clarenden Press.

ING India Gazette (Calcutta). (1833). Wed. Aug. 28 1833, III, 859; Fri, Sep. 6 1833, III, 867.

Kayal, J. R., Saginaram, Singh, O. P., et al. (2003) Aftershock of the 1999 Chamoli earthquake and seismotectonic structure of the Garhwal Himalaya. *Bulletin of the Seismological Society of America, 93*, 109–117. https://doi.org/10.1785/0119990139.

Khattri, K. N., Chander, R., Gaur, V. K., & Sarkar, I. (1989). New seismological results on the tectonics of the Garhwal Himalaya. *Proceedings of the Indian Academy of Sciences. Earth and planetary sciences, 98*(1), 91–109.

Mahesh, P., Gupta, S., Rai, S. S., & Sarma, P. R. (2012). Fluid driven earthquakes in the Chamoli Region, Garhwal Himalaya: evidence from local earthquake tomography. *Geophysical Journal International, 191*, 1295–1304. https://doi.org/10.1111/j.1365-246X.2012.05672.x.

Malik, J. N., Naik, S. P., Santiswarup, S., & Okumora, K. (2016). Paleoseismic evidence of the 1505 CE (?) and 1803 CE earthquakes from the foothill zone of the Kumaun Himalaya along the Himalayan frontal thrust (HFT). *India, Tectonophysics,* 714–715. https://doi.org/10.1016/j.tecto.2016.07.026

Morell, K. D., Sandiford, M., Rajendran, C. P., et al. (2015). Geomorphology reveals active décollement geometry in the central Himalayan seismic gap. *Lithosphere, 7*(3), 247–256.

Mugnier, J. L., Gajurel, A., Huyghe, P., et al. (2013). Structural interpretation of the great earthquakes of the last millennium in the central Himalaya. *Earth-Science Reviews, 127*, 30–47.

Munshi, R. N. (1911). *The history of the Kutb Minar (Delhi)* (p. 94). Fort Printing Press.

Oldham, T. (1883). A catalogue of Indian earthquakes from the earliest to the end of 1869. *Memoirs of the Geological Survey of India, 19*(3), 1–53.

Oldham, R. D. (1899). Report of the great earthquake of 12th 1897. Memoirs of the Geological Survey of India, reprinted: 1981, Geol. Surv. India, Calcutta, p. 379.

Parija, M. P., Kumar, S., & Tiwari, V. M., et al. (2021). Coulomb Stress Modeling and Seismicity in the Western Himalaya, India since 1905: Implications for the incomplete ruptures of the Main Himalayan Thrust. *Tectonics, 4*0, e2020TC006204. https://doi.org/10.1029/2020TC006204.

Peña, F., Lourenço, P. B., Mendes, N., & Oliveira, D. V. (2010). Numerical models for the seismic assessment of an old masonry tower. *Engineering Structures, 32*, 1466–1478.

Piddington, H. (1804). Bengal occurrences for October 1803. *Asiat Ann Reg, 6*(35), 57–58.

Rajendran, C. P., Sanwal, J., & John, B., et al. (2019). Footprints of an elusive mid-14th century earthquake in the central Himalaya: Consilience of evidence from Nepal and India. *Geological Journal, 54*, 2829–2846. https://doi.org/10.1002/gj.3385.

Rajendran, C. P., John, B., & Anandasabari, K., et al. (2018). On the paleoseismic evidence of the 1803 earthquake rupture (or lack of it) along the frontal thrust of the Kumaun Himalaya. *Tectonophysics, 722*, 227–234.

Rajendran, K., Parameswaran, R., & Rajendran, C. P. (2017). Seismotectonic perspectives on the Himalayan arc and contiguous areas: Inferences from past and recent earthquakes. *Earth-Science Reviews, 173*, 1–30. https://doi.org/10.1016/j.earscirev.2017.08.003.

Rajendran, C. P., Rajendran, K., Sanwal, J., & Sandiford, M. (2013). Archaeological and historical database on the medieval earthquakes of the central Himalaya: Ambiguities and inferences. *Seismological Research Letters, 87*, 1098–1108.

Rajendran, C. P., & Rajendran, K. (2005). The status of central seismic gap: A perspective based on the spatial and temporal aspects of the large Himalayan earthquakes. *Tectonophysics, 395*, 19–39.

Raper, F. V. (1810). Narratives of a survey for the purpose of discovering the resources of the Ganges. *Asiat Research, 11*, 446–563.

Robert, X., van der Beek, P., & Braun, J., et al. (2009). Assessing Quaternary reactivation of the Main Central thrust zone (central Nepal Himalaya): New thermochronologic data and numerical modelling. *Geology, 37*, 731–734.

Satyabala, S. P., & Bilham, R. (2006). Surface deformation and subsurface slip of the 28 March 1999 $Mw = 6.4$ west Himalayan Chamoli earthquake from InSAR analysis. *Geophysical Research Letters, 33*, L23305. https://doi.org/10.1029/2006GL027422.

Schelling, D., & Arita, K. (1991). Thrust tectonics, crustal shortening and the structure of the Far-Eastern Nepal Himalaya. *Tectonics, 10*, 851–862.

Seeber, L., & Armbruster, J. (1981). Great detachment earthquakes along the Himalayan arc and long-term forecasting. In D. W. Simpson & P. G. Richards (Eds.), *Earthquake prediction-an international review* (Vol. 4, pp. 259–277). Maurice Ewing Series, Am. Geophys. Union.

Sharma, V. D. (2001) *Delhi and its neighbourhood*. New Delhi: Archaeological Survey of India (p. 161).

Srivastava, P., & Mitra, G. (1994). Thrust geometries and deep structure of the outer and lesser Himalaya, Kumaon and Garhwal (India): Implications for evolution of the Himalayan fold-and-thrust belt. *Tectonics, 13*, 89–109.

Stevens, V. L., & Avouac, J.-P. (2015). Interseismic coupling on the main Himalayan thrust. *Geophysical Research Letters, 42*, 5828–5837. https://doi.org/10.1002/2015GL06484.

Szeliga, W., Hough, S., Martin, S., & Bilham, R. (2010). Intensity, Magnitude, Location, and Attenuation in India for Felt Earthquakes since 1762. *Bulletin of the Seismological Society of America, 100*, 570–584. https://doi.org/10.1785/0120080329.

Valdiya, K. S. (1980). Geology of the Kuma^un Himalaya. Wadia Inst. Him. Geol, Dehra Dun (p. 289).

Wobus, C., Heimsath, A., Whipple, K., & Hodges, K. (2005). Active out-of-sequence thrust faulting in the central Nepalese Himalaya. *Nature, 434*(7036), 1008–1011. https://doi.org/10.1038/nature03499.

Xu, W., Bürgmann, R., & Li, Z. (2016). An improved geodetic source model for the 1999 Mw 6.3 Chamoli earthquake, India. *Geophysical Journal International, 205*(1), 236–242. https://doi.org/10.1093/gji/ggw016.

Yin, A. (2006). Cenozoic tectonic evolution of the Himalayan orogen as constrained by along-strike variation of structural geometry, exhumation history, and foreland sedimentation. *Earth-Science Reviews, 76*(1), 1–131.

Zilio, D.L., van Dinther, Y., & Gerya, T., et al. (2019). Bimodal seismicity in the Himalaya controlled by fault friction and geometry. *Nature Communications, 10*, 48. https://doi.org/10.1038/s41467-018-07874-8

Chapter 9
Kangra 1905

9.1 Introduction

The April 4, 1905, Kangra earthquake (M_w 7.8) that occurred in the early hours of the morning killed more than 20,000 people in the Kangra Valley and destroyed nearly 100,000 dwellings. This earthquake tops in terms of causalities and extent of damage among the 20th-century earthquakes in India. Most of the original information on this earthquake is collated in the reports of Charles Middlemiss (1905, 1910), of the Geological Survey of India, who travelled extensively, largely on foot, throughout the affected areas, to make the assessments of damage. The isoseismal maps presented in the 1910 report by Middlemiss identify nineteen locations of intensity VIII, nineteen of intensity IX, and eight of intensity X (Rossi-Forel). One highlight of this report, which continues to remain a topic of discussion, is the dual centers of high intensity around Kangra (X) and Dehra Dun (VIII), separated by ~200 km (Fig. 9.1).

The geometry of the causative fault of the Kangra earthquake has been variously interpreted. It has long been regarded as a pure and single thrust fault rupture that spanned the width of the plate boundary for ~300-km, implying its origin on the MHT (Seeber & Armbruster, 1981). According to Yeats and Lillie (1991), the earthquake could have propagated on a blind thrust and the deformation at shallow depth was accommodated through folding. A model by Gahalaut et al. (1994) also proposed a thrust fault, wherein the SW-edge of the inferred rupture terminated near the Jwalamukhi Thrust (JMT; Fig. 9.1b). Chander (1988) considers that the earthquake was associated with a sequential rupture involving two spatially distant segments. The work of Wallace et al. (2005), wherein the data from the Great Trigonometrical Survey of India (GTS) was combined with observations from a 2001 GPS survey, is the latest in the effort to constrain the rupture parameters of this earthquake.

Although there are no references to surface rupture in the reports of Middlemiss (1910), some recent papers argue that the Kangra earthquake was sourced on a 60-km-long, right-lateral fault called the Kangra Valley Fault (KVF) (Malik et al., 2015; Sahoo & Malik, 2017; Fig. 9.2). Multiple earthquakes are reported to have occurred

C. P. Rajendran and K. Rajendran, *Earthquakes of the Indian Subcontinent*,
GeoPlanet: Earth and Planetary Sciences,
https://doi.org/10.1007/978-981-16-4748-2_9

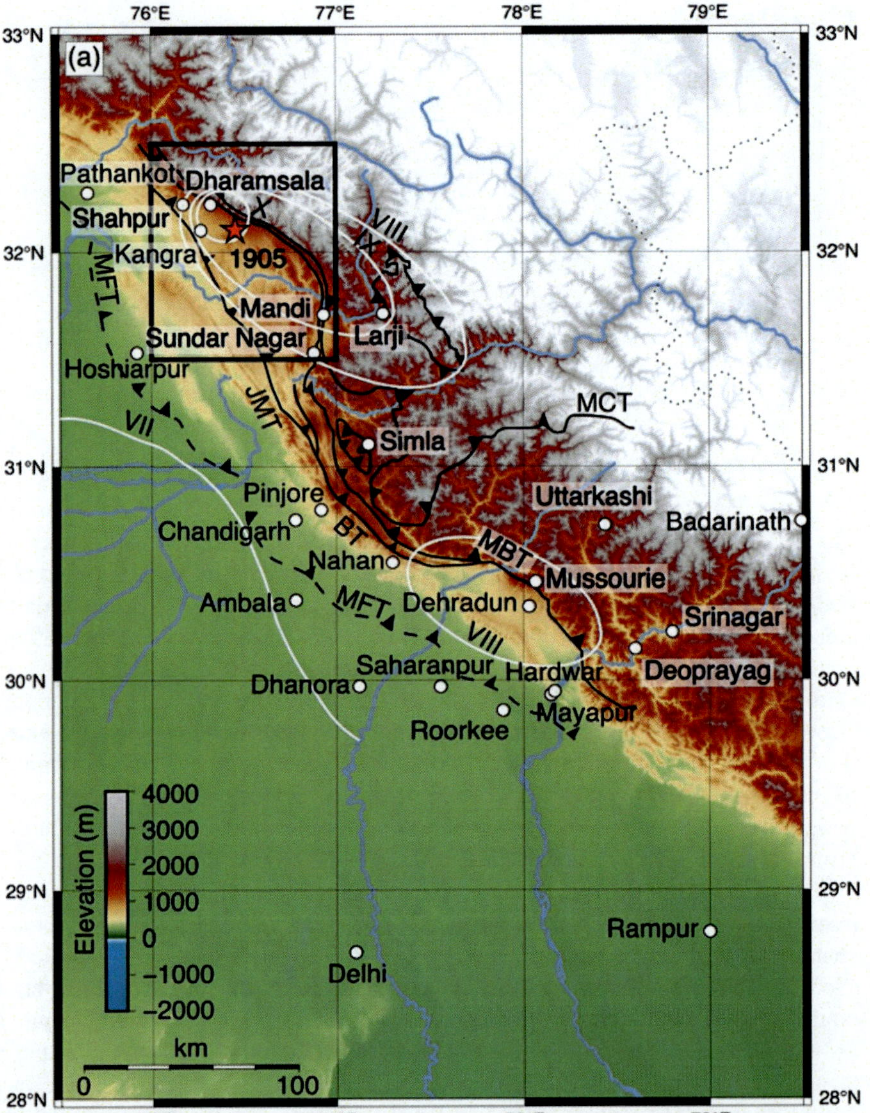

Fig. 9.1 **a** Area showing the location of the Kangra earthquake and the isoseismals on Rossi-Forel scale (after Middlemiss, 1910). Traces of the Jwalamukhi Thrust (JMT) and Bilaspur Thrust (BT) are also shown along with other structures (MCT, MBT and MFT). The epicenter is marked in the highest intensity zone. The area in the box is shown in Fig. 9.1b (modified from Singh et al., 2012)

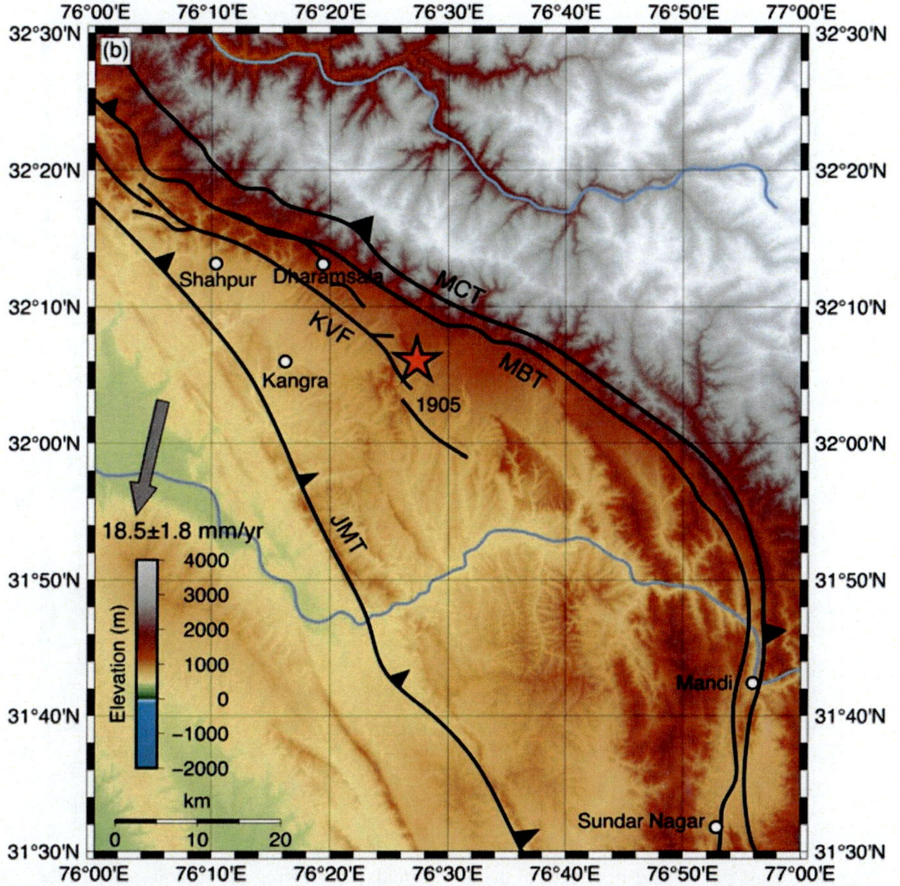

Fig. 9.1 **b** Location of the 1905 earthquake, trace of the Kangra Valley Fault (KVF) and the GPS vector from Stevens and Avouac (2015)

on the KVF since 2500 BCE, with a coseismic dextral component from the 1905 earthquake, a based on the interpretation of the sedimentary section on the fault trace. However, the geodetic strain model by Szeliga and Bilham (2017) shows that any movement identified on the KVF can only be secondary, and that if indeed it had moved in 1905, the total slip could only have been < 0.6 m, accounting for less than 2% of the total moment).

Geodetic models suggest convergence of 14 ± 1 mm/yr on the MFT in northwest Himalaya, and the inter-seismic strain on the detachment fault will ultimately be relieved by 1905- or 1934-type earthquake (Banerjee & Burgmann, 2002). But the available geological evidence is quite ambiguous on the previous earthquakes in the region. A thousand-year (1050 ± 150) recurrence interval is suggested for the KVF, assuming that it is the primary host structure (Malik et al., 2015). However, for such long interseismic intervals, Szeliga and Bilham (2017) suggests 3 to 4 times

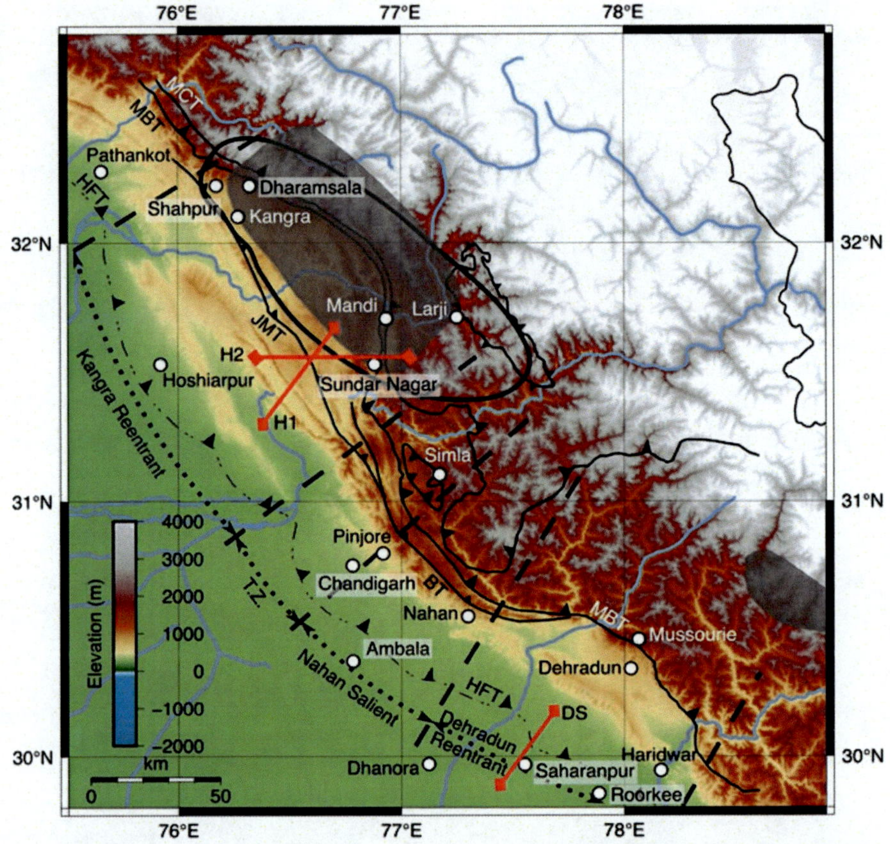

Fig. 9.2 Topography and structures in the NW Himalaya, showing the sinuous trace of the MBT giving rise to reentrants and salients (concave and convex features, respectively), with respect to the foreland. The sub-Himalayan belt is also marked by the presence of longitudinal intermontane depressions (duns better developed in the Kangra and Dehradun re-entrants). TZ marks a major transition zone. JMT: Jwalamukhi Thrust; BT: Bilaspur Thrust (adapted from Singh et al., 2012). H1, H2, and DS mark the transects as discussed in Prasad et al. (2011). The MSK VIII contour (solid line) is adapted from Ambraseys and Douglas (2004). The shaded patch shows the highest intensity zone from Middlemiss (1910)

larger slip than the highest observed for the 1905 earthquake. They also note that the strike-slip dominated KVF fault could fail independent of such earthquakes. Considering all the varied views, this chapter reviews the observations on the 1905 Kangra earthquake and its significance in the overall context of the seismicity and tectonics of the Himalaya.

9.2 Epicentral Parameters

The reports of Middlemiss (1905, 1910) contain the macroseismic location of the Kangra earthquake based on the highest intensities. The epicentral intensity distribution given by Middlemiss shows two regions, about 200 km apart, at Kangra and Dehra Dun, with maximum intensity values of X and VIII, respectively, on the Rossi–Forel (RF) scale. Concerns have been raised about the values assigned based on the initial surveys and the interpretation of the higher isoseismals (e.g., Molnar, 1987). Hough et al. (2005) re-evaluated the intensity values and observed that shaking from the mainshock rupture was overpredicted, due to site amplification within the river valleys. Further, the broad zone of high residuals found near Dehradun was attributed to a triggered aftershock.

The magnitude of the Kangra earthquake was estimated from the intensity reports, with some uncertainties. The first account written within months of the earthquake, Middlemiss had defined only the isoseismals exceeding intensity VII. The second account published five years later provides additional details about the higher intensity observations, but without any revision to the first set of contours. A review of the various magnitude estimates by Ambraseys and Bilham (2000) suggests that the surface-wave magnitude, estimated between 7.5 and 8.6, was inferred from 15 Milne-Shaw trace amplitudes. Similarly, the intensity magnitude estimates (7.5 > Mi > 7 to 8 > Mi > 7.5) by Bilham (1995) are based on calibration constants derived for continental areas such as North America and Australia, and Rossi-Forel (RF) intensities.

The earthquake was recorded by 66 recording stations around the world, based on which Szirtes (1909) estimated its epicenter at 32.10°N, 76.30°E, close to the town of Kangra. Gutenberg and Richter (1954) adopted the epicenter at 33.00°N, 76.00°E. A magnitude of 8.6 was estimated by Duda (1965), based on amplitudes of P, S, and L phases recorded by the Wiechert seismograph at Uppsala. Kanamori and Abe (1979) suggested M_w 8.2 was based on 14 maximum-trace amplitudes of surface-waves and a global calibration formula. Ambraseys and Bilham (2000) used 18 observation points and provided a surface-wave magnitude of Ms = 7.83 ± 0.18. From a reappraisal of the intensity data, they concluded that the intensity values reported by Middlemiss (1905, 1910) must have been inflated by excessive reporting from urban centers, and possibly biased, due to the limitations of the RF scale. The revised value was estimated as less than Ms 7.8. Although some modern studies continue to consider the Kangra event as a great earthquake, we follow the re-estimated magnitude of 7.8 and hence it is not regarded as a great earthquake in our discussion.

9.3 Damage Reports and Other Post-Seismic Observations

Detailed accounts of damage caused by the Kangra earthquake are available from the reports of Middlemiss (1905 and 1910). The most damaging effects were reported from an area comprising Shahpur and Dharamsala. The tract between these two locations was littered with intensely damaged houses and many had reportedly become mere heaps of bricks. With highest number of casualties. the town of Dharamsala suffered the most severe loss of lives, and the military and civil staff were reduced to about one-half by deaths. Another center of major damage was Kangra, where the devastation was reported to be total, with not a single house standing. Based on these reports, the area encompassing Dharamsala and Kangra is considered to constitute the zone of X in RF intensity scale,

Middlemiss (1910) makes some remarkable statements that impinge on the question of surface rupture, for which there is no reported evidence. The first-hand information presented by him reads like this: "… not a single railway has recorded any damage to the track, not a single road or path has been deflected, raised or lowered, no rivers or streams have changed their courses or been temporarily dammed up—except as due directly to landslips from slopes of such steepness that they might as easily have occurred after a heavy torrential rain-storm". Therefore, according to him, it is highly probable that the source of Kangra earthquake "was at some considerable depth and of comparatively simple shape,"—a conclusion based on the shape of isoseists that represents an elongated tract with the isoseismals of highest intensity distributed within the Dharamsala-Kangra-Kulu area. The freshly "reopened" fissure, marked by a vertical separation of strata (of limestone and slates) along a steeply inclined plane to the southwest, was the only feature that may vaguely be related to surficial manifestation of movement at depth, but there was no mention of any co-seismic displacements.

The distant impact of the 1905 earthquake on the Gangetic alluvial plain was manifested in the form of ground fissures and liquefaction features. Middlemiss (1905) reports about a stream bed at Ambala (located in the Punjab plains), which was previously dry, but was covered by water after the earthquake. Here the subsoil water was reported to have been forced out in five-feet-high jets, leaving little cones of sand of 1 to 0.5 feet high. From Dhanora (Karnal District, Punjab) near the Western Jhelum Canal bank, fifteen springs of water were reported to have "burst out from the earth" along a 100-feet-long stretch in the north–south direction. Several parallel fissures (the largest was 10 feet long) were reported from the bank of Solani stream, near the town of Roorkee, from which sand and water were poured out. Several crater-like hollows were also seen in this area, which most probably represent what could be termed as 'sandblows.'

Ground-level changes following the Kangra earthquake have been a subject of discussion, and the data has been used by various authors to model the rupture parameters (Bilham, 2001 and references therein). Triangulation surveys carried out between 1860 and 1880 (Walker, 1863, 1873) provide limited geodetic coverage along the northern edge of the Gangetic Plain. Fortuitously, a 1904, first-order leveling line from Saharanpur to Dehra Dun and Mussoorie was reoccupied after the earthquake. Measurements along this line revealed a change in elevation of ~15 cm for Dehradun, relative to Mussoorie, 15 km to its north. This observation and data on uplift formed the basis for several rupture models, despite reservations on the data quality (Chander, 1988, 1989; Gahalaut & Chander, 1999). Presenting the basis for sources of errors, Bilham (2001) had concluded that the elevation changes of 13.5 cm used in these analyses is not well founded and thus ruling out any co-seismic deformation near Dehradun, consistent with the magnitude estimate of Ms = 7.8.

9.4 Tectonic Framework

The epicentral area near Kangra forms a part of the foreland-fold and thrust belt within the Sub-Himalaya, the regional tectonic setting of which has been discussed by many workers (Yeats & Lillie, 1991; Powers et al., 1998; Singh et al., 2012; Cortés-Aranda et al., 2018 (Fig. 9.2)). Two important entities, "reentrants" and "salients" dominate the structural fabric of the NW Himalaya. These are the Kangra and Dehradun reentrants, where the sub-Himalaya is wider and the "Nahan Salient", where it gets narrower. Within the Nahan Salient, the Tertiary formations are present as imbricate thrusts while a series of broad synclinal basins containing alluvial deposits marks the Kangra reentrant. An analogous synclinal valley occupies the Dehra Dun area east of Kangra. In the Kangra region of the Sub-Himalaya, the MFT runs straight, in a NW–SE direction, while the surface trace of the Main Boundary Thrust (MBT) assumes a sinuous trend. Between the Kangra Re-entrant and the Nahan Salient, the lateral variations are gentle, due to the possible presence of a transition zone (Singh et al., 2012). Spatial correspondence of these topographic/tectonic features, with the dual zones of intensity maxima of the 1905 earthquake has been suggested, with a possible tectonic connection.

There are two major contractional structures that are somewhat parallel to the MBT, Jwalamukhi (JMT) and Bilaspur Thrusts (BT), associated with upper crustal tectonic processes and may represent out-of-sequence thrusts (Dey et al., 2016). During the Late Quaternary Period, thrusts located at a distance up to 50 km from the active front have accommodated out-of-sequence deformation within the external Kangra Reentrant. For example, the Jwalamukhi Thrust deforms fluvial terraces of ca. 10 ka with a mean slip rate of 7.5 mm/yr (Dey et al., 2016; Thakur et al., 2014). Long-term shortening rate of the Kangra Reentrant has been estimated to be 14 ± 2 mm/yr from balanced cross-sections (Powers et al., 1998). GPS measurements suggest a slip rate of 14 ± 1 mm/yr along the ductile part of the MHT (Banerjee & Burgmann, 2002). Consistency between these estimates makes these Late Quaternary slip rates useful proxies for long-term estimates of slip. These estimates are also supported by the documented steady convergence between India and Eurasia over the geological time scale (e.g., DeMets et al., 2017).

9.5 Geological Constraints on Active Tectonics of the Kangra Reentrant Region

Steepened longitudinal river profiles, back tilted fluvial terraces, and stratigrahic offsets within Kangra reentrant indicate vigorous Holocene activity on the Jwalamukhi Thrust (JMT) and other thrusts in the region (Dey et al., 2016). Large-scale sediment aggradation and removal observed within the valley are attributed to the depositional and erosional processes related to the glacial activity (80–20 ka). Morphological and structural data combined with cosmogenic radionuclide dating of fluvial terraces provide some constraints on the uplift rates across the JMT, which is regarded as an out-of-sequence thrust. The longitudinal profiles of the rivers across this structure exhibit sharp knickpoints, coinciding with the inferred position of the ramp, indicative of high uplift rates. The estimated shortening rates range between 5.6 ± 0.8 and 7.5 ± 1.1 mm/year since 10 ka, considered to be much higher than what was previously estimated as 3.5 to 4.2 mm/year (Thakur et al., 2014).

The current geodetically constrained shortening rate in the Kangra reentrant ranges from 13.3 ± 1.7 mm to 14 ± 2 mm/year (Banerjee & Burgmann, 2002; Stevens & Avouac, 2015). These geological constraints allow Dey et al. (2016) to conclude that the JMT accommodates ~40–60% of the total convergence across the NW Himalaya since the Holocene. Much of the Holocene shortening is equally distributed over the JMT and other sub-Himalayan thrusts, including the MFT in the Kangra reentrant. Such shortening is cited as a trigger for initiating faulting on the JMT, located within the toe of the Himalayan orogenic wedge where long-term variations in sedimentary load are pertinent. Changes in sediment load (aggradation) in the intermontane valleys or near the front of the wedges are known to enhance the taper and trigger foreland propagation of deformation to maintain the critical taper (Dahlen, 1990). This seems to be a simple first-order mechanical explanation for the propagation of the out-of-sequence thrusts, and inherent in this mechanism is the time-dependent behavior of such thrust systems.

9.5.1 Structure Inferred from Shallow Seismic Reflection Profiles

In the backdrop of the above discussion on the foreland propagation of orogenic wedges, the results of shallow seismic reflection survey across this region provide some relevant database to conceptualize the evolution of the landforms defined as recess and re-entrant (Fig. 9.2; Prasad et al., 2011). These profiles indicate that the Himalayan detachment plane (décollement) in the Kangra recess occurs at 6–8 km depth above the thin Neoproterozoic Vindhyan strata. The reflection data also indicates that the Vindhyan strata is thinner in the Kangra recess area, compared to the Nahan salient, located to its southeast (Fig. 9.3a, b). Thus, the width of the Lesser Himalayan thrust belt and the existence of the Kangra recess could be related to the

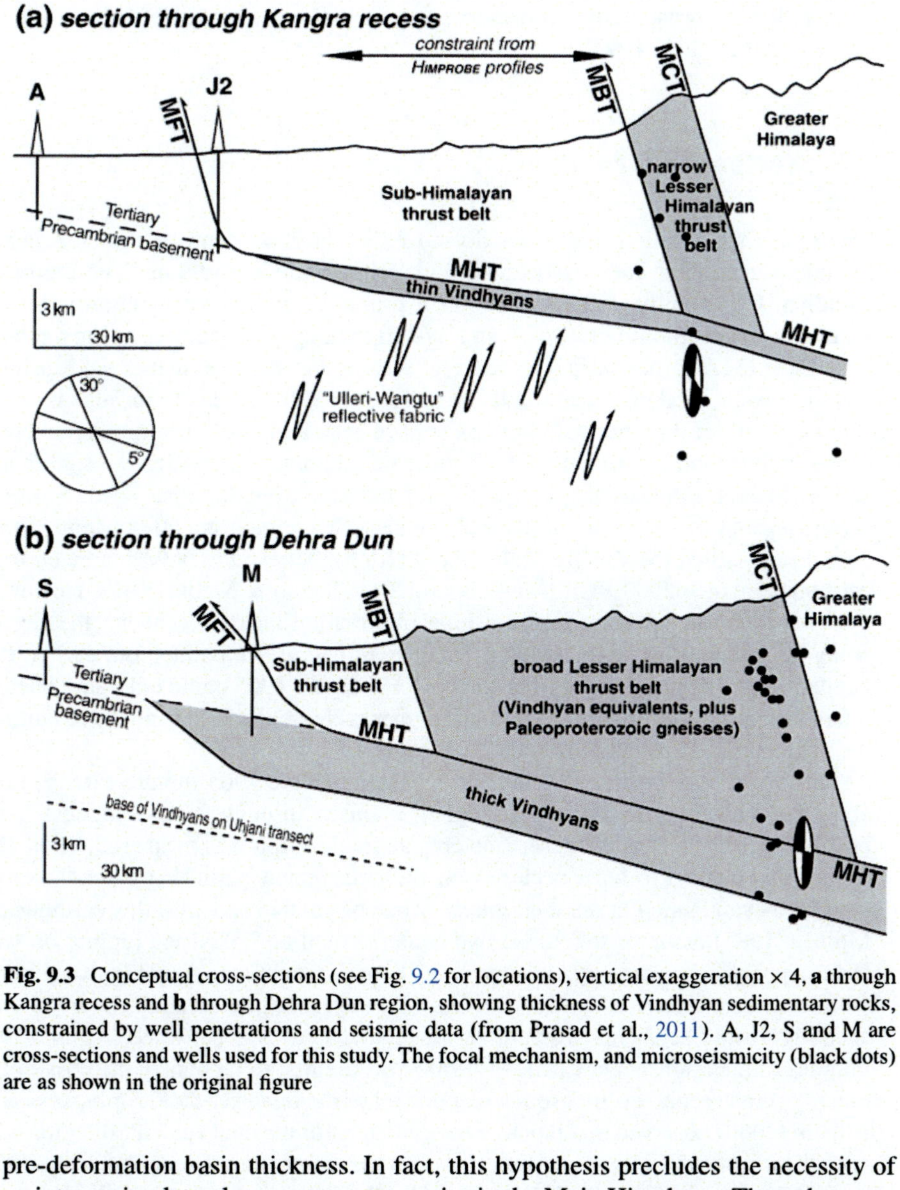

Fig. 9.3 Conceptual cross-sections (see Fig. 9.2 for locations), vertical exaggeration × 4, **a** through Kangra recess and **b** through Dehra Dun region, showing thickness of Vindhyan sedimentary rocks, constrained by well penetrations and seismic data (from Prasad et al., 2011). A, J2, S and M are cross-sections and wells used for this study. The focal mechanism, and microseismicity (black dots) are as shown in the original figure

pre-deformation basin thickness. In fact, this hypothesis precludes the necessity of an intervening lateral ramp or promontories in the Main Himalayan Thrust between Kangra recess and Nahan salient, as suggested by Powers et al. (1998). In the absence of a lateral ramp, the alternate explanation is that the widening of the thrust wedge in Kangra recess is affected by the reduced local thickness of the subducted Vindhyan strata (Proterozoic sedimentary rocks distributed in the northern part of the Indian craton). This inference is supported by some of the analog models of fault-bend-folds wherein it is shown that sediment thickness can act as a determining factor controlling the width of a thrust wedge to preserve volume balance during deformation (e.g., Macedo & Marshak, 1999).

9.6 Surface Deformation

One dominant question that is usually raised in the discussion on the Himalayan earthquakes is about their association with primary surface rupture. The report of Middlemiss (1910; page 350) rules out a primary rupture, but mentions about a secondary coseismic structure, a fresh NW–SE striking scarp, located at Barwar lake near Larji (31.712°N, 77.251°E; see Fig. 9.2 for location), which is said to be a "fresh movement along an old fault". Szeliga and Bilham (2017) surmise that the rupture length of this "normal" fault must have been less than a few hundred meters with slip, less than 2 m, and that Middlemiss might not have considered it significant enough to do an extensive mapping of this feature. However, this observation assumes significance in the backdrop of the recent papers that argue for a 60-km-long, right-lateral fault called the Kangra Valley Fault (KVF), believed to have generated four earthquakes since 2500 BCE (Malik et al., 2015; Sahoo & Malik, 2017). However, there is a contradiction between the original observation and the recent one, the former being normal faulting and the latter, being right lateral. As pointed out earlier, the geodetic data suggest that the total slip on KVF during 1905 could only have been < 0.6 m, accounting for < 2% of the total moment released, which cannot account for the primary slip (Szeliga & Bilham, 2017).

Seeber and Armbruster (1981) had suggested that the 1905 rupture extended for about 300 km, from the area of the highest intensity through the southeastern zone of high intensity where the coseismic change in elevation was measured. Thus, the rupture is believed to have occurred on the detachment plain that extends below the Himalayan front. On the dual intensity patches associated with this earthquake, Molnar (1987) assumes that those two centers could be linked via rupture on two fault segments, separated by aseismic creep between them, or a single fault segment, with less strain release around Dehra Dun. By suggesting sequential rupture of two spatially distant segments wherein an associated fault was activated within a few minutes of the mainshock, Chander (1988) was the first to recognize the possibility of a triggered second event. From a reappraisal of the intensity data, Ambraseys and Bilham (2000) observed that much of the evidence for the dual epicentral region was an artifact of how the macroseismic observations were interpreted. The idea of the second epicentral tract around Dehra Dun has also been challenged initially by many others, as discussed below.

9.6.1 Interpretations on the Two Epicentral Tracts

The spatial gap between the two intensity VIII contours has been widely discussed, most importantly whether it is an artifact of sparse damage reports from the intervening region (e.g., Molnar, 1987). These studies asserted that the observed absence of intense damage in the intervening region between the VIII contour closures was real. The re-evaluated MSK data also suggested that the intensity map by Middlemiss (1910) reflects the actual picture of damage distribution with an intervening low intensity between Kangra and Dehra Dun (Hough et al., 2005). Besides identifying a high-intensity zone, apparent evidence for a second epicentral tract near Dehra Dun was perceived from the results of a pre-1905 geodetic survey (Middlemiss, 1910). A first-order leveling line from Sharanpur to Dehra Dun, extended northward to Mussoorie in 1904, had been resurveyed immediately after the earthquake. Bilham (2001) has used these measurements to suggest an apparent elevation difference of 15 cm of the town of Dehra Dun (and the Siwalik ranges) relative to Mussoorie.

Several authors used the resurvey data as key evidence to support the idea that the 1905 rupture reached the second high intensity zone near Dehra Dun (e.g., Gahalaut & Chander, 1992, 1997). However, a re-evaluation of original raw data indicated that the leveling data reported by Middlemiss (1910) was riddled with systematic errors originating from temperature or humidity-induced rod-length changes (Bilham, 2001). These findings question the possibility of a co-seismic tectonic uplift in the Dehra Dun region. That the 1905 rupture did not extend far eastward and was limited to about 200 km is also consistent with re-evaluated of the magnitude of Ms 7.8 $\pm$ 0.2 (Ambraseys & Bilham, 2000).

The possibility of a triggered earthquake in the second high intensity zone near Dehra Dun is a topic that is being debated (Bilham, 2001). Following up on this suggestion, Hough et al. (2005) pointed out that the circularity of the residual 1905 intensity plot, located slightly west of Middlemiss' intensity VII outlier (29.0°N, 78.7°E with an uncertainty of $\pm$ 0.5°; see Fig. 9.2a) is indicative of a triggered earthquake. Although instrumental recordings are sparse, some wave arrivals related to the earthquakes do exist. These include two P-wave arrivals identified on the record from the observatory in Colaba (Bombay/Mumbai, India) in the first two minutes of the earthquake, and two S-wave arrivals separated by 7–8 min from a recording from Gottingen and Leipzig (Germany). Interestingly, the record from Leipzig indicates that the sS-S time of the second event is larger than the sS-S time of the first event, which proves that the second earthquake originated from a deeper source (Hough & Bilham, 2008; Hough et al., 2005). It is concluded that this remotely triggered earthquake could be comparable to a magnitude of $M_w \leq 7$ and occurred at about 30 km depth.

9.6.2 Mechanism of the Kangra Earthquake

As indicated earlier, the 1905 earthquake was initially considered as a plate boundary event that ruptured the MFT (Seeber & Armbruster, 1981; Ni & Barazangi, 1984). Later, this has been interpreted as a blind thrust (Yeats & Lillie, 1991). Using the data from the Great Trigonometrical Survey of India (GTS marks established between 1849 and 1852) and measurements from a 2003 GPS survey, Wallace et al. (2005) have attempted to constrain the rupture parameters. The resultant model indicates the rupture extending along a NE dipping plane that is ~100 km long, striking 155° ± 5 with a reverse-slip of 2–8 m and a sinistral slip of −1 to + 2 m. The geodetic result corresponds with the area and strike of the MSK intensity VIII as a proxy for rupture area. The SW edge of the inferred 1905 rupture corresponds to the prominent Jwalamukhi Thrust, where the rupture is believed to have terminated (Gahalaut & Chander, 1992). With a steady convergence rate of 14–20 mm/yr in this part of the Himalaya (Banerjee & Bürgmann, 2002), the Kangra earthquake occurred when a slip deficit of 7–9 m was already stored. The earthquake might have released half of or most of the accumulated slip, depending on the amount on its actual co-seismic slip.

Although no primary surface rupture was reported for the Kangra earthquake, as mentioned earlier, recent papers by Malik et al. (2015) and Sahoo and Malik (2017) argue that it was sourced on a 60-km-long, right-lateral fault called the Kangra Valley Fault (KVF). This structure reportedly hosted four earthquakes since 2500 BCE including the 1905 earthquake. Further, it is argued that the dextral slip on the 60-km-stretch on the KVF is sufficient to account for the strain release and the magnitude of the 1905 earthquake. This proposal contradicts the predominantly reverse slip fault modeled from geodetic data (Wallace et al., 2005). It may also be mentioned that the morphological indicators in the region, like lateral stream offsets, terrace elevation changes, and deformed alluvial fan surfaces, however, show a right-lateral fault shift.

The latest to have re-analyzed the geodetic data, Szeliga and Bilham (2017) obtained shear strain magnitudes and maximum contraction direction and compared the results with those predicted from dextral displacement on the KVF and thrust slip on the subsurface. The dextral slip on a 60-km-long fault (with a depth of either 10 or 15 km) lacks consistency with the actual strain release during the 1905 earthquake. The geodetic inversion allows only up to 0.6-m lateral slip on the KVF, accounting for not more than 2% of the total moment, equivalent to a $M_w \geq 6.8$. Their strain models thus suggest that any movement on the KVF is subsidiary to the primary faulting associated with the Kangra earthquake. The primary displacement is attributed to Jwalamukhi Thrust, which must have accommodated a subsurface coseismic dip slip of 1.1 ± 0.25 m. Dextral slip as little as zero or as much as 0.6 m must have occurred on the KVF along its length of ~190 km or about 100 km.

It is relevant to note here that Malik et al. (2015) used the idea of slip partitioning and the analogy of kinematic relationship between the Karakoram fault and the Kashmir Himalayan frontal arc (Kundu et al., 2014). This model asserts that the motion between India and Southern Tibet is oblique, and the slip is partitioned between the dextral motion on the Karakoram fault and the "predominantly" arc-normal motion along the frontal arc. The geodetic model by Szeliga and Bilham (2017) for the Kangra region indicates that convergence occurs only on the Jwalamukhi Thrust, whose azimuth is almost normal to the GPS vector, and therefore may not be optimal for slip partitioning as it happens on the Karakoram fault.

9.7 Some Outstanding Questions

Like many other Himalayan earthquakes, the Kangra earthquake had also its share of surprises and there are two attributes that stand out: one, it had two centers of distantly located high intensity, and two, its reported lack of surface rupture. It has long been regarded as a blind thrust event, with no surface rupture. It is likely that the rupture did not extend to the MFT, and thus created no surface rupture. It is also possible that the earthquake was sourced on the Jwalamukhi Thrust, an out-of-sequence thrust, paralleling the Main Boundary Thrust. Even so, the lack of a surface rupture is intriguing.

Our previous understanding of the causative structure needs to be revisited following the report of coseismic slip on a 60-km-long dextral fault (KVF) that traverses close to the region of highest shaking intensity. However, the geodetic slip models predict only nominal slip on this fault (KVF), considered insufficient for the size of the Kangra earthquake. Whether there is any significant slip partitioning along the KVF and if this fault has the potential to generate large/great earthquakes are issues that need further investigations.

An important question is if the strike-slip faults associated with the Himalayan thrust systems have the potential to generate large earthquakes. The recent history of earthquakes in the Himalaya does not suggest large earthquakes sourced on dextral faults, but admittedly, the interval for which reliable information exists is not long enough to rule out that possibility. It is likely that the dextral faults serve as passive partners in the genesis of great earthquakes and thus may share part of the slip, as it might have happened in the case of the Kangra earthquake. GPS strain models could possibly suggest the nature of strain partitioning on such faults and how they would affect the overall slip budget along and across the plate boundary.

The Kangra earthquake was also associated with a secondary meizoseismal area around Dehra Dun, which has been widely discussed. Hough et al. (2005) consider this as the effect of a triggered earthquake, with its hypocentral depth ranging from 30 to 50 km and postulated to have occurred on the subducting Indian plate. An analogy can be drawn with the Gorkha (Nepal) earthquake of M_w 7.8 that triggered an aftershock of M_w 7.2 at the eastern terminus of its rupture zone, but both the earthquakes had similar mechanisms consistent with the geometry of the thrust. In this context one has to also visualize how the strike-slip faulting scenario for the Kangra earthquake would work, as the second epicenter is far beyond the currently mapped extent of the KVF.

There is much uncertainty about the previous occurrences of earthquakes within the Kangra source zone, as the available geological evidence is ambiguous. Geodetic models suggest convergence of 14 ± 1 mm/yr on the MFT in northwest Himalaya and the inter-seismic strain on the detachment fault will ultimately be relieved by 1905- or 1934-type earthquakes. Paleoseismic studies deduced a 1050 ± 150-year-recurrence interval for slip on the KVF, but for such long inter-event gaps, Szeliga and Bilham (2017) suggest that the slip should be at least 3–4 times larger than the highest, estimated for the 1905 event.

Overall, the studies in the Kangra intermontane basin bring out the role of the out-of-sequence thrusts like the Jwalamukhi Thrust in accommodating part of the tectonic strain and the possibility that they are as active in terms of generating major earthquakes as the MFT. The Kangra region continues to be seismically active with spatially clustered occurrence of earthquakes located immediately northeast of the 1905 epicenter. An event with a peak magnitude of M_w 5.4 near Dharamshala in 1986 stands out in that cluster (Ram et al., 2005). All of these observations lead to a conclusion that in the absence of a lateral ramp (if that hypothesis holds up), the Kangra segment may host ruptures larger than the 1905 Kangra earthquake.

References

Ambraseys, N., & Bilham, R. (2000). A note on the Kangra Ms = 7.8 earthquake of 4 April 1905. *Current Science, 79*, 45–50.

Ambraseys, N., & Douglas, J. J. (2004). Magnitude calibration of north Indian earthquakes. *Geophysical Journal International, 159*, 165–206.

Banerjee, P., & Bu¨rgmann, R. (2002). Convergence across the northwest Himalaya from GPS measurements. *Geophysical Research Letter, 29*(13), 1652. https://doi.org/10.1029/2002GL015184.

Bilham, R. (1995). Location and magnitude of the 1833 Nepal earthquake and its relation to the rupture zones of contiguous great Himalayan earthquakes. *Current Science, 69*, 101–127.

Bilham, R. (2001). Slow tilt reversal of the Lesser Himalaya between 1862 and 1992 at 78° E, and bounds to the southeast rupture of the 1905 Kangra earthquake. *Geophysical Journal International, 144*, 713–728.

Chander, R. (1988). Interpretation of observed ground level changes due to the 1905 Kangra earthquake, northwest Himalaya. *Tectonophysics, 149*, 289–298.

Chander R. (1989). On applying the concept of rupture propagation to deduce the location of the 1905 Kangra earthquake epicenter. *Journal of Geological Society of India, 33*, 150–158.

Cortés-Aranda, J., Vassallo, R., Jomard, H., Pousse-Beltrán, L., Astudillo, L., Mugnier, J.-L., Jouanne, F., Malik, M., & Carcaillet, J. (2018). Late quaternary out-of-sequence deformation in the innermost Kangra Reentrant, NW Himalaya of India: Seismic potential appraisal from 10Be dated fluvial terraces. *Journal of Asian Earth Sciences, 158*, 140–152.

Dahlen, F. A. (1990). Critical taper model of fold-and-thrust belts and accretionary wedges. *Annual Review Earth Plant Science, 18*, 55–99.

DeMets, C., Calais, E., & Merkouriev, S. (2017). Reconciling geodetic and geological estimates of recent plate motion across the Southwest Indian Ridge. *Geophysical Journal International, 208*, 118–133.

Dey, S., Thiede, R. C., Schildgen, T. F., Wittmann, H., Bookhagen, B., Scherler, D., & Strecker, M. R. (2016). Holocene internal shortening within the northwest Sub-Himalaya: Out-of sequence faulting of the Jwalamukhi Thrust, India. *Tectonics, 35*, 1–21.

Duda, S. (1965). Secular seismic energy release in circum-Pacific belt. *Tectonophysics, 2*, 409–452.

Gahalaut, V. K., Chander, R. (1992). On the active tectonics of Dehradun region from observations of ground elevation changes. *Journal of Geological Society of India, 39*, 61–68.

Gahalaut, V. K., & Chander, R. (1997). Evidence for an earthquake cycle in the NW outer Himalaya near 78°E longitude from precision levelling operations. *Geophysical Research Letters, 24*, 225–228.

Gahalaut, V. K., & Chander, R. (1999). Geodetic evidence for accumulation of earthquake generating strains in the NW Himalaya near 75.5°E longitude. *Bulletin Seismological Society of America, 89*, 837–843.

Gahalaut, V. K., Gupta P. K., Chander, R., & Gaur, V.K. (1994). Minimum norm inversion of observed ground elevation changes for slips on the causative fault during the 1905Kangra earthquake. *Proceedings of Indian Academy Science* (Earth Planet Science), *103*, 401–411.

Gutenberg, B., & Richter, C. F. (1954). *Seismicity of earth and associated phenomenon* (2nd ed.). Princeton University Press.

Hough, S., & Bilham, R. (2008). Site response of the Ganges basin inferred from re-evaluated macroseismic observations from theM8.1 Shillong 1897, M7.8 Kangra 1905 and 1934Nepal M 8.1 earthquakes. *Journal of Earth System Science, 117*, 773–782.

Hough, S. E., Bilham, R., Ambraseys, N., & Feldl, N. (2005). Revisiting the 1897 Shillong and 1905 Kangra earthquakes in northern India: Site response, Moho reflections and a triggered earthquake. *Current Science, 2*, 1632–1638.

Kanamori, H., & Abe, K. (1979). *Journal of Geophysical Research, 84*, 6131–6139.

Kundu, B., Yadav, R. K., Bali, B. S., Chodhury, S., & Gahalaut, V. K. (2014). Oblique convergence and partitioning in the NW Himalaya: Implications from GPS measurements. *Tectonics, 33*. https://doi.org/10.1002/2014TC003633.

Macedo, J., & Marshak, S. (1999). Controls on the geometry of fold–thrust belt salients. *Geological Society of America Bulletin, 111*, 1808–1822.

Malik, J. N., Santiswarup Sahoo, S., & Satuluri, K. O. (2015). Active fault and paleoseismic studies in Kangra Valley: evidence of surface rupture of a great Himalayan 1905 Kangra Earthquake (M_W 7.8), Northwest Himalaya, India. *Bulletin of the Seismological Society of America* 2015, *105* (5), 2325–2342.

Middlemiss, C. S. (1905). Preliminary account of the Kangra earthquake of 4 April 1905. *Geological Society of India Memoirs, 32*(Pt. 4), 258–294.

Middlemiss, C. S. (1910). The Kangra earthquake of 4th April 1905. *Memoirs Geological Survey of India, 38*, 409p.

Molnar, P. (1987). The distribution of intensity associated with the 1905 Kangra earthquake and bounds on the extent of rupture. *Journal of the Geological Society of India, 29*, 221–229.

Ni, J., & Barazangi, M. (1984). Seismotectonics of the Himalayan collision zone: Geometry of the underthrusting Indian plate beneath the Himalaya. *Journal of Geophysical Research, 80*, 1142–1163.

Powers, P. M., Lillie, R. J., & Yeats, R. S. (1998). Structure and shortening of the Kangra and Dehra Dun Reentrants, Sub-Himalaya India. *Geological Society American Bulletin, 110*, 1010–1027.

Prasad, B. R., Klemperer, S. L., Vijaya Rao, V., Tewari, H. C., & Khare, P. (2011). Crustal structure beneath the Sub-Himalayan fold–thrust belt, Kangra recess, northwest India, from seismic reflection profiling: Implications for Late Paleoproterozoic orogenesis and modern earthquake hazard. *Earth and Planetary Science Letters, 308*(1–2), 218–228.

Ram, V. S., Kumar, D., & Khattri, K. N. (2005). The 1986 Dharamsala earthquake of Himachal Himalaya – estimates of source parameters, average intrinsic attenuation and site amplification functions. *Journal of Seismology, 9*, 473–485. https://doi.org/10.1007/s10950-005-1418-x

Sahoo, S., & Malik, J. N. (2017). Active fault topography along Kangra Valley fault in the epicentral zone of 1905 Mw 7.8 earthquake NW Himalaya India. *Quaternary International, 462*, 90–108.

Seeber, L., & Armbruster, J. (1981). Great detachment earthquakes along the Himalayan arc and long-term forecasting. In D. W. Simpson & P. G. Richards (Eds.), *Earthquake prediction-an international review*. Maurice Ewing Series, American Geophysical Union, 4, 259–277.

Singh, T., Awasthi, A. K., & Caputo, R. (2012). The sub-Himalayan fold-thrust belt in the 1905 Kangra earthquake zone: a critical taper model perspective for seismic hazard analysis. *Tectonics, 31* (6).

Stevens, V. L., & Avouac, J.–P. (2015). Interseismic coupling on the main Himalayan thrust. *Geophysics Research Letter, 42*, 5828–5837. https://doi.org/10.1002/2015GL06484.

Szeliga, W., & Bilham, R. (2017). New constraints on the mechanism and rupture area for the 1905 Mw 7.8 kangra earthquake, northwest himalaya: New constraints on the mechanism and rupture area for the 1905 Mw 7.8 kangra earthquake. *Bulletin of the Seismological Society of America, 107*, 2467–2479.

Szirtes, S. (1909). Publ. Bur. Centr. Assoc. Int. Seismol. Ser. A., Strassburg, 13–14.

Thakur, V. C., Joshi, M., Sahoo, D., Suresh, N., Jayangondapermal, R., & Singh, A. (2014). Partitioning of convergence in Northwest Sub-Himalaya: Estimation of late quaternary uplift and convergence rates across the Kangra reentrant North India. *International Journal of Earth Science, 103*, 1037–1056.

Walker, J. T. (1873). *Account of the operations of the great trigonometrical survey of India, Principal Triangulation III, 1879*. Survey of India.

Walker, J. T. (1863). Tables of Heights in Sind, the Punjab, NW Provinces and Central India Determined by the Great Trigonometrical Survey of India Trigonometrically and by Spirit Levelling Operations to May 1862, Ambala to Dehra Dun (pp. 112–113). Public Works Department Press, Calcutta.

Wallace, K., Bilham, R., Blume, F., Gaur, V. K, & Gahalaut, V. (2005). Surface deformation in theregion of the 1905 Kangra Mw = 7.8 earthquake in the period1846–2001. *Geophysical Research Letter, 32*, L15307.

Yeats, R. S., & Lillie, R. J. (1991). Contemporary tectonics of the Himalayan frontal fault system. Folds, blind thrusts and the 1905 Kangra earthquake. *Journal of Structural Geology, 13*, 227–233.

Chapter 10
Kashmir 2005

10.1 Introduction

The M_w 7.6 Kashmir (Muzaffarabad) earthquake struck northern Pakistan and adjacent India on 8 October 2005, with its epicenter located ~19 km northeast of Muzaffarabad and 95 km north–northeast of Islamabad, Pakistan (Fig. 10.1). This shallow focus earthquake (~10 km; Avouac et al., 2006) impacted areas on both sides of the Line of Control (LoC) between India and Pakistan. Centered near the city of Muzaffarabad, it also impacted nearby Balakot in Khyber Pakhtunkhwa and some areas of Jammu and Kashmir. Deadliest to strike South Asia since the 1935 Quetta earthquake, the 2005 event led to an estimated death toll of 87,350. About 138,000 people were injured, and over 3.5 million were rendered homeless (Hussain et al., 2009). The earthquake also triggered thousands of landslides in an area of >7500 km^2, adding to the fatalities, destroying roads, and disrupting communications (Owen et al., 2008). In the Jhelum Valley, south of Hattian (Fig. 10.1), the earthquake activated a huge rock avalanche called the Hattian Bala slide (Schneider, 2009). The slide created a natural dam blocking the waterways of the tributaries of the Jhelum River and destroyed the village. Aside from the magnitude of the earthquake and its shallow focus, the widespread occurrences of landslides were attributed to the geology of the area, climatologic and geomorphologic conditions, mudflows, widening of the roads without stability assessment, and heavy rainfall that followed the earthquake (Mahmood et al., 2015). Considering the level of damage and destruction, the Kashmir earthquake is among the most devastating event to have occurred along the Himalayan arc during the documented history. Although not matching with the size of the 1905 (Kangra, M_w 7.8) or the 1934 (Bihar-Nepal, M_w 8.2) events, this earthquake surpassed them both, in its damage potential.

The meizoseismal area of the earthquake is close to a northwest-rending belt of high microseismicity, called the Indus-Kohistan Seismic Zone (IKSZ). The earthquake had nucleated in this tectonically active zone that was identified first by

C. P. Rajendran and K. Rajendran, *Earthquakes of the Indian Subcontinent*,
GeoPlanet: Earth and Planetary Sciences,
https://doi.org/10.1007/978-981-16-4748-2_10

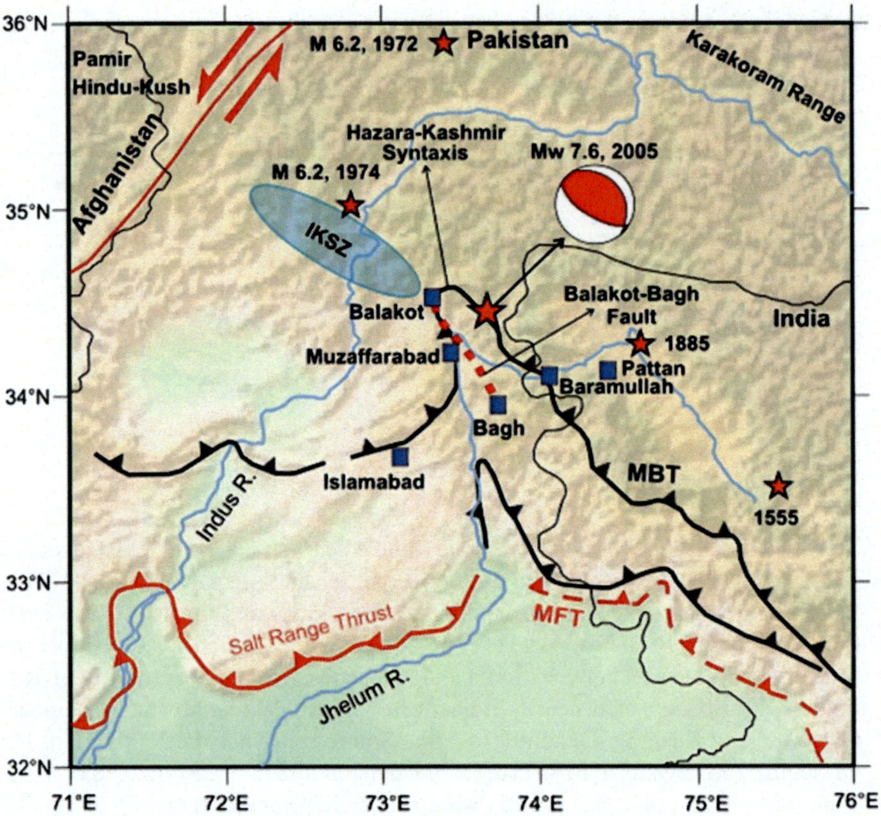

Fig. 10.1 Structural map of the Kashmir Himalaya showing the location and focal mechanism of the 2005 M_W 7.6 Kashmir earthquake (Harvard CMT), Indus Kohistan Seismic Zone (IKSZ) and the trend of the Balakot-Bagh Fault (BBF, Kaneda et al., 2008) (after Rajendran et al., 2017)

Armbruster et al. (1978). This is also the first modern-day event to have generated a significant coseismic surface rupture. With its source on a sub-Himalayan out-of-sequence thrust, the earthquake generated a 70-km-long rupture and vertical offset of about 7 m (Avouac et al., 2006; Chini et al., 2011; Kaneda et al., 2008; Pathier et al., 2006). The damage is reportedly severe along the rupture and was concentrated on the hanging wall side. The earthquake was followed by a series of aftershocks, some of which were damaging. Nearly 500 aftershocks that occurred between 8 October 2005 and 20 March 2006 formed a tight cluster between Balakot and Muzaffarabad (Rehman et al., 2016).

Sourced near the western syntaxial bent (the Hazara-Kashmir Syntaxis) along the Balakot-Bagh fault, known also as the Muzaffarabad fault (Fig. 10.1), the Kashmir event stands apart from the earthquakes that originate on the Main Himalayan Thrust (MHT). First, it was not hosted by any major thrust systems, but by an independent fault rooted on the MBT. Second, it is the first modern-day large Himalayan earthquake to occur on an out-of-sequence thrust, which generated remarkable coseismic surface rupture and fault offsets. It was also a demonstration that such shallow, out-of-sequence thrust faults can cause immense damage, surpassing even that of some of the previous great earthquakes. The 2005 earthquake is one of few modern-day Himalaya earthquakes whose coseismic deformation has been studied extensively using geodetic, seismologic (e.g., Avouac et al., 2006; Parsons et al., 2006); and geologic tools (Kondo et al., 2008). This chapter is a review of the mechanism of the 2005 Kashmir earthquake and its implications for the seismic hazard in Kashmir and elsewhere in the Himalaya.

10.2 Structural Setting and Background Seismicity

The Kashmir earthquake took place in the high mountainous area of the northwest Himalaya, near the syntaxial bend, ~200 km north of the Salt Range Thrust (SRT), a zone of complex tectonics (see Figs. 10.1 and 10.2). Here the physiography and the geological structures including the MBT takes the shape of a hair-pin-curve, forming the Kashmir-Hazara Syntaxis (Avouac et al., 2006; Wadia, 1931). In Pakistan to its west, the northwest-trending Himalaya arc abruptly changes its strike in a east–west direction. The SRT, an equivalent to the Main Frontal Thrust (MFT), marks the southern margin of the folded mountain chains here. To the north of the SRT lies the Sub-Himalaya, composed mainly of folded and faulted Mesozoic and Tertiary sedimentary rocks together with the foredeep molasse sediments.

The Main Boundary Thrust (MBT), which locally displaces the Late Quaternary strata in the Indian and Nepal Himalaya (e.g., Nakata, 1989), is considered as mostly inactive in the NW Himalaya. Three active structures rooted in the Main Himalayan Thrust (MHT) have been mapped in this region. These are the Medlicott–Wadia Thrust (MWT; Thakur et al., 2010) and the Main Frontal Thrust (MFT; Powers et al., 1998), which together absorb much of the shortening across the Himalayan belt. Defined by a known belt of seismicity, this area is noted for active faulting, inferred primarily from the tectonically emergent morphological features including vertical dislocation of the Pleistocene fan surfaces at the foot of the Muzaffarabad anticlinal ridge. For example, the 16-km-long, northeast-dipping active thrust named the Tanda fault that, cuts the fluvial terrace surfaces, is regarded as evidence of Late Quaternary faulting. The northern part of the extended Tanda fault mostly runs along the northwest-trending Murree fault, a local name for the MBT, which separates the Sub-Himalayan rocks from the Lesser-Himalayan rocks (Fig. 10.2).

Morphotectonic analyses by Vassallo et al. (2015) show that the Late Quaternary deformation in this part of northwestern India occurs both by in-sequence and out-of-sequence thrusting. From the morphological studies of the local drainage and river terraces, a shortening rate of 11.2 ± 3.8 mm/yr for the MWT during the last ~14 ka, in contrast to the shortening rate of 9.0 ± 3.2 mm/yr is estimated for the MFT during the last ~24 ka. This implies a total Late-Quaternary shortening rate

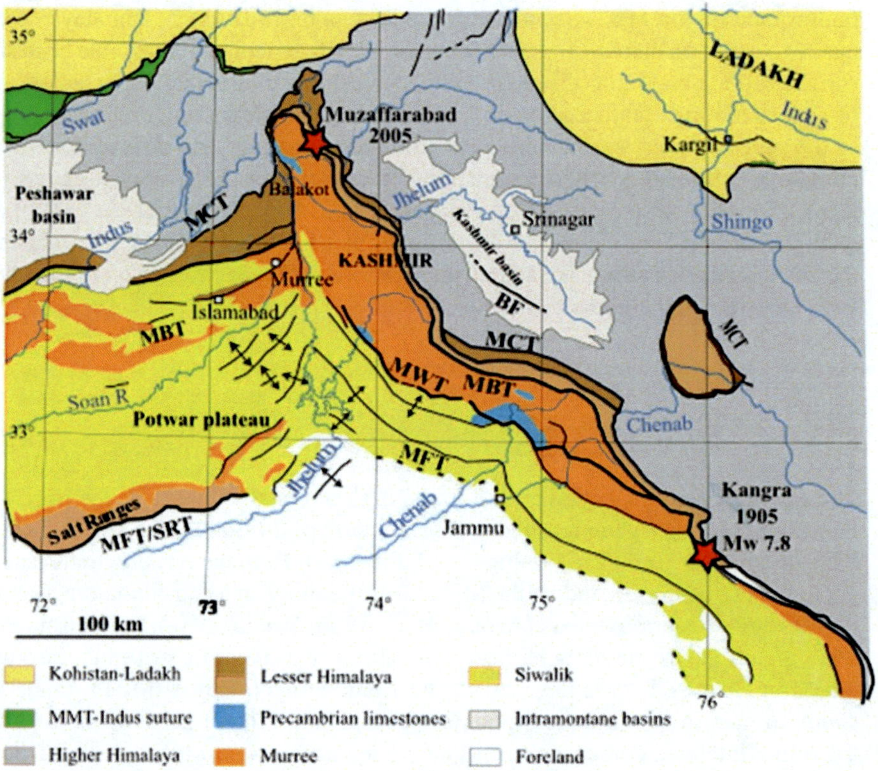

Fig. 10.2. Simplified structural map of the Northwestern Himalayan front in the syntaxial region BF is the Balapur fault (after Vassallo et al., 2015); the red stars mark the location of the 2005 and 1905 earthquakes. The structures: MCT, MBT, MFT, SRT and MWT are discussed in the text

between 13.2 and 27.2 mm/yr for the MHT, which is somewhat consistent with geodetically estimated rates. This assessment also agrees with the shortening rate of 11.2 ± 3.8 mm/yr, which is more than half of the regional shortening rate measured by geodesy (Schiffman et al., 2013; Jade et al., 2014; Jouanne et al., 2014). It is also believed that while the MWT is an emergent thrust, the MFT behaves as a blind thrust that manifests at the surface as a frontal anticline.

The 2005 earthquake is located within the Kashmir-Hazara Syntaxis – an area generally marked by moderate to high-level seismicity (e.g., Quitmeyer et al. 1979). Located approximately 19 km northeast of Muzaffarabad, the event is the most damaging to have occurred here. The September 3, 1972 (M 6.2) event, followed by a series of aftershocks, as listed in some catalogs, was the previous major event known from this region (Fig. 10.1). Another moderate event, known as the Pattan earthquake, occurred on December 28, 1974 (Ambraseys et al., 1981). The source of the 2005 earthquake is linked neither with the MFT nor the Main Himalayan Thrust (MHT), but with a group of three active faults or fault segments within the Sub-Himalaya, collectively called the Balakot-Bagh fault (BBF) (Kaneda et al., 2008; Fig. 10.1). The association with a non-MHT fault is what makes the Kashmir earthquake unique, and it gives an opportunity for a fresh assessment of the status of the MCT and MBT in the NW Himalaya and their roles in earthquake generation.

The MCT and MBT are considered as inactive in the Hazara syntaxis, and are assumed to be passive structural markers, which are displaced by the younger faults such as the Balakot-Bagh, thrust. The documented seismic activity of the Indus-Kohistan seismic zone, including the Pattan 1974 earthquake (M_w 6.2) and distribution of the aftershocks from the 2005 earthquake, which were clustered around Balakot, suggest the presence of active faults in this area (e.g., Jouanne et al., 2014). Such high concentration of seismic activity is regarded as an indication that the area, northwest of the Balakot-Bagh, is characterized by several active thrusts. As they are not associated with any clear morphological signatures, Jouanne et al. (2011) interpreted them as an immature structure linked to the northwestward lateral propagation of the Balakot-Bagh thrust. According to this interpretation, the Balakot thrust is an out-of-sequence thrust corresponding to a ramp, dipping 30°NE, connected to a flat, which is assumed to be as an equivalent of the Main Himalayan Thrust (MHT) of the central Himalaya.

Although the BBF is considered to be a part of the Himalayan arc based on its geometry, there is a difference in its shortening rate as compared to the MFT. At similar longitudes to the south, the Salt Range Thrust (SRT), the E-W trending equivalent of the MFT takes up N-S shortening of at least 9–14 mm/yr (Baker et al., 1988). Combining the 10–30 ka slip history observed from the terrace morphology and the 2005 coseismic slip, Kaneda et al. (2008) argue that horizontal shortening rate across the BBF is only 7%–27% of the 15–20 mm/yr slip across the Himalaya. With such low convergence rates and estimated millennium-long recurrence intervals, the BBF cannot be considered as the main player in accommodating the Himalayan convergence across the Kashmir Himalaya. However, the potential for future devastating surface rupturing earthquakes north of the frontal thrusts are not ruled out. Such active out-of-sequence thrusts found in the Kashmir Himalaya may also occur to most parts of the Himalaya (Avouac et al., 2006).

10.3 Coseismic Deformation and Source Parameters

Coseismic deformation of the Kashmir earthquake has been studied extensively using geodetic, and seismic data and other field observations (Avouac et al., 2006; Pathier et al., 2006; Yan et al., 2013). With some minor variations in terms of rupture geometry and slip distribution, all the models generally agree that the ~70 km long, NE dipping fault that failed primarily by thrust mechanism, with a minor component of right-lateral slip. The rupture models are consistent with the seismic moment tensor solutions determined from the modeling of long-period surface waves. Avouac et al. (2006) were the first to propose a rupture model using correlation of ASTER images (acquired on November 14, 2000, and October 27, 2005) to measure the ground deformation in the epicentral area. A clear discontinuity extending for ~75 km could be traced in the offset field. Despite the 5-year-long interval between the two images, the correlation was reportedly good, except at locations where major landslides had disrupted the surface profiles. The image highlighted some sharp discontinuities, with deformations localized within narrow zones, only a few hundred meters wide. This was a clear indication that the rupture had reached the surface, confirmed by field investigations by Hussain et al. (2009), who also mapped a folded scarp near the northern termination zone of the rupture near Balakot.

According to the rupture model of Avouac et al. (2006), the fault trace is remarkably linear and follows the northeastern flank of the valley for about 30 km north of Muzaffarabad along the previously mapped Tanda fault (Fig. 10.3). The Tanda fault trace is more irregular where it joins the Muzaffarabad fault, a feature that attributed to the roughness in the orography of the region. It was noted that the fault trace makes a 'V' shape across the upper Jhelum River valley, and based on this geometry, the dip angle was inferred to be about 10°. The maximum amplitude of the horizontal slip vector was 7.15 ± 0.4 m, estimated at a location about 10 km northwest of Muzaffarabad. Along the straight fault segment of the Tanda fault the horizontal slip was reported as nearly constant, around 4 ± 0.8 m. The rupture is nearly pure dip–slip as the azimuth of horizontal slip motion is on average N41°E, almost perpendicular to the general strike of the fault trace (138°E).

The northern end of the mapped surface rupture is located ∼3 km north of Balakot. Kaneda et al. (2008) traced the northwest-trending surface rupture that extends from immediately north of Balakot to the northwest of Bagh. The 70-km-long surface rupture, with a vertical separation of ~7 m is not attributed to the MFT or the MBT, but to three fault segments within the Sub-Himalaya, collectively called the Balakot-Bagh Fault (BBF). The rupture zone exhibits northward bifurcation near its northern end, which shows geomorphic evidence for branching of the fault. Near Muzaffarabad, the fault marks the contact of the Precambrian limestone with shale on the northeast and the Miocene Murree Formation (clastic deposits) to the southwest. Towards southeast, the fault is located entirely within the Murree Formation (Hussain et al., 2009). To the southeast, the surface rupture passes through mountainous terrain, where extensive earthquake-induced landslides have obscured most of the surface rupture. The highest fault scarp along the entire surface rupture was mapped on this mountainous section, and with a vertical separation of 7.05 ± 0.35 m (Kaneda et al., 2008). The detailed description of the various segments of the surface rupture along the BBF enabled the site selection for paleoseismologic investigations (Kondo et al., 2008).

Pathier et al. (2006) used sub-pixel correlation of ENVISAT SAR images to show discontinuity of displacement across an almost continuous 80 km-long NE-SW trending fault trace, which was mapped with an accuracy of ~600 m. By comparing pre- and post-event images of 60 cm resolution, it was possible to find evidence for a segmented rupture along the inferred fault trace. A left-step of ~1.5 km was considered as evidence of fault segmentation and as a transition zone that connects with the ~55-km-long straight portion of the fault, named as Muzaffarabad-Bagh segment. The northwestern Muzaffarabad-Balakot segment, where larger displacements are

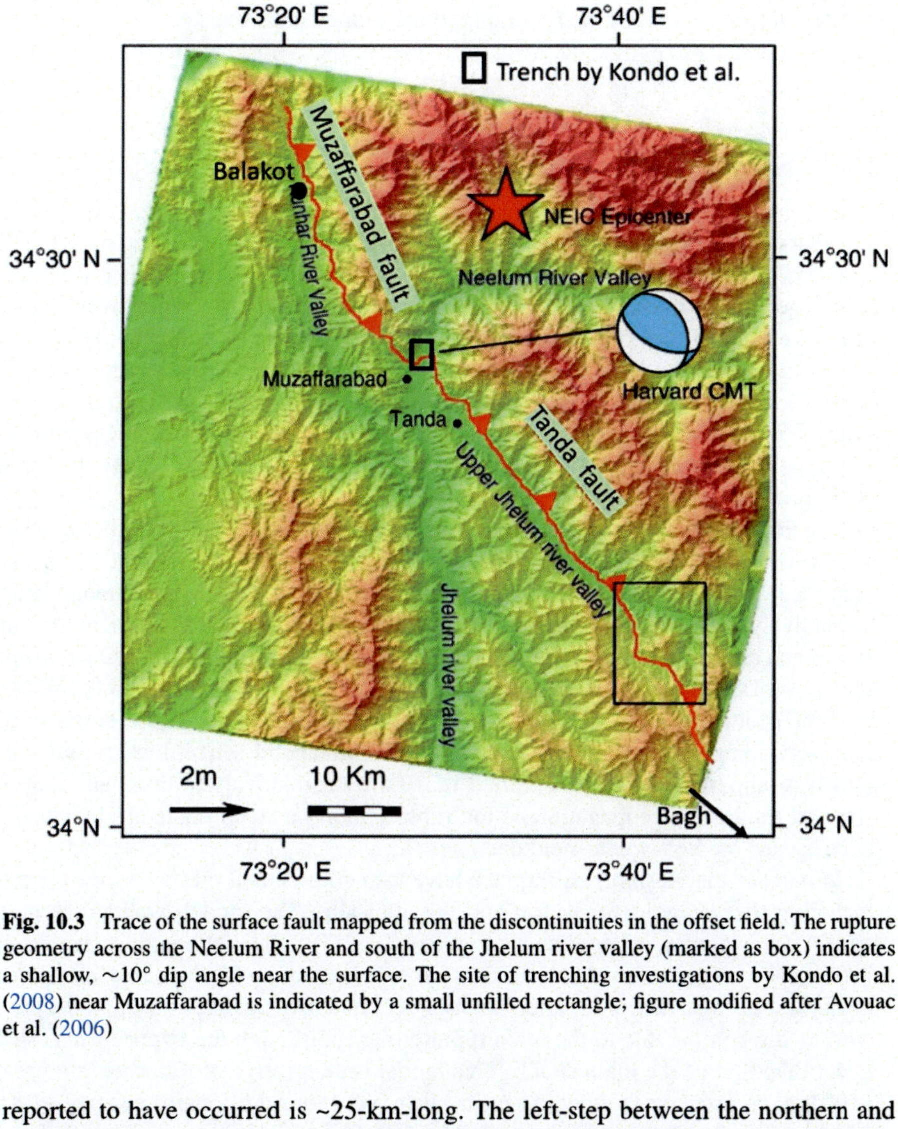

Fig. 10.3 Trace of the surface fault mapped from the discontinuities in the offset field. The rupture geometry across the Neelum River and south of the Jhelum river valley (marked as box) indicates a shallow, ~10° dip angle near the surface. The site of trenching investigations by Kondo et al. (2008) near Muzaffarabad is indicated by a small unfilled rectangle; figure modified after Avouac et al. (2006)

reported to have occurred is ~25-km-long. The left-step between the northern and southern segments align with the North–South Jhelum Valley southward of Muzaffarabad, regarded as the outcome of the current left-lateral motion of the Jhelum fault.

Chini et al. (2011) analyzed the surface features by comparing the pre- and post-images of both the Muzaffarabad and Balakot areas and reported the conspicuous nature of the fault trace at several places. The exposures were prominent wherever the fresh white surface of the dolomitic limestone was exposed on the upthrown, northeast side of the fault plane. The newly formed coseismic scarps coincided with sharp topographic changes, considered as an indication of repeated fault movements.

The left-lateral motion of the Jhelum fault mentioned is perhaps an outcome of such movements.

10.4 Source Parameters

The Harvard CMT solution shows a northeast-dipping fault plane striking N133°E, with a rake of 123° and a dip angle of 40°. The USGS focal mechanism suggested a fault geometry of 133°, 140°, and 29° (strike, rake, and dip, respectively) for the causative fault. The finite source model from teleseismic waveforms (0.01–1 Hz) by Avouac et al. (2006) is analogous to that of Harvard CMT. By using various dip angles between 25° and 40°, the polarity of the P and S wave first motions for this solution were best adjusted for a dip angle of 29°, consistent with the USGS solution. From the fault geometry inferred from the fault trace, and the best fitting dip angle, the hypocentral depth was estimated to be 11 km.

The fault geometry proposed by Avouac et al. (2006) included two fault segments, a 60-km-long southern segment striking 320°, and a 15-km-long northern segment striking 343°, approximately coinciding with the Tanda and the Muzaffarabad faults, respectively. Modelling suggested slip on a single planar fault segment striking 108°E and dipping 31° to the northeast just above the hypocenter, involving multiple segments of the BBF. The model by is analogous to the USGS solution and consistent with the geometry of the BBF. The best-fitting model shows a simple source with a relatively compact high-slip zone spanning the Tanda and Muzaffarabad faults and mostly up-dip of the nucleation point (Fig. 10.4). The relatively compact and shallow slip and the up-dip propagation of the rupture along a steep plane are believed to have caused the heavy near-field damage.

Models of the Kashmir earthquake have also commented on its various aspects, including the wedge deformation and the after-slip. The model by Bendick et al. (2007), for example, is based on a combination of GPS data (although sparse, as the 2001 array was reoccupied only in October and November 2005), surface displacements and the distribution of aftershocks. Focal mechanisms indicate slip on a plane with its dip comparable to the main rupture plane, although the hypocenter is shallower than that of the main shock. This model is suggestive of a surface rupture of ~100 km, on a reverse fault, with a mean slip of ~5.1 m. Additionally, a slip of ~1.8 m on a blind wedge thrust, the total moment of 2.86×10^{20} Nm consistent with that of the Harvard CMT is also indicated. More than 75% of the aftershocks occurred in a cluster located ~30 km southwest of the strike of the main rupture, which follows the trend within the Indus-Kohistan seismic zone identified by previous workers (e.g., Seeber & Armbruster, 1979).

Based on the modeled displacement, the distribution of aftershocks, and historical seismicity. Bendick et al. (2007) infer the presence of an active blind wedge extending WNW from the MBT and a cascade of slip that propagated to the wedge. Bendick et al. (2007) also point out that wedge deformations have not previously been recognized in the coseismic geodetic data, and their complex fault geometry is not conducive

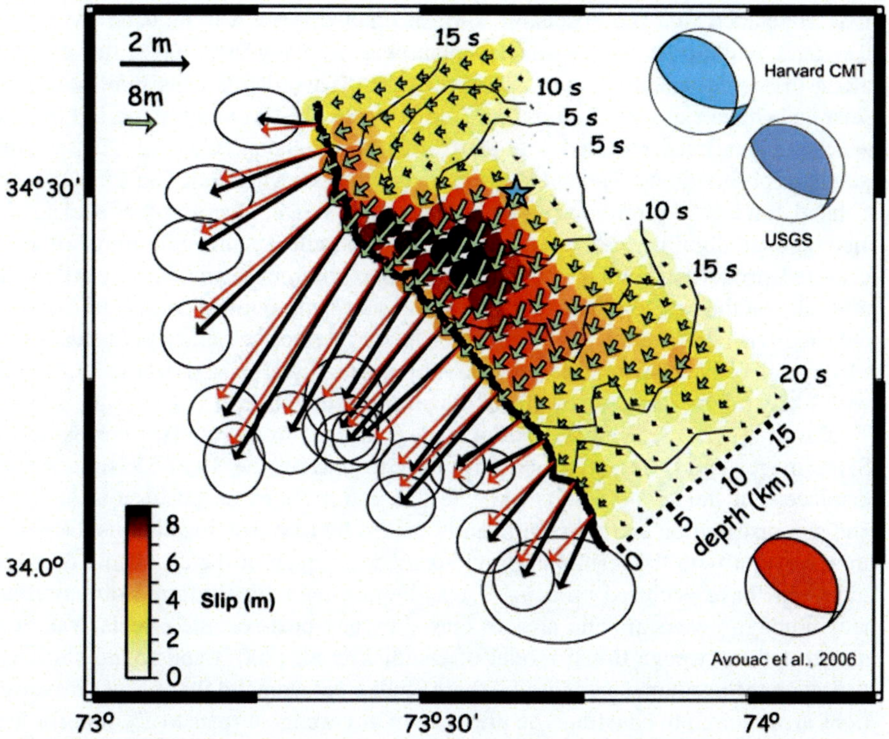

Fig. 10.4 Focal mechanism and slip distribution for the 2005 M_w 7.6 Kashmir earthquake from seismic waveforms and surface slip distribution. Harvard CMT and USGS solutions are also shown. Black, red, and green arrows, respectively, are measured and theoretical horizontal slip vectors along the trace of the surface fault and slip vectors on the fault plane at depth. The star marks the location of the nucleation points on the fault plane, assumed to coincide with the USGS epicenter (34.493°N,73.629°E). (After Avouac et al., 2006)

to model the dislocation geometry at depth without additional inputs. From these arguments it is quite clear that the wedge-thrust model proposed for this region requires more clarity and validation. There are also some limitations with the data. For example, Yan et al. (2013) noted that the Bendick et al.'s data were obtained along the Indus Valley, located northwest of the rupture. Further, this is a region of many blind thrusts and some of them might have been activated during the 1974 Pattan earthquake.

Models based on the post-2005 earthquake GPS data suggest that post-seismic displacement and spatial and temporal distribution of aftershocks were probably induced by an after-slip along a décollement, north of the ramp affected by the main shock (Jouanne et al., 2011). This mechanism was active at least during the first weeks following the main shock, followed by another phase of after-slip and viscous relaxation. The displacement along this 10° north-dipping décollement, connected to the ramp, was reported to be 30.8 cm between November 2005 and August 2006.

It is concluded that the coseismic surface displacement was induced by slip on the ramp, while post-seismic displacement was probably induced by the after-slip along the dé collement located NE of the ramp. This idea is consistent with post-seismic behavior observed elsewhere, for example, the three-month period of post-seismic relaxation associated with the Chi-Chi earthquake (Hsu et al. 2002). Using geodetic observations, L-band Advanced Land Observing Satellite (ALOS), and C-band Envisat interferometric synthetic aperture data, Wang and Fialko (2014) also reported dominance of after-slip. This model indicates that maximum after-slip occurred primarily downdip of the area of maximum coseismic slip. Further, the after-slip on the northern half of the fault was more pronounced compared to that on the southern half. Overall, the observed pattern of surface velocities indicates several early years of post-seismic deformation dominated by after-slip on the fault plane, with likely minor contributions from the poro-elastic rebound.

Yan et al. (2013) used 23 sets of data measured from subpixel correlation of SAR images and differential interferometry to retrieve the 3-D coseismic surface displacement field. The results indicate that estimates of coseismic displacement and slip distribution are not significantly biased by any post-seismic displacement. In agreement with the results of previous models, much of the coseismic displacement may have occurred on a ~40° NE-dipping fault. The inferred direction and magnitude of coseismic slip are not biased by any post-seismic effects. Yan et al. also tested the wedge thrust model of Bendick et al. (2007) and found some slip deficit on the northern segment. All these studies recommend that further investigations are required to constrain the presence of any wedge thrusts to account for their contribution while interpreting the measured displacement field.

10.5 Return Period of Earthquakes in the Region

Avouac et al. (2006) reported that the average slip on the fault patch ruptured by the Kashmir earthquake is ~4.2 m. They also observed that if the geodetically determined shortening rate of ~14 mm across the range were to be accommodated by the repetition of 2005-type earthquakes, the segment-specific return period of such events along the specific segment of the arc would be about 300 yrs. It has been argued that the interseismic strain accumulation in this region is generally dominated by thrust-normal compression (7–8 mm/yr), radial to the Himalayan arc (Bendick et al., 2007). Further, it is noted that the dominant stress regime in the western syntaxis is SW-NE-shortening and convergence rates here are at least a factor or two smaller than in the central arc and that the recurrence interval for the 2005-type events is 530–720 yr. However, historical catalogue of the region does not show many such events, and thus, it is likely that the shortening may be accommodated by less frequent, but significantly larger events like the 1555 CE earthquake (Avouac et al., 2006).

Near Muzaffarabad, the BBF fault takes a bend to form a lateral ramp and strikes almost E-W. The trenches excavated at this location by Kondo et al. (2008) revealed fault strands that cut through Late Holocene fluvial deposits. The first trench excavated was 11-m-long and 4-m-deep, across the 2-m-high scarp. The second one, excavated a year later, located just east of the first trench was 20-m-long and 8-m-deep. The trench sections exposed two generations of faulting, from the recent earthquake and its penultimate event. The penultimate event on the Balakot-Bagh fault, weakly constrained by the radiocarbon ages, is temporally placed between ca. 300 and 2200 yr B.P. The fact that that the 500-yr-old fort in Muzaffarabad had remained intact until the 2005 earthquake can be used as a proxy to fix the minimum age of the event.

The segment of the Himalaya to the southeast of the 2005 source has generated large earthquakes in the past (Ambraseys & Jackson, 2003). A review of the historical records suggests two earthquakes that stand out in the region; the May 30, 1885 (M_w 6.3) and the 1555 (M_w 7.6) events. These earthquakes mostly damaged the northern parts of the Kashmir Valley, including the capital city of Srinagar, Baramulla, and Pattan (Sana et al., 2018). The 1885 earthquake is reported to have caused 3400 causalities and damaged prominent buildings including the twin temples of Pattan. Despite its low magnitude, the unusual rain that lasted till the end of May might have contributed to the triggering of landslides and excessive soil liquefaction, mostly reported from the district of Baramulla (Sana & Nath, 2016). Joshi and Thakur (2016) have reported damage to medieval period temples in the Chamba region (32° 30′N 76° 30′ E), attributed to the 1555 earthquake. Temples of Pattan are also reported to have been damaged by this earthquake. Based on the pattern of damage, it is suggested that both these earthquakes might have occurred on a hinterland structure, possibly the Balapur Fault, in the Kashmir basin (Madden et al., 2010; Shah, 2015). An earthquake of unknown size in the mid-ninth century has also been reported from the region (Urooj et al., 2021). This earthquake had triggered a huge landslide that blocked the narrow gorge of the Jhelum River near Baramullah (Fig. 10.1), where it exits the Kashmir Valley. The landslide impounded a lake, flooding the capital Srinagar.

Bilham and Wallace (2005) argue that since the 1555 earthquake, strain equivalent to slip >6 m has probably accumulated in the 2005 rupture area. From the observations in the trench, Kondo et al., (2008) estimate a slip rate of ~3 mm/yr, which makes the BBF a significant structure, but that alone could not be solely absorbing the convergence over 10 mm/yr across the Indo-Asian boundary at this longitude (e.g., Bettinelli et al., 2006). The range of ages obtained by Kondo et al. (2008) is very broad, and the size of the 2005 earthquake and the recurrence behavior do not seem to catch up with the estimated strain, and there is much uncertainty on the renewal rates. For example, the younger range of penultimate age obtained from the trench suggests either aperiodicity or clustering of events, with inter-event intervals longer than 2000 years. On the other hand, the older end of the age range would place the average renewal interval greater than 2000 years.

As discussed initially, the Kashmir earthquake reactivated a landslide south of Hattian. Some researchers have suggested that during the Holocene, two landslides may have occurred on Dana Hill, which had supposedly left two scars, one of which was the cause of the Hattian slide (Dunning et al., 2007). The evidence for previous landslides could be an additional input in developing the chronology of the previous 2005-like earthquakes. This would suggest a significantly longer recurrence interval than the range of 500–900 yr suggested for the large surface-breaking earthquakes along the MFT (e.g., Bilham & Wallace, 2005). This issue needs to be addressed through further trenching excavations and by developing better chronological constraints.

10.6 The 2005 Kashmir Earthquake and its Regional Context

The Kashmir event is an exceptional Himalayan earthquake from the tectonic point of view because of its association with an out-of-sequence thrust and its spectacular surface rupture. Perhaps, the dominant view that the crustal shortening is localized at the Main Frontal Thrust (MFT), where the earthquakes release most of the inter-seismic strain (Lavé & Avouac, 2000) does not hold in this case. The 2005 event demonstrates how the shortening can be distributed across the width of the orogenic belt, with active out-of-sequence thrusts hosting large earthquakes (Wobus et al., 2003; Hodges et al., 2004; Morell et al., 2015). In terms of its magnitude and surface slip, there are few global examples that could be compared with the 2005 Kashmir earthquake. One could see an analogy with the 1999 M_w 7.6 Chi-Chi (Taiwan) earthquake, which was also attributed to an out-of-sequence thrust that produced a surface slip of ~10 m (Chen et al., 2001).

Apart from the BBF, there are other mapped faults such as the Balapur Fault (BF, Fig. 10.2), which offsets the Quaternary alluvial terraces and holds potential for future earthquakes. For instance, the trenches excavated across the Balapur fault exposed evidence of 2 to 4 surface rupturing events in the latest Quaternary (Madden et al., 2010). Results indicate that the 40-km-long BF with a vertical separation of about 13 m is a low-slip-rate fault and poses seismic hazard to the city of Srinagar and the Kashmir Valley in general. These findings suggest that the BBF, the host structure of the 2005 earthquake is not an exception and that analogous out-of-sequence structures that can generate large ~M_w 7.5 earthquakes could be present elsewhere in the Kashmir region.

There are arguments that the 2005 earthquake might load the neighboring faults and trigger more earthquakes. Hough et al. (2009) argue that with its modest energy release, this earthquake could have done little to dissipate the accumulated stress in its neighborhood; it certainly has not dissipated the stresses already stored near the 1555 rupture zone. Based on the Coulomb stress changes mapped on optimally oriented faults, Parsons et al. (2006) showed increased stress northwest and southeast of the rupture, where large earthquakes have occurred in 1555 and 1885. The convergence rate, as per GPS measurement, is about 11 mm (Schiffman et al., 2013), and this means that a potential slip of about 5.5 m has already been accumulated here since the 1555 earthquake, which is enough to drive a large or great earthquake. Whether the 2005 earthquake may have served to enhance the existing stresses on the nearby active faults is a case for future investigations. The 2005 earthquake created unprecedented damage, both coseismically and through secondary effects, due to its shallow source on an out-of-sequence thrust. This example underscores the role of out-of-sequence thrusts in the seismogenesis along the Himalaya arc. These issues are important in the discussions of seismic hazard assessment of the Kashmir and NW Himalaya that need to be addressed in future investigations.

References

Ambraseys, N., & Jackson, D. (2003). A note on early earthquakes in northern India and southern Tibet. *Current Science, 84*, 570–582.

Ambraseys, N., Lensen, G., Moinfar, A., & Pennington, W. (1981). The Pattan (Pakistan) earthquake of 28 December 1974: field observations. *14*, 1–16.

Armbruster, J., Seeber, L., & Jacob, K. H. (1978). The Northwestern termination of the Himalayan Mountain Front: Active tectonics from microearthquake. *Journal of Geophysical Research, 83*, 269–282.

Avouac, J. P., Ayoub, F., Leprince, S., Konca, O., & Helmberger, D. V. (2006) The 2005, Mw 7.6 Kashmir earthquake: Sub-pixel correlation of ASTER images and seismic waveforms analysis. *Earth Planet Science Letter, 249* (3), 514–528.

Baker, D. M., Lillie, R. J., Yeats, R. S., Johnson, G. D., Yousuf, M., & Zamin, A. S. H. (1988). Development of the Himalayan frontal thrust zone: Salt Range Pakistan. *Geology, 16*(1), 3–7.

Bendick, R., Bilham, R., Khan, M. A., & Khan, S. F. (2007). Slip on an active wedge thrust from geodetic observations of the 8 October 2005 Kashmir earthquake. *Geology, 35*, 267–270. https://doi.org/10.1130/G23158A.1

Bettinelli, P., Avouac, J.-P., & Flouzat, M. et al. (2006). Plate motion of India and interseismic strain in the Nepal Himalaya from GPS and DORIS measurements. *Journal of Geodesy 80*, 567–589.

Bilham, R., & Wallace, K. (2005). Future Mw > 8 earthquakes in the Himalaya: Implications from the 26 Dec 2004 Mw = 9.0 earthquake on India's eastern plate margin. *Geological Survey of India, Special Publications, 85*, 1–14.

Chen, W. S., Huang, B. S., & Chen, Y. G. et al. (2001). 1999 Chi-Chi earthquake: a case study on the role of thrust–ramp structures for generating earthquakes. *Bulletin Seismological Society of America, 91*, 986–994.

Chini, M., Cinti, F. R., & Stramondo, S. (2011). Co-seismic surface effects from very high resolution panchromatic images: The case of the 2005 Kashmir (Pakistan) earthquake. *Natural Hazards and Earth Systems Sciences, 11*(931–943), 2011.

Dunning, S. A., Mitchell, W. A., & Rosser, N. J. (2007). The Hattian Bala Rock Avalanche and associated landslides triggered by the Kashmir earthquake of 8 October 2005. *Engineering Geology, 93*, 130–144. https://doi.org/10.1016/j.enggeo.2007.07.00

Hodges, K. V., Wobus, C., Ruhl, K., Schildgen, T., & Whipple, K. (2004). Quaternary deformation, river steepening, and heavy precipitation at the front of the Higher Himalayan ranges. *Earth and Planetary Science Letters, 220*, 379–389. https://doi.org/10.1016/S0012-821X(04)00063-9

Hough, S., Bilham, R., & Bhat, I. (2009). *Kashmir Valley Mega-Earthquakes: Estimates of the magnitudes of past seismic events foretell a very shaky future for this pastoral valley., 97*, 42–49.

Hsu Y. J., Bechor, N. Segall, P. Yu, S. B. Kuo, L. C., & Ma, K. F. (2002). Rapid afterslip following the 1999 Chi-Chi, Taiwan earthquake. *Geophysical Research Letter, 29*, 1754-

Hussain, A., Yeats, R. S., & Lisa, M. (2009). Geological setting of the 8 October 2005 Kashmir earthquake. *Journal of Seismology, 13*(3), 315–325.

Jade, S., Mukul, M., Gaur, V. K., Kumar, K., Shrungeshwar, T. S., Satyal, G. S., Dumka, R. K., Jagannathan, S., Ananda, M. B., Kumar, P. D., & Banerjee, S. (2014). Contemporary deformation in the Kashmir-Himachal, Garhwal and Kumaon Himalaya: Significant insights from 1995–2008 GPS time series. *Journal of Geodesy, 88*, 539–557.

Joshi, M., & Thakur, V. C. (2016). Signatures of 1905 Kangra and 1555 Kashmir Earthquakes in Medieval Period Temples of Chamba Region, Northwest Himalaya. *Seismological Research Letter, 87*(5), 1150–1160.

Jouanne, F., Latif, M., Majid, A., Kausar, A., Pecher, A., & Mugnier, J. L. (2011). Current shortening across the Himalayas: Quantification of inter-seismic deformation in Nepal and first results of postseismic deformation in Pakistan after the 8th October earthquake. *Journal of Geophysical Research, 116*, B07401.

Jouanne, F., Awan, A., Pêcher, A., Kausar, A., Mugnier, J. L., Khan, I., Khan, N. A., & Van Melle, J. (2014). Present-day deformation of northern Pakistan from Salt Ranges to Karakorum Ranges. *Journal of Geophysical Research, Solid Earth.*

Kaneda, H., Nakata, T., Tsutsumi, H., Kondo, H., Sugito, N., Awata, Y., Akhtar, S. S., Majid, A., Khattak, W., Awan, A. A., & Yeats, R. S. (2008). Surface rupture of the 2005 Kashmir, Pakistan, earthquake and its active tectonic implications. *Bulletin of the Seismological Society of America, 98*(2), 521–557.

Kondo, H., Nakata, T., Akhtar, S. S., Wesnousky, S. G., Sugito, N., Kaneda, N., Tsutsumi, H., Khan, A. M., Khattak, W., & Kausar, A. B. (2008). Long recurrence interval of faulting beyond the 2005 Kashmir earthquake around the northwestern margin of the Indo-Asian collision zone. *Geology, 36*, 731–734. https://doi.org/10.1130/G25028A.1

Lavé, J., & Avouac, J. P. (2000). Active folding of fluvial terraces across the Siwaliks Hills, Himalayas of central Nepal. *Journal of Geophysical Research, 105*, 5735. https://doi.org/10.1029/1999JB900292

Madden, C., Trench, D., Meigs, A., Ahmad, S., Bhat, M. I., & Yule, J. D. (2010). Late Quaternary shortening and earthquake chronology of an active fault in the Kashmir. *Seismological Research Letter, 81*(2), 346.

Mahmood, I., Qureshi, S. N., Tariq, S. et al. (2015). Analysis of landslides triggered by October 2005, Kashmir Earthquake. PLOS Currents Disasters. https://doi.org/10.1371/currents.dis.0bc3ebc5b8adf5c7fe9fd3d702d44a99.

Morell, K. D., Sandiford, M., Rajendran, C. P., Rajendran, K., & Sanwal, J. (2015). Geomorphology reveals active décollement geometry in the central Himalayan seismic gap. *Lithosphere, 7*(3), 247–256.

Nakata, T. (1989). Active faults of the Himalaya of India and Nepal. In Malinconica, L.L., Jr., & Lillie, R.L. (Eds.),Tectonics of the Western Himalayas: Geological Society of America Special Paper 232, pp. 243–264.

Owen, L. A., Kamp, U., Khattak, G. A., Harp, E. L., Keefer, D. K., & Bauer, M. A. (2008). Landslides triggered by the 8 October 2005 Kashmir earthquake. *Geomorphology, 94*, 1–9.

Parsons, T., Yeats, R. S., Yagi, Y., & Hussain, A. (2006). Static stress change from the 8 October, 2005M _ 7:6 Kashmir earthquake. *Geophysical Research Letters, 33*, L06304. https://doi.org/10.1029/2005GL025429

Pathier, E., Fielding, E. J., Wright, T. J., Walker, R., Parsons, B. E., & Hensley, S. (2006). Displacement field and slip distribution of the 2005 Kashmir earthquake from SAR imagery. *Geophysical Research Letters, 33*, L20310.

Powers, P. M., Lillie, R. J., & Yeats, R. S. (1998). Structure and shortening of the Kangra and Dehra Dun Reentrants, Sub-Himalaya, India. *Geological Society of America Bulletin, 110*, 1010–1027.

Quittmeyer, R. C., Farah, A., & Klaus, H. J. (1979). The seismicity of Pakistan and its relation to surface faults, Lamont-Doherty Geols Contribution No. 2667, pp. 271–284.

Rajendran, K., Parameswaran, R., & Rajendran, C. P. (2017). Seismotectonic perspectives on the Himalayan arc and contiguous areas: Inferences from past and recent earthquakes. *Earth-Science Reviews, 173*, 1–30.

Rehman, K., Qadri, S. M. T., Aamir, A., & Asghar, A, Ahmed. (2016). Analysis of the devastating Kashmir earthquake 2005 aftershocks. *Arabic Journal of Geoscience, 9*, 379.

Sana, H., & Nath, S. K. (2016). Liquefaction potential analysis of the Kashmir valley alluvium. *NW Himalaya Soil Dynamics and Earthquake Engineering, 85*(11–18), 2016.

Sana, H., Bhat, F. A., & Sana, S. (2018). The ancient temples of Kashmir turned from marvel to ruin by earthquakes? A case study of the Pattan twin temples (A.D. 883–902). *Seismological Research Letter, 90* (1) 358–365.

Schiffman, C., Bali, B. S., Szeliga, W., & Bilham, R. (2013). Seismic slip deficit in the Kashmir Himalaya from GPS observations. *Geophysical Research Letters, 40*, 5642–5645.

Schneider, J. F. (2009). Seismically reactivated Hattian slide in Kashmir, Northern Pakistan. *Journal of Seismology, 13*, 387–398.

Seeber, L., & Armbruster, J. (1979). Seismicity of the Hazara arc in northern Pakistan: decollement vs. basement faulting, in Geodynamics of Pakistan, A. Farah and K. A. DeJong (Editors), Geological Survey of Pakistan, Quetta, 131–142.

Shah, A. A. (2015). Kashmir Basin Fault and its tectonic significance in NW Himalaya, Jammu and Kashmir India. *International Journal Earth Science (geol Rundsch)*. https://doi.org/10.1007/s00531-015-1183-1

Thakur, V. C., Jayangondaperumal, R., & Malik, M. A. (2010). Redefining Medlicott-Wadia's main boundary fault from Jhelum to Yamuna, an active fault strand of the main boundary thrust in northwest Himalaya. *Tectonophysics, 489*, 29–42.

Urooj, M., Bilham, R., Bali, B. S., & Ahmed, S. I. (2021). Suyyas's flood: Numerical models of Kashmir's Medieval Megaflood and Ancient Lake Kerewa drainage events. *Earth Science System and Society, 1*, 10040. https://doi.org/10.3389/esss.2021.10040.

Vassallo, R., Mugnier, J.-L., Vignon, V., Malike, M. A., Jayangondaperumal, R., Srivastava, P., Jouanne, F., & Carcaillet, J. (2015). Distribution of the late-quaternary deformation in Northwestern Himalaya. *Earth Planet Science Letter, 411*, 241–252.

Wadia, D. N. (1931). The syntaxis of the northwest Himalaya: Its rocks, tectonics and orogeny. *Rec. Geological Survey of India, 65*(2), 189–220.

Wang, K., & Fialko, Y. (2014). Space geodetic observations and models of postseismic deformation due to the 2005 M 7.6 Kashmir (Pakistan) earthquake: Relaxation due to Kashmir Earthquake. *Journal of Geophysical Research, 119*, 7306–7318. https://doi.org/10.1002/2014JB011122.

Wobus, C. W., Hodges, K. V., & Whipple, K. X. (2003). Has focused denudation sustained active thrusting at the Himalayan topographic front? *Geology, 31*, 861–864. https://doi.org/10.1130/G19730.1

Yan, Y., Pinel, V., Trouve, E., Pathier, E., Perrin, J., Bascou, P., & P and. Jouanne, F,. (2013). Coseismic displacement field and slip distribution of the 2005 Kashmir earthquake from SAR amplitude image correlation and differential interferometry. *Geophysical Journal International, 2013*(193), 29–46.

Chapter 11
Nepal-Bihar 1934

11.1 Introduction

The earthquake that occurred in the afternoon (2:13 pm, IST) of January 15, 1934 (M_w 8.2), is one of the strongest among the twentieth century Himalayan events (Fig. 11.1). It resulted in a huge death toll and caused extensive destruction to buildings and infrastructure both in the central-eastern Nepal and the adjacent parts of India. The official death toll was reported to be 8000 in Nepal and about 7000 in India, but the unofficial sources put the Indian figure near 25,000. The ground movement lasted about five minutes in the central tract of Bihar, and the shock was perceived over almost the whole of northern and central India (e.g., Dunn et al., 1939; Rana, 1935, 2013). The earliest of the Himalayan great twentieth century earthquakes, it was the first to give a glimpse of the unprecedented scale of ground deformation in a riverine landscape that could emanate from shaking intensity.

Although the concept of plate tectonics and its relation to the Himalayan orogenesis was far from being included in the geological lexicon at the time of the Nepal-Bihar earthquake, the rudiments of modern earthquake science were already in place, thanks to a popular published work on the impact of the great 1897 Shillong earthquake (see Chap. 14). In the modern seismological studies, the 1934 earthquake evolved as a classic example that typified the generally accepted model of seismogenesis in the Himalaya. In essence, these tectonic models proposed that the great Himalayan earthquakes ($M > 8.0$) nucleate on the ramp at the northern end of décollement (e.g., Seeber & Armbruster, 1981). It was hypothesized that their ruptures would propagate southward toward the Main Frontal Thrust (MFT) at the mountain front, which would either break the surface or terminate as a blind thrust (see Fig. 7.6).

The Nepal-Bihar earthquake, originally regarded as a blind thrust event, has been considered as classic example of great earthquakes originating on the MHT with their ruptures failing to reach the surface. This was the prevailing view until Sapkota et al. (2013) reported a surface rupture which they attributed to the 1934 earthquake. This interpretation has later been challenged by Wesnousky et al. (2017a, b) and

C. P. Rajendran and K. Rajendran, *Earthquakes of the Indian Subcontinent*, GeoPlanet: Earth and Planetary Sciences,
https://doi.org/10.1007/978-981-16-4748-2_11

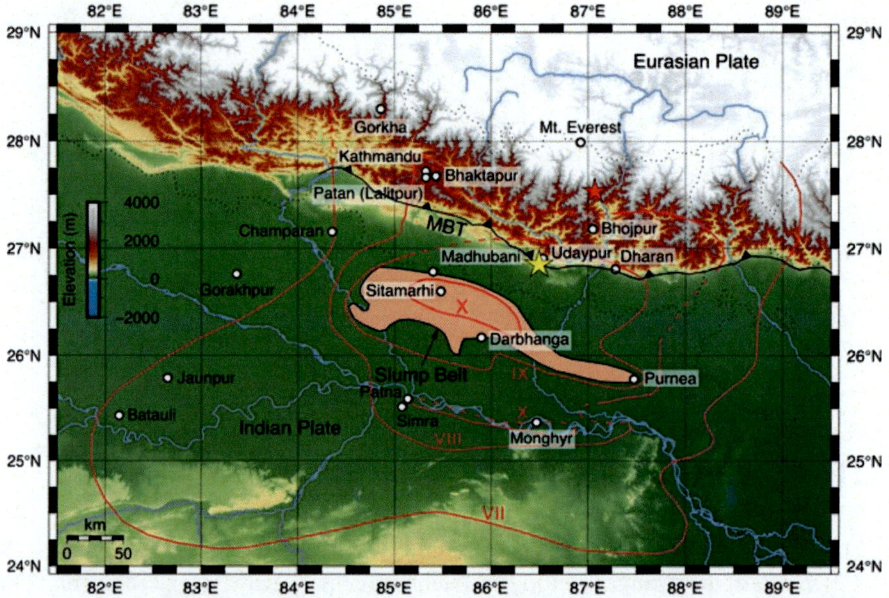

Fig. 11.1 Location of the 1934 earthquake and the intensity contours (Mercalli scale). The yellow and red stars denote the epicentral locations by Seeber and Armbruster (1981), and Chen and Molnar (1977), respectively; the slump belt as reported in Dunn et al. (1939) is shown in beige. (Modified after Seeber & Armbruster, 1981)

no fresh evidence to substantiate the surface rupture has been reported since these publications. The rupture zone of the Nepal-Bihar earthquake is one of the few sites in the Himalaya where distinct paleoseismological evidence of previous faulting episodes is exposed (Lavé et al., 2005). The 2015 Nepal earthquakes (Chap. 12) have aroused a renewed interest in the seismogenesis of this part of the Himalaya, and observations presented in many recent papers provide insights on the rupture mechanism of large and great earthquakes in this segment. This chapter presents an evaluation of the work that has been done on various aspects of this earthquake, particularly its rupture extent. We start this chapter with a brief historical perspective to highlight how this earthquake became a major milestone in earthquake science of the Himalaya.

11.2 Historical Perspective

The interest in the Nepal-Bihar earthquake permeated even the top Indian political leadership, including Gandhi, who visited Bihar after the earthquake and wrote that it was "providential retribution for India's failure to eradicate untouchability". The reference was about the practice of ostracizing a group of people regarded as "untouchables" resulting in segregation and persecution from the people regarded as "higher" caste. This prompted Rabindranath Tagore, a distinguished poet, and a key figure in Indian renaissance, to retort in a letter that "physical catastrophes must have origin in physical facts". Tagore was perturbed by the irrationality in Gandhi's statement, even though he was in total agreement on the issue of untouchability. "I find it difficult to believe it. But if this be your real view on the matter, I do not think it should go unchallenged," Tagore sent the statement to Gandhi on January 28, 1934, with a request to publish it in his newspaper. Tagore's statement appeared in *Harijan*, on February 16, 1934, under the title, 'The Bihar Earthquake'.

The earthquake thus generated popular discourses on the contemporary understanding of earthquake generation along with pseudoscience, as reflected in the exchanges between Tagore and Gandhi. Marcussen (2017) provides an insightful historical perspective of this earthquake and shows how such discourses in the aftermath of the earthquake led to a popular debate by the defenders of science against those who were spreading pseudoscience and superstitions. To ally the rising panic among the public due to any rumors, the local authorities had to resort to circulating pamphlets explaining the scientific cause of earthquakes. In one such pamphlet, it is stated that the cause of the earthquake was the gradual movement of the Indian Peninsula towards the Himalaya, and this was squeezing up the Gangetic plain. Considering the infancy of the continental drift theory, in the 1930s, this statement could be considered as one of the earliest references to this concept in India. It was used by the administrators, possibly advised by the Geological Survey of India (GSI), to avert the concerns of the public.

However, this earthquake was not free of scientific controversies. For example, the disagreements on the location of the epicenter of the Nepal-Bihar earthquake continued for many years. The GSI team, who located the epicenter within the Bihar Plains, was not quite aware of the damage intensity in the central-eastern part of the Himalaya in Nepal. J.B. Auden's brief visit to Kathmandu (Nepal) did not help in filling this communication gap, which was complicated by the fact that there was an official ban on foreigners from visiting other parts of Nepal unless waived in individual cases. The publication of a book, "*Nepal ko Mahabukhampa*"—"Great Earthquake of Nepal" in Nepali, published by Brahma Shumsher Rana, a senior officer of the Nepalese Army in 1935 (Rana, 1935, republished in 2013), that included many details, was unavailable to the GSI team, who were working on the earthquake from the Indian side at that time. Interestingly, Nobuji Nasu, the Japanese seismologist from the Earthquake Research Institute in Tokyo, who visited the damage area in Bihar, offered an explanation that appeared much closer to the truth, though it was not accepted by the GSI. Nasu (1935) suggested that the shock originated beneath the Himalaya, and that the formation of the slump belt was due to the presence of particularly weak sediments to the south. Thus, the earthquake, despite the self-evident regional manifestations of its effects and intense field investigations by the experienced workers, was mired in controversies from the very beginning. Though later studies solved the issue of its epicenter location, it is interesting that the controversies on this earthquake refuse to die even to this day. Currently, it is more about the surface rupture and the debate is whether this earthquake ruptured the Main Frontal Thrust (MFT) or if it ended up as a blind thrust.

11.3 Intensity, Field Observations and Estimates of Epicenter

Damage from the Nepal-Bihar earthquake was the most pervasive in two parallel belts: the Bihar Plains of India and in the Kathmandu Valley and its eastern part. The GSI conducted detailed field investigations in the Bihar Plains, mostly focusing on the delineation of intensity values, using a modified version of the Rossi-Forel scale, and focusing on local building styles and soil characteristics. However, these investigations were confined to Bihar and the neighboring areas in India because of the restrictions placed on foreigners entering Nepal. The survey also did not cover the mountainous region of the eastern Nepal as those conducting the work were unaware of the earthquake impact in that region. The detailed report by the GSI was published in 1939 (Dunn et al., 1939). One of the members of the GSI team who conducted a reconnaissance survey in Nepal, J. B. Auden also submitted a report to the Government of Nepal. However, as he was unsuccessful in obtaining damage information beyond Kathmandu, much of the damage in Nepal remained under-reported for a long time to the outside world.

Ground shaking was extremely severe in the alluvial plains of northern Bihar on the Indian side, comprising an area of about 46,600 km^2 (Dunn et al., 1939). The most affected areas on the Indian side include the towns of Bettiah, Motihari, Sitamarhi, Madhubani, Muzaffarpur, Darbhanga, and Monghyr (Munger) in the state of Bihar, forming what was described as a "slump belt", which also experienced severe liquefaction at some locations (Figs. 11.1 and 11.2). The slump belt is reported to have extended for ~ 320 km, from Purnea (25.7771°N: 87.4753°E) in the east to Champaran (27.1543°N, 84.3542°E) in the west in (Bihar, India). This coseismically generated "slump belt", which displayed wide field of liquefaction features and ground subsidence in the northern Bihar, persuaded Dunn et al. (1939) to conclude that the epicenter was under the Gangetic alluvial plains of Bihar, and not in the Himalaya. This conclusion earned credence when Gutenberg and Richter (1954), also accepted it as the favored epicentral location. They also inferred a strike-slip sense of fault movement along a west-northwest direction from the distribution of isoseismals. Later, Seeber and Armbruster (1981) argued that this event originated on the low-angle (3°–5°) plane of detachment beneath the Lesser- and Sub-Himalaya and suggested that the rupture extended far south of the Himalaya beneath the alluvial plains and located the epicenter about 50 km north of the slump belt (Fig. 11.1).

While the early understanding of the earthquake was acquired mostly relying on the observations in the report published by Dunn et al. (1939), the contemporary account (Rana, 1935 in Nepali), which had remained inaccessible and unnoticed for a long time, became an additional source of information. Rana's report provided

Fig. 11.2 Ground failure at Sitamarhi (see Fig. 11.1 for location) in the 'slump belt' formed during the 1934 earthquake (after Bilham, 2019)

extensive details from eastern Nepal, describing that the destruction was most intense in the eastern districts. He reported that in the areas east of Kathmandu, nearly half of the population was killed, and that most of the houses were destroyed. Damage to monuments was quite pronounced in Gorkha, Bhaktapur (located 80 km NW and 11 km east of Kathmandu, respectively) and Patan (also called Lalitpur, located 9 km south of Kathmandu), a center of many ancient buildings including temples and palaces. Landslides were reported from Udayapurgadhi, Dharan and Bhojpur, within the eastern districts of Udaipur, Sunsari and Bhojpur of Province No. 1 in Nepal. Rana used the Sanskrit word '*patala*' (meaning hell) to qualify the level of destruction in some of these sites.

Landslides were reported not only from the mountainous parts of central Nepal, but also from the southern, lower Himalaya. The ground deformation in the form of fissures, some of them as wide as 3–4 m and as deep as 20–30 m, was widespread in the region. Thus, a zone of severe damage (IX and X on the Rossi-Forel scale) that included both ground failure and soil liquefaction in the Kathmandu Valley was recognized as a remarkable feature of this earthquake (Rana, 1935). Using these cues derived from ground reports, Chen and Molnar (1977) identified the epicentral location in central-east Nepal (27.55°N 87.07°E), approximately 10 km south of Mount Everest (Fig. 11.1).

11.4 Source Parameters, Geometry and Rupture

Following the generally accepted tectonic models, the great Himalayan earthquakes of moment magnitude (M_w) > 8.0 originate on the décollement, or the Main Himalayan Thrust (MHT), and their ruptures propagate southward to the Main Frontal Thrust (e.g., Seeber & Armbruster, 1981). The January 15, Nepal-Bihar earthquake was considered a typical event that exemplifies the rupture on the MHT that propagates to the southern plains but falls short of reaching the surface. The view that no coseismic surface rupture was generated during the earthquake was questioned by Sapkota et al. (2013). Geomorphological data and paleoseismological age models were used to suggest that a nearly 150 km-long surface rupture was formed along the Main Frontal Thrust fault in Nepal, between 85.8° and 87.3°E longitudes. Interestingly, trenching investigations at a location close to their site (86°E, 27°N) by Wesnousky et al. (2019) have revealed no evidence for a surface rupture that could be attributed to the 1934 Nepal-Bihar earthquake, as discussed later in this chapter (see Figs. 11.3 and 11.4).

In the absence of any mapped surface expression of primary faulting, the geometry of the causative fault had to be based primarily on the intensity reports and observations of ground failures. Based on the intensity reports, the width of the fault plane could be about 100 km, although its northern edge remains ambiguous (Pandey & Molnar, 1988). Chen and Molnar (1977) calculated a seismic moment of 1.1×10^{21} dyne cm, assuming a low angle thrust plane dipping 20° and slip of 5.4 m, with a rupture area of 130–50 km^2, corresponding to $M_w = 8.0$. Ambraseys and Douglas (2004) revised the magnitude as M_w 8.11, assuming rupture dimension of 150×80 km^2 and slip of 5 m. However, using a shallower and more probable dip of 5°

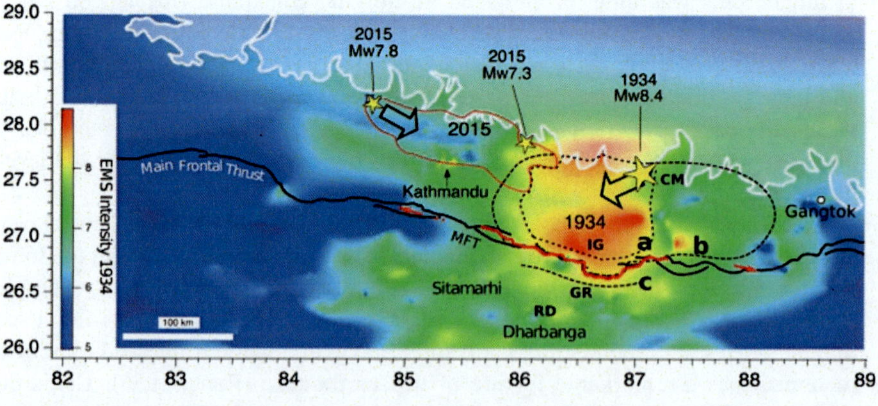

Fig. 11.3 Intensities observed in the 1934 earthquake (Ambraseys & Douglas, 2004; Martin & Szeliga, 2010). Various epicentral locations estimated for the earthquake are indicated by letters: RD—Dunn et al. (1939), GR—Gutenberg and Richter (1949), IG-ISC/GEM, CM—Chen and Molnar (1977). Three 1934 rupture scenarios, a, b, and c are discussed by Bilham (2019). Large unfilled arrows show inferred rupture propagation directions of the 1934 and 2015 earthquakes (after Bilham, 2019)

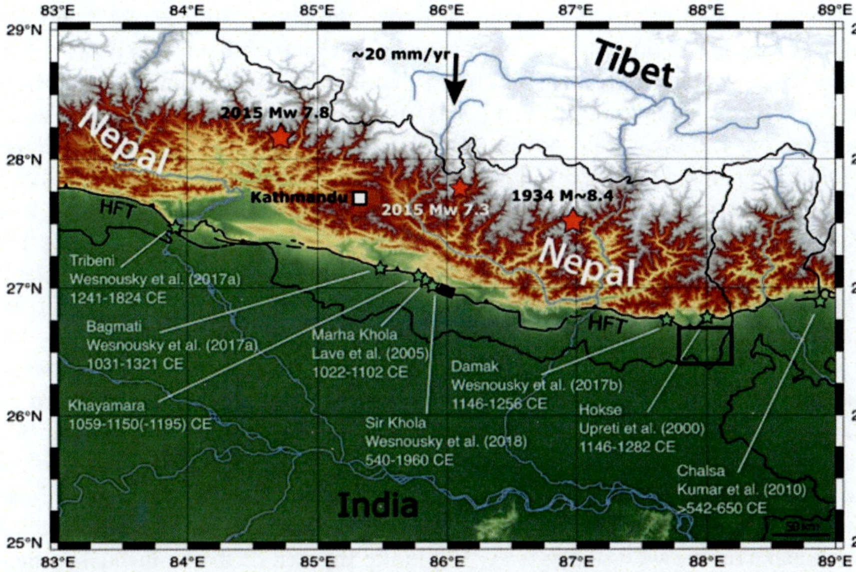

Fig. 11.4 Location of various trenches excavated along the MFT of eastern Nepal Himalaya. Sites are tagged with the age of the last large surface displacement and citations., Red stars show locations of the two large earthquakes of 2015. Locations of seismic reflection studies marked as open square shows the location of Bhadrapur High (Duvall et al., 2020); filled-black square near Sir Khola indicates the location of seismic lines by Almeida et al. (2018); the Charnath trench excavated by Rizza et al. (2019) is located to the east of Sir Khola. (Modified after Wesnousky et al., 2019)

(consistent with recent focal mechanisms and the current knowledge of the décollement geometry), the seismic moment from the earthquake was calculated as 4.1×10^{28} dyne cm, and its magnitude was projected to be M_w 8.4 (Molnar & Deng, 1984). However, Bilham (2019) argues that, for a rupture area of $130 \times 100\ km^2$, the mean slip could have been ~10 m. Thus, the slip estimate of the 1934 earthquake continues to be in the realm of speculation, still devoid of objective constraints.

The large intensities felt in the northern Bihar compelled Seeber and Armbruster (1981) to consider that the rupture had extended onto the plains, possibly as a blind thrust. Apart from the rupture directivity, the exceptionally wide zone of liquefaction in the Bihar Plains is now attributed mainly to high ground motions caused by basin resonance (Hough & Bilham, 2008). The re-measurement of the first order spirit leveling lines in northern Bihar after the earthquake had indicated subsidence of 1.1 m in the region (Bomford, 1937; Burrard, 1934). This was attributed to the intense and spatially widespread lateral spreading and soil liquefaction rather than elastic deformation (Hough & Bilham, 2008). As verified from the residents in Jaunpur and Batauli, the 1934 liquefaction of the eastern Uttar Pradesh was likely to have extended further into the parts of eastern Uttar Pradesh (see Fig. 11.1 for locations).

Several models have been proposed to explain the rupture from the 1934 earthquake. Singh and Gupta (1980) suggested that the rupture occurred as a combination of thrust as well as strike-slip faulting (strike direction 280°; dip 26°; slip angle 40°). Using the region of maximum shaking intensity and subsidence as proxies, Bilham (1995) had suggested that the rupture propagated from east to west for about 160 km. Hough and Bilham (2008) argue that a 130–160 km-long eastward-propagating rupture would not be feasible with the relocated epicenter (Chen & Molnar, 1977) and the observed MSK data unless it crosses over to Sikkim, which had experienced only low shaking. If the rupture was to extend westward to the Kathmandu Valley, their model predicted greater damage than what was observed. Assuming an along-strike distance of 160 km and an appropriate scaling relationship (for a magnitude of M_w 8.1), the western and eastern edges of the rupture were constrained to be lying between 85.5 ± 0.2°E and 87.0 ± 0.2°E.

Bilham (2019) proposed three possible rupture scenarios, with a caveat that the lack of sufficient data inhibits an unequivocal explanation. In one model, the rupture extent is bounded in the west by the 2015 earthquake rupture at 86.1° and to the east by an apparent reduction in intensity near 87.2°. In this scenario, the rupture zone measures 100 × 85 km^2, corresponding to a mean slip of 17 m for an estimated magnitude of M_w 8.4. In an alternate scenario, the rupture overlaps with the 2015 earthquake on the west and extends 100 km to the east of the mainshock, a model that requires a bi-lateral rupture that reduces the mean slip to 8 m. In the third option, the southern edge of the rupture extends to the south of the Himalaya Frontal Thrust, requiring the fault to be a blind thrust. The third possibility gains support in the backdrop of reports of the existence of several blind thrusts and subsurface anticlines, which extend 1–10 km south of the Main Himalaya Thrust (e.g., Almeida et al., 2018; Duvall et al., 2020; Yeats & Thakur, 2008). These three scenarios are indicated as a, b, and c in Fig. 11.3. In this context, it is important to review the evidence that argues for a surface rupture from the 1934 earthquake (e.g., Sapkota et al., 2013), and those that refute them (Wesnousky et al., 2017a, b; 2019), which will be discussed next in some detail.

11.5 Debates on the 1934 Surface Rupture: Evidence from Paleoseismology

The Main Frontal Thrust (MFT) of the central Nepal has been a locus of trenching excavation for more than a decade (e.g., Bollinger et al., 2014; Kumar et al., 2001, 2006, 2010; Lavé et al., 2005; Rizza et al., 2019; Sapkota et al., 2013;: Wesnousky et al., 2017a, b, 2019; see Fig. 11.4 for the locations of trenches). Uncovering the potential evidence for past large/great earthquakes has been a common goal in all these studies. Paleoseismological investigations initiated along the MFT in the central Himalaya and their study of the Black Mango fault displayed evidence of two large surface rupture earthquakes in the past 650 years (Kumar et al., 2001).

What set the stage for a series of investigations in the Nepal Himalaya was the report of a surface-breaking earthquake in 1100 CE from a site called Marha Khola in the eastern part of the central Nepal, southeast of Kathmandu (Lavé et al., 2005). The only surface-breaking event from this trench for which these authors could find any evidence was the 1100 CE event (M_w ~ 8.8), thus ruling out any surface breaks from any earthquakes including the 1934 event. The 1100 CE event is believed to have produced a surface displacement of ~17 (+5/–3) meters with a lateral extent of 240 kms or more. This was the first-reported geological evidence that suggested that a great earthquake, larger than the 1934 event, had indeed occurred in the MFT of eastern Nepal, with potential for future analogous events (e.g., Wesnousky et al., 2019, Wesnousky, 2020).

The idea that the 1934 earthquake occurred on a blind thrust was challenged by Sapkota et al. (2013), by interpreting the findings from a trench at Sir Khola on the MFT. A subsequent study by Bollinger et al. (2014) on a section along the Ratu Nadi River, ~10 km east of Sir Khola, reported on an uplift event postdating 1021–1166 CE. Two surface ruptures, each with a plausible throw of about 4–5 m, were attributed to the 1934 Bihar-Nepal earthquake and the historically described catastrophic 1255 CE event. Subsequently, Mishra et al. (2016) reported that the 1255 event was a giant earthquake that ruptured 800 km of the eastern Himalayan arc between 85.87° and 93.76°E longitudes. This finding reinforced the argument that the Himalaya plate boundary has the potential to generate giant earthquakes of M_w ~ 9.0 with rupture lengths as large as ~800 km.

A re-examination of the section at Sir Khola, however, had raised questions on the surface ruptures attributed to both the 1934 and the 1255 CE earthquakes (Wesnousky et al., 2017a, b, 2018). The radiocarbon age brackets used were the same, but the revised temporal bounds overlap with an earthquake around 1100 CE, rather than the historical candidate of 1255 CE. The alternate interpretation also highlights that the 1255 event is more likely to have ruptured the western part of the Nepal MFT. Wesnousky et al. (2019) also used their observations from the Khayarmara trench, located to the left of the trench by Lave et al. (2005) to buttress their alternate interpretation of the earthquake chronology. The stratigraphy, structure, and radiocarbon data from Khayarmara indicate an event between 1060 and 1195 CE. In summary, the dates of the ancient earthquakes fall between calendric intervals of 1146–1256, 1022–1102, and 1221–1262 CE on the eastern, central, and western segments, respectively.

Trenching investigations by Rizza et al. (2019), at the outlet of Charnath River east of Sir Khola was the latest to report a surface scarp, at a site located 20 km east of Ratu/Sir Khola sites. However, according to the authors' admission, the interpretation of the stratigraphy of the sections at the Charnath site is not straightforward as the key structural inputs like fault offsets are obfuscated by the post-earthquake aggradation processes, which have blurred the most recent surface rupture. The age constraints of the interpreted events based on the wedges in front of the scarps are reported as ambiguous. Further, they note that vertical separations across the scarps produced by the most recent displacements are of similar size (Bagmati ~ 7 m; Marha Khola ~ 7 m; Damak (~5.5 m). Thus, they conclude that the ~250 km section of the MFT must have ruptured around ~1100 CE. This study, therefore, by itself may not be able to resolve the issues concerning the surface rupture reported from the other sites.

As witnessed in 1934, the Gangetic Plains have the potential to amplify seismic energy and generate huge liquefaction fields. The sedimentary sections within these sites of liquefaction most likely expose event horizons representing previous earthquakes. Sukhija et al. (2002) have deciphered evidence for past events from the Bihar Plains. Using multiple exposures within the slump belt of the Bihar Plains, Rajendran et al. (2016) identified different generations of liquefaction episodes as proxies of medieval earthquakes. Liquefaction features generated by successive earthquakes were dated at 829–971, 886–1090, 907–1181, 1130–1376, 1112–1572, 1492–1672, 1733–1839, and 1814–1854 CE. Of these, the earliest liquefaction event bracketed between 829 and 1181 CE, may correlate with the great earthquake of ~1100 CE, recognized in trench sections across the frontal thrust in central eastern Nepal. The sedimentary signatures of the 1833 and 1934 and the 1988 (Udaipur Ms 6.6 Bihar-Nepal border) earthquakes overlap in the sedimentary sections and thus discriminating each of these events become difficult. The destruction to a medieval monument in the eastern Uttar Pradesh may be helpful to constrain the minimum elapsed time prior to the 1934 earthquake. The Mughal bridge near Jaunpur (Fig. 11.1) completed in 1569 was severely damaged in 1934; seven of its arches had to be rebuilt, and the bridge is still in use.

11.6 The 1934 Earthquake: Some Outstanding Questions

As discussed, at the beginning of this chapter, the lack of surface rupture from the 1934 earthquake has been a point of discussion for a long time. In fact, the slump-belt and the extensive liquefaction in the Ganga plains and the missing rupture have led to the formulation of the idea that the great Himalayan thrust earthquakes tend to be blind (Seeber & Armbruster, 1981). There was a change in perception when Sapkota et al. (2013) published evidence that argued in favor of a surface rupture. Subsequently, Wesnousky et al. (2017a, b) interpreted the stratigraphy of the same trench differently and ruled out the possibility of a coseismic rupture. From the additional information obtained from other sites, Wesnousky et al. (2019) argue that the sections expose evidence for an earthquake between 1059 and 1195 CE, but none during the intervening period. Neither have they observed any record of any surface rupture associated with the great 1934 earthquake.

While the question of surface rupture on the MFT from the 1934 earthquake remains debated, several studies have shown that the forward portion of the Ganga foreland cannot be considered as un-deformed. One of the original proposals on the rupture mechanism of the 1934 earthquake suggests that the southern edge of the rupture had extended to the south of the Himalaya Frontal Thrust onto the Ganga foreland as a blind thrust. It is interesting to address this question in the light of some of the studies that predict the southward extension of a regional deformation in front of the MFT (e.g., Yeats & Thakur, 2008). It is suggested that the MFT is extending farther south into the Indo-Gangetic Plains, as an emergent thrust or an anticline.

This segment, located 15 km to the south of the MFT, is named as the Piedmont Thrust.

Geophysical evidence from the Nepal-Bihar segment also suggests the possibility for the existence of blind thrusts. For example, multichannel seismic lines used for petroleum exploration in Nepal also traced blind thrusts in the Terai (plains region) south of the MFT (Bashyal, 1998). These studies imply that the emergent foreland structure south of the MFT could be a regional structure forming beyond the mountain front. However, the Bihar slump belt does appear to lie on the eastern extension of the purported Piedmont Thrust. The studies of Almeida et al. (2018) also confirm the existence of frontal ramps in central Nepal. They present two pre-stack depth migrated seismic profiles from the Bardibas region of Nepal, where the trench sites are located. They could image the deformation associated with displacement along the youngest frontal ramps of the MFT system, locally called the Bardibas and Patu Thrusts, and infer that the depth to the decollement beneath these structures is ~2 km. They show that the southernmost fault ramp, the Bardibas thrust, is blind and that the deformation reaches the surface as a folding of fluvial sediments.

In another study at sites located further east that used seismic reflection data, Duvall et al. (2020) have presented evidence for a foreland blind thrust in the Ganga plains at ~37 km south of the MFT. They report that the Bihar slump belt and Bhadrapur high fall between the 26°–27° latitude (Fig. 11.4), but it is not clear from their analysis if this structure follows the NW–SE trend of the Piedmont Thrust. It was also reported that the deformation zones beneath the Bhadrapur topographic high were formed because of subsurface propagation of the Main Himalayan Thrust (MHT) onto the foreland basin as an outer frontal thrust. They estimated a cumulative slip of ~100 m, accumulated in <0.5 Ma, over a minimum slipped area of ~780 km^2. These observations demonstrate that Himalayan ruptures may pass under the present-day trace of the MFT as blind faults and would remain inaccessible to trenching. It is also to be recognized that paleoseismic studies may underestimate the Holocene convergence. Whether the 1934 earthquake rupture is indeed accommodated in these step-out subsurface structures and if the slump belt is a structurally controlled surface manifestation need to be tested against these new findings from the Ganga foreland basin.

References

Almeida, V., Hubbard, J., Liberty, L., Foster, A., & Sapkota, S. N. (2018). Seismic imaging of the main frontal thrust in Nepal reveals a shallow décollement and blind thrusting. *Earth and Planetary Science Letters, 494*, 216–225.

Ambraseys, N., & Douglas, J. J. (2004). Magnitude calibration of north Indian earthquakes. *Geophysical Journal International, 159*, 165–206.

Bashyal, R. P. (1998). Petroleum exploration in Nepal. *Journal of Nepal Geological Society, 18*, 19–24.

Bilham, R. (2019). Himalayan earthquakes: A review of historical seismicity and early 21st century slip potential. In: P. J. Treloar, & M. P. Searle (Eds.), *2019. Himalayan tectonics: A modern synthesis. Geological society* (Vol. 483, pp. 423–482). London:Special Publications.

Bilham, R. (1995). Location and magnitude of the 1833 Nepal earthquake and its relation to the rupture zones of contiguous great Himalayan earthquakes. *Current Science, 69*, 101–127.

Bollinger, L., Sapkota, S. N., Tapponnier, P., Klinger, Y., Rizza, M., Van Der Woerd, J., Tiwari, D. R., Pandey, R., Bitri, A., & Bes de Berc, S. (2014). Estimating the return times of great Himalayan earthquakes in eastern Nepal: evidence from the Patu and Bardibas strands of the main frontal thrust. *Journal of Geophysical Research Solid Earth, 119*(9), 7123–7163.

Bomford, G. (1937). Leveling in Bengal and Bihar, 93–97. In H. J. Couchman (Ed.), *Survey of India geodetic report, 1936. Geodetic Branch Survey of India* (p. 97). Dehra Dun,.

Burrard, S. (1934). Ground levels in Bihar in relation to the earthquake of January 15 1934. *Nature* (London), 582–583.

Chen, W. P., & Molnar, P. (1977). Seismic moments of major earthquakes and the average rate of slip in central Asia. *Journal of Geophysical Research, 82*, 2945–2969.

Dunn, J. A., Auden, J. B., Ghosh, A. M. N., & Wadia, D. N. (1939). The Bihar–Nepal Earthquake of 1934. Mem. Geol. Soc. India, 73. Survey of India, Calcuta (391 pp).

Duvall, M., Waldron, J. W. F., Godin, L., & Najman, Y. (2020). Active strike-slip faults and an outer frontal thrust in the Himalayan foreland basin. *PNAS, 117*, 17615–17621.

Gutenberg, B., & Richter, C. F. (1949). *Seismicity of the earth and associated phenomena*. Princeton University Press.

Gutenberg, B., & Richter, C. F. (1954). *Seismicity of earth and associated phenomenon* (2nd ed.). Princeton University Press.

Hough, S., & Bilham, R. (2008). Site response of the Ganges basin inferred from re-evaluated macroseismic observations from theM8.1 Shillong 1897, M7.8 Kangra 1905 and 1934Nepal M 8.1 earthquakes. *Journal of Earth System Science* 117, 773–782.

Kumar, S., Wesnousky, S., Rockwell, T. K., Briggs, R. W., Thakur, V. C., & Jayangondaperumal, R. (2006). Paleoseismic evidence of great surface rupture earthquakes along the Indian himalaya. *Journal of Geophysical Research, 111*, B03304. https://doi.org/10.1029/2004JB003309.

Kumar, S., Wesnousky, S. G., Rockwell, T. K., Ragona, D., Thakur, V. C., & Seitz, G. G. (2001). Earthquake recurrence and rupture dynamics of Himalayan frontal thrust, India. *Science, 294*, 2328–2331.

Kumar, S., Wesnousky, S. G., Jayangondaperumal, R., Nakata, T., Kumahara, Y., & Singh, V. (2010). Paleoseismological evidence of surface faulting along the northeastern himalayan front, India: Timing, size, and spatial extent of great earthquakes. *Journal of Geophysical Research-Solid Earth*, 115

Lave, J., Yule, D., Sapkota, S. N., Basant, K., Madden, C., Attal, M., & Pandey, R. (2005). Evidence for a great medieval earthquake (~1100 A.D.) in the central himalayas. *Nepal Science, 307*, 1302–1305.

Marcussen, E. (2017). Explaining the 1934 Bihar-Nepal Earthquake: The Role of Science, Astrology, and "Rumours". In G. J. Schenk (Ed.), *Historical disaster experiences, transcultural research—Heidelberg studies on Asia and Europe in a global context* (p. 241–266).

Martin, S., & Szeliga, W. (2010). A catalog of felt intensity data for 589 earthquakes in India, 1636–2008. *Bulletin of the Seismological Society of America, 100*(2), 562–569.

Mishra, R. L., Singh, I., Pandey, A., Rao, P. S., Sahoo, H. K., & Jayangondaperumal, R. (2016). Paleoseismic evidence of a giant medieval earthquake in the eastern himalaya. *Geophysical Research Letters, 43*, 5707–5715.

Molnar, P., & Deng, Q. (1984). Faulting associated with large earthquakes and the average rate of deformation in Central and Eastern Asia. *Journal of Geophysical Research, 89*(B7), 6203–6227.

Nasu, N. (1935). The great Indian Earthquake of January 15, 1934. *Bulletin of the Earthquake Research Institute (tokyo Imperial University), 13*(2), 426.

Pandey, M. R., & Molnar, P. (1988). The distribution of intensity of the Bihar-Nepal earthquake of 15 January 1934 and bounds on the extent of the rupture zone. *Journal of Nepal Geological Society, 5*, 22–44.

Rajendran, C. P., John, B., Rajendran, K., & Sanwal, J. (2016). Liquefaction record of the great 1934 earthquake predecessors from the north Bihar alluvial plains of India. *Journal of Seismology, 20*, 733–745. https://doi.org/10.1007/s10950-016-9554-z.

Rana, B. S. (1935). *Nepal Ko Maha Bhukampa (The Great Earthquake of Nepal)* (pp. 1–250). Jorganesh Press.

Rana, B. S. (2013). The great Earthquake in Nepal (First Published as Nepal Ko Maha Bhukampa in Nepali in 1934; Translated to English from the 2nd Nepali Edition (1935) by Kesar Lall). Ratna Pustak Bhandar, Kathmandu, Nepal (136 p).

Rizza, M., Bollinger, L., Sapkota, S. N., Tapponnier, P., Klinger, Y., Karakaş, Ç., et al. (2019). Post earthquake aggradation processes to hide surface ruptures in thrust systems: The M8.3, 1934, Bihar-Nepal earthquake ruptures at Charnath Khola (Eastern Nepal). *Journal of Geophysical Research: Solid Earth, 124*, 9182–9207.

Sapkota, S., Bollinger, L., Klinger, Y., et al. (2013). Primary surface ruptures of the great himalayan earthquakes in 1934 and 1255. *Nature Geoscience, 6*, 71–76.

Seeber, L., & Armbruster, J. (1981). Great detachment earthquakes along the himalayan arc and long-term forecasting. In D. W. Simpson & P. G. Richards (Eds.), *Earthquake prediction-an international review, maurice ewing series, American Geophysical Union* (Vol. 4, pp. 259–277).

Singh, D. D., & Gupta, H. K. (1980). Source dynamics of two great earthquakes of the Indian subcontinent; the Bihar-Nepal earthquake of January 15, 1934 and the Quetta earthquake of May 30, 1935. *Bulletin of the Seismological Society of America, 70*, 757–773.

Sukhija, B. S., Rao, M. N., Reddy, D. V., Nagabhushanam, P., Kumar, D., Lakshmi, B. V., & Sharma, P. (2002). Palaeoliquefaction evidence of prehistoric large/great earthquakes in north Bihar, India. *Cur Sci, 83*, 1019–1025.

Wesnousky, S. G. (2020). Great pending himalaya earthquakes. *Seismological Research Letters, XX*, 1–9.

Wesnousky, S. G., Kumahara, Y., Chamlagain, D., & Neupane, P. C. (2019). Large himalayan frontal thrust paleoearthquake at Khayarmara in Eastern Nepal. *Journal of Asian Earth Sciences*

Wesnousky, S. G., Kumahara, Y., Chamlagain, D., Pierce, I. K., Karki, A., & Gautam, D. (2017a). Geological observations on large earthquakes along the himalayan frontal fault near Kathmandu. *Nepal Earth Planetary Sciences Letters, 457*, 366–375.

Wesnousky, S. G., Kumahara, Y., Chamlagain, D. K., Pierce, I. K., Reddy, T., Angster, S., & Giri, B. (2017b). Large paleoearthquake timing and displacement near Damak in eastern Nepal on the Himalayan frontal thrust. *Geophysical Research Letters*. https://doi.org/10.1002/2017GL074270

Wesnousky, S. G., Kumahara, Y., Nakata, T., Chamlagain, D., & Neupane, P. (2018). New observations disagree with previous interpretations of surface rupture along the himalayan frontal thrust during the great 1934 Bihar-Nepal earthquake. *Geophysical Research Letters, 45*, 2652–2658.

Yeats, R. S., & Thakur, V. C. (2008). Active faulting south of the himalayan front: Establishing a new plate boundary. *Tectonophysics, 453*, 63–73.

Chapter 12
Gorkha (Nepal) 2015

12.1 Introduction

The April 25, 2015, M_w 7.8 earthquake (local time 11:56 a.m.) that occurred in Barpak Village of Gorkha District, about 80 km northwest of Kathmandu, Nepal, is the largest modern-day earthquake in the Himalaya. Its epicenter was placed east-southeast of Lamjung, 77 km southwest of Kathmandu, and hypocentral depth was ~15 km (USGS). This earthquake caused extensive damage and destruction in the region, killing nearly 9000 people and injured about 21,952 (Ohsumi et al., 2016). The earthquake was followed by strong aftershocks on 25 April (M_w 6.7), 26 April (M_w 6.9) and 12 May 2015 (M_w 7.3). The largest aftershock of May 12 was located ~75 km north-east of Kathmandu, near the Nepal-China border (Fig. 12.1). This event also caused several casualties, leaving at least 172 people dead and another 3470 injured. According to various reports, a total of about 5,000,000 buildings and houses were partially damaged, and about half of them had collapsed. The earthquake caused ground deformation leading to land subsidence, ground fissures, slope failures, and liquefaction in the Kathmandu Valley (e.g., Kobayashi et al., 2015; Yamada et al., 2016). It also triggered avalanches near Mount Everest (~60 km NNE of Kathmandu), killing several climbers who were camping out. In an avalanche in the Langtang Valley about 50 km north of Kathmandu, 250 people went missing. Numerous small and large landslides followed the earthquake totalling to ~3600, mostly from central Nepal, and some of them affected the hydro-electrical projects in the region. The worst natural disaster to strike Nepal since the 1934 earthquake, the impact of the Gorkha event was minimal in the Indo-Gangetic Plains.

As the first modern-day large earthquake in the Central Himalaya that also triggered a large aftershock, the 2015 event attracted global attention. With the rich and diverse database generated through international collaborations, this earthquake is undoubtedly among the best-studied Himalayan earthquakes. With a different style of rupture and energy directivity, the 2015 event differed from the 1934 earthquake, as manifested by its nominal impact in the Gangetic plains. However it seems to share

C. P. Rajendran and K. Rajendran, *Earthquakes of the Indian Subcontinent*,
GeoPlanet: Earth and Planetary Sciences,
https://doi.org/10.1007/978-981-16-4748-2_12

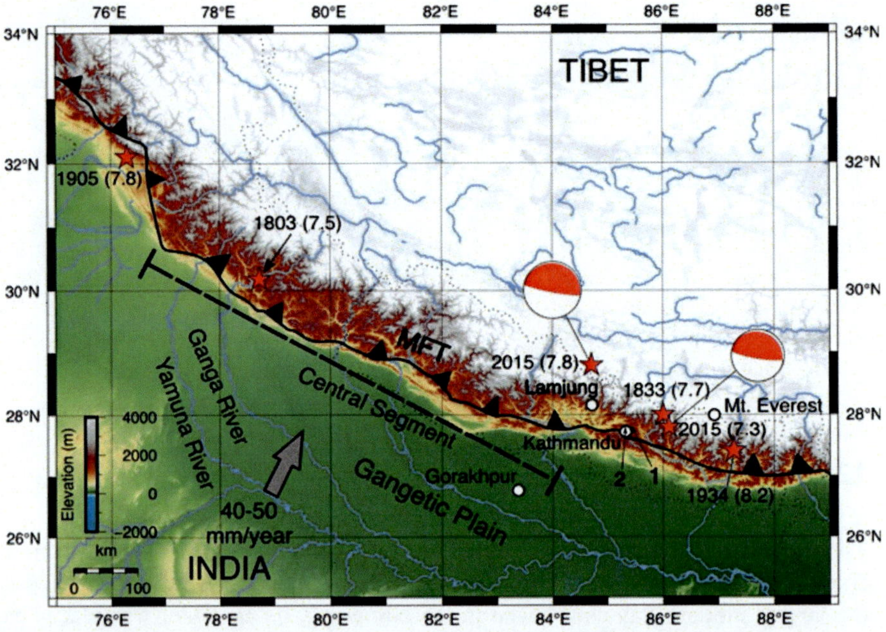

Fig. 12.1 Locations of the 1803, 1833, 1934 and 2015 earthquakes, also showing the fault plane solutions for the 2015 earthquakes (Harvard CMT). Numbers 1 and 2 mark the locations of Manahara and Bagmati River basins, and the liquefaction sites mentioned in the text

some similarities with the 1905 Kangra event, particularly in the way it triggered a large aftershock about 15 min after the mainshock. It also shares similarities with the 1833 (Nepal) and the 1803 (Garhwal) ruptures, both of which are believed to have propagated southward along the basal detachment, but arrested mid-way (e.g., Mugnier, 2013). Quite notably, like these earthquakes the 2015 earthquake did not generate any primary surface rupture.

The Gorkha earthquake, the largest to have occurred in the central Himalaya, was also concurrently recorded by GPS (Avouac et al., 2015), SAR, global broadband stations (Fan & Shearer, 2015; Lindsey et al., 2015), strong motion (Grandin et al., 2015), and local seismic networks (Adhikari et al., 2015). Thus, unlike the 1803 and 1833 events, this earthquake could be investigated in greater detail, using a broad spectrum of data and advanced analytical tools. For example, the observation that the high mountain range subsided by about 60 cm due to the elastic extension north of the region of maximum southward slip was possible through space geodesy (Elliot et al., 2016). Such exceptional opportunities helped to investigate the interplay between seismic deformation and the role of orogenic wedges in developing the topography of the Nepal Himalaya. The data generated from this earthquake was used to further understand the geometry of the MHT.

12.2 Damage Assessment and Intensity

Intensity estimates for modern-day earthquakes are based on the spatial pattern of damage sustained by built structures, whereas for the historical earthquakes, archival information is the only source. The 2015 earthquake occurred at a time when several seismic recording stations were in place, providing data to generate shaking intensity maps. Thus, the damage characteristics could be quantified to understand how a large Himalayan earthquake impacts the landscape and the built-environment, practically in a real-time mode. There were also several post-earthquake surveys that documented the damage. The mainshock destroyed a large number of buildings and infrastructure in the urban and rural areas in central Nepal and triggered numerous landslides and rock falls, blocking roads, and hampering rescue and recovery activities (Goda et al., 2015). For example, among the 1440 buildings in the Barpak Village (close to the epicenter), 1250 collapsed completely and the rest sustained serious structural damage (Gautam, 2017). Although there were many reports on ground failures, damage to buildings and heritage structures, there were also some surprises. For example, the Gorkha Town, though located ~20–30 km south of the epicentre experienced relatively low intensity. Several single and multi-storied brick masonry buildings here survived, with only minor cracks on their inner walls.

Parameswaran et al. (2015) conducted a road-survey, starting from Gorakhpur (India) towards Kathmandu and other affected areas using descriptions by Murakami and Kagami (1991). Using the Modified Mercalli Intensity Scale (MMIS) score out of XII, they reported the maximum intensity of VII–VIII at Kathmandu and Bhaktapur. Although the interior regions were not covered in detail, the overall damage distribution pattern within the valley was reported to be quite random. The damaged buildings were mostly the stone/brick masonry structures with no seismic detailing, whereas the RC buildings with seismic designs generally survived. Strong motion data from the KATNP station, located in the Durbar Square, Bhaktapur Town and reported PGA of about 150–170 and 70–80 cm/s^2, respectively, for the M_w 7.8 mainshock and the M_w 7.3 aftershock (see Fig. 12.2 for locations) (Goda et al. (2015)

A detailed account of damage to structures is provided by Yamada et al. (2016) who conducted building damage surveys along with microtremor measurements around the Kathmandu Valley. The impact in the valley was reported as mixed, and only less than 5% of the total number of buildings had collapsed. Outside of the Kathmandu Valley, especially to its east, the impact was severe. More than 40% buildings had collapsed and the severity of impact was attributed to the eastward rupture directivity. Based on microtremor observations, they attributed the severity of damage to the rupture characteristics, and the seismic performance of buildings, rather than to the local site conditions.

Most of the intensity surveys had also focused on the extent of damage to the ancient buildings and temples in Kathmandu and the nearby areas. The Durbar Square at Bhaktapur town, located 13 km east of Kathmandu, an important heritage site and a cultural center, was heavily damaged. One of the temples in the square had lost its roof, and one of its subsidiary units, called the Vatsala Devi temple,

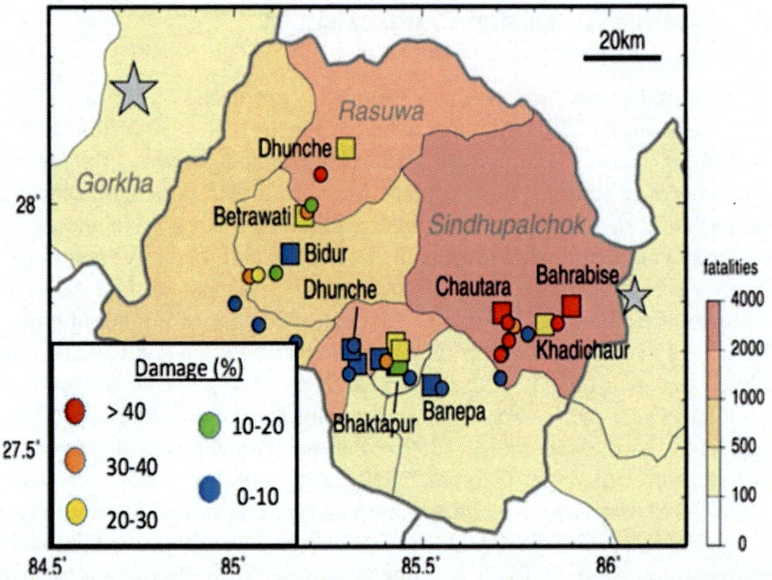

Fig. 12.2 Summary of the damage survey showing the percentage of totally collapsed buildings at the sites visited (squares) and visually inspected from the vehicle (circles). The grey stars show the epicenter of the mainshock and the major aftershock. The background color code indicates the fatalities in each district (after Yamada et al., 2016)

known for its sandstone walls and gold-topped pagodas, was partially destroyed. A few of these long-lived structures have been affected also by past earthquakes. It has been documented that many of the temples in this town were impacted by the 1833 and 1934 earthquakes (Basukala et al., 2014). Another important monument at the center of Kathmandu, the nine-storied, ~63 m-high-minaret called Dharahara was completely destroyed by the 2015 earthquake. What used to be a twin-tower, built in 1832, this structure had partially collapsed during the 1833 and 1934 earthquakes and was restored as a single tower. The *Punva Mahadev* (a part of the Pashupatinath Temple complex), presumably more than 200 years old, withstood the ground shaking, although their walls had developed shear cracks, and the decorative metallic pieces on the top of the structure were damaged (Fig. 12.3a, b).

The 2015 earthquake caused liquefaction and ground deformation in many parts of the Kathmandu Valley. Sandblows, lateral spreads, subsidence of ground, and slope failures were reported from various parts of central Nepal, including Kathmandu (e.g., Goda et al., 2015; Parameswaran et al., 2015). Subsidence of walkways, fissures, slope failures, and liquefaction, were observed at many locations, mostly within the Kathmandu basin. As in the case of some of the previous historical earthquakes sourced in the east-central Nepal Himalaya (e.g., the 1833 and 1934), the 2015 earthquake had also shaken the northern Bihar Plains. However, ground deformation including liquefaction in the Nepal and Indian Plains was nominal in contrast to the impact of the 1934 event. Some of the sites near Madhubani in the Bihar Plains

Fig. 12.3 Damage to heritage structures at **a** remains of Dharahara tower. **b** Punva Mahadev temple and **c** Sandblows at Ramkot near Kathmandu

and in Chitwan District in the Nepal Plains are reported to have been liquefied in a smaller scale. The features near Madhubani were of 3–5 cm wide circular-shaped 'blisters'. Streams of very fine sand that were emplaced made a clear distinction with the surrounding flood plain alluvium. The contrast between the 1934 and 2015 earthquake in their impact on the southern plains can be explained as being due to directivity effect (Grandin et al., 2015).

It was also reported that the location of ground failures induced by liquefaction in the Kathmandu Valley was limited to margins of the basin, possibly due to the effects of edge amplification. Such spatial preference is attributed to edge effects, which results in ground motion amplification and change in the frequency content at the edges of the basin (Gautam, 2017). These patterns are caused by the constructive interference of the shear waves propagating vertically through basin sediments and horizontally propagating surface waves, diffracted into the basin. In the basin interior, the ground shaking generated rather low cyclic stress ratio (CSR), which reportedly was not enough to cause liquefaction. The low intensity shaking seemed just enough to trigger the liquefaction but not strong enough to fully weaken the liquefiable materials, exhibiting "marginal" liquefaction. Additionally, the frequency and the intensity (low PGA of 0.16 g) of the shaking appeared to be less than it would be expected for an event of the same magnitude (Maharjan, 2017).

It was observed that at most liquefaction sites, the sand was ejected to agricultural fields forming veneers or sheets of sand (Fig. 12.3c). The liquefaction features were mostly observed along the floodplain area of the Manahara and Bagmati River basins. Some sand boils (when the liquefied sand reaches and spreads out, they are called sand blows or sand boils) were formed in agricultural fields, as observed in the floodplains. Of all the liquefaction sites, Ramkot, located to the western margin of the Kathmandu Valley, exhibited locally extensive sand boils as explored by Rajendran (2021) (Fig. 12.4). Findings of paleoliquefaction features implied previous earthquakes that had impacted the Kathmandu region, most likely the 1833 and 1934 earthquakes.

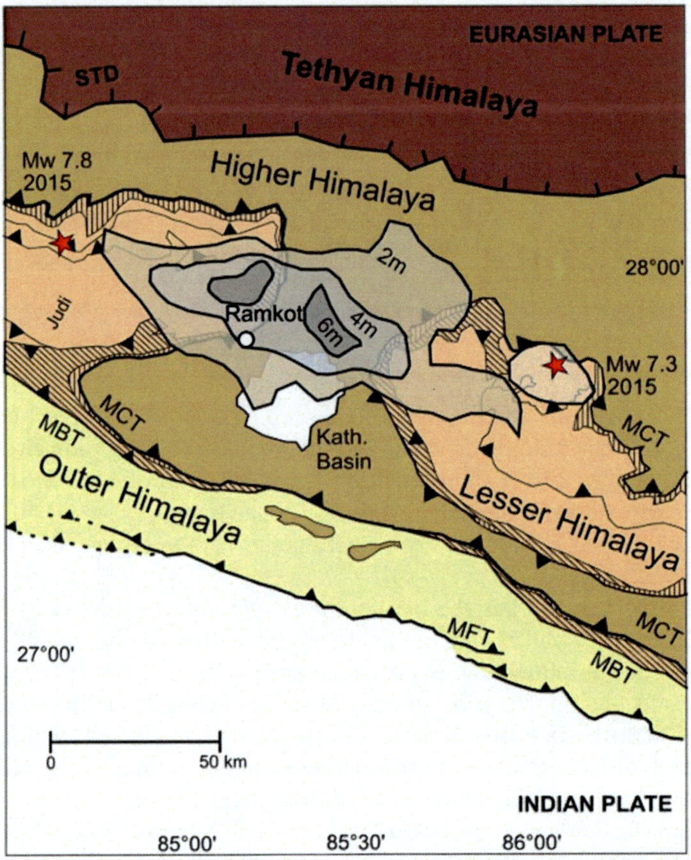

Fig. 12.4 A simplified geological map of the Nepal Himalaya compiled from various sources by Mugnier et al. (2017). Locations of 2015 April–May earthquakes are shown along with coseismic slips (transparent grey patches) inferred from structural models by Hubbard et al. (2016)

12.3 Structural Setting of the Nepal Himalaya

The mechanism of wedge deformation and its role in the evolution of the steep front of the high range remain debated, as we have discussed in Chap. 7. Recall that the geometry of the Himalaya orogenic wedge has been described as a stack of thrust sheets that sole into a main basal décollement or a mid-crustal reflector. The mid-crustal ramp along the MHT (Pandey et al., 1995; Lave and Avouac, 2001); ramp over-thrusting and underplating associated with duplex development of the Himalayan wedge (e.g., Bollinger et al., 2004) and the active out-of-sequence thrusting at the front of the high Himalaya (Wobus et al., 2005) are among the various mechanisms proposed. Receiver function analysis from the Hi-CLIMB (Himalayan–Tibetan Continental Lithosphere during Mountain Building) experiment had revealed a low-velocity zone (LVZ) interpreted as the signature of the MHT. It is almost horizontal and is located at 10–15 km depth below Nepal, but it is reported to attain a depth of about 40 km beneath southern Tibet (Nabelek et al., 2009).

Mugnier et al. (2017) provide a review of the crustal-scale deformation of the central Nepal Himalaya and its important structures (see Fig. 12.4). The MHT, locally imaged by geophysical data (e.g., Nabelek et al., 2009), projects at the surface along the frontal part of the Himalaya. This structure is referred to as the MFT or the HFT (e.g., Schelling & Arita, 1991). On the MHT, which absorbs approximately 20 mm/yr of convergence in Nepal (Lavé & Avouac, 2000), a crustal ramp has been deduced from balanced cross-sections and models based on earthquake data (e.g., Pandey et al., 1995).

All the major thrust systems are expressed in the central Nepal Himalaya. Here, the MCT is passively folded, and it depicts a regional antiform, north of Kathmandu and a synform beneath Kathmandu. Following Mugnier et al. (2017), the geometry of the MHT in the Nepal Himalaya can be described by its basic features. These include a southern frontal ramp, a flat detachment beneath the Siwalik and Lesser Himalayan zones, a crustal ramp cutting through the upper crust of the Indian plate, and a lower flat that extends far to the north beneath the Tibetan Plateau. In addition to this classically described geometry, some minor ramps probably affect the flat segments. The discussion of active faults in the Nepal Himalaya also highlight the right-lateral strike-slip faults with a thrust component that reactivate the roof thrust of the antiformal duplex. Other active structures affecting the Tibetan Plateau fall under the category of normal faults, which extend to the High Himalaya (Armijo et al., 1986). All these faults are seismogenic, but their long-term motions remain very small (a few tens of meters during thousands of years). With their locations above the ductile MHT, they do not significantly influence the seismic cycle of the brittle MHT (Avouac et al., 2001).

12.4 Surface Deformation

The M_w 7.8 earthquake was located to the west of the 1934 earthquake source, on the eastern part of the central Himalayan segment, thus leaving much of the postulated gap un-filled (Fig. 12.1). Extensive post-2015 field surveys found no evidence for coseismic surface rupture along the MFT (e.g., Goda et al., 2015). Source models that explain the origin, rupture propagation, and aftershock pattern show south-eastward propagation of the rupture with no significant component relayed to the MFT (e.g., Avouac et al., 2015; Lindsey et al., 2015; Parameswaran & Rajendran, 2017).

Focussing on the areas that showed maximum deformation on the InSAR interferometry image, Parameswaran et al. (2015) explored the region east of the 2015 epicenter, where broad zones of uplift and subsidence were juxtaposed in the close proximity. Subtle ground cracks with minor displacements that were spatially consistent with the coseismic deformation observed on the InSAR imagery (Lindsey et al., 2015) were reported from an agricultural field near Phujel, located about 20 km south of the epicenter (see Fig. 12.5 for location). These subtle features were consid-

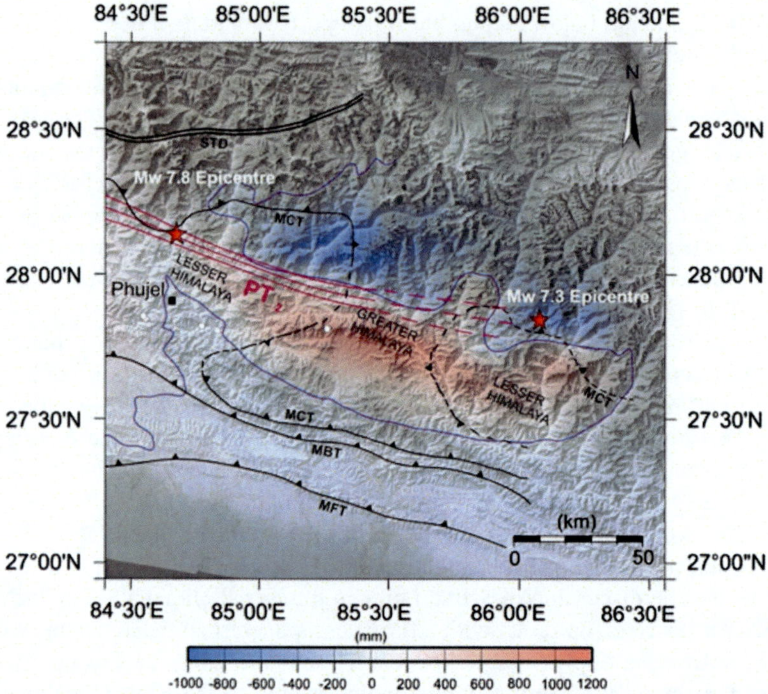

Fig. 12.5 Unwrapped InSAR data from Lindsey et al. (2015) over central Nepal showing the sum of vertical displacements from the April 25 earthquake and its major aftershock. Thin purple line: zero-change contour. Trace of PT_2 (pink lines) (Wobus et al., 2006) follows the zero-change contour (thin purple line) between the major zones of uplift (red; up to 1.08 m) and subsidence (blue; up to −0.80 m); adopted from Parameswaran et al. (2015)

ered as the likely surface expressions of splay thrusts. They spatially correspond to the physiographic break named NPT2, considered as an out-of-sequence thrust that separates the southern regions of moderate elevation from those of higher elevations in the north (Hodges et al., 2004; Wobus et al., 2003, 2006). However, extensions of this feature mapped in Phujel have not been reported from any later surveys and thus, any spatial correspondence of the subtle features with the postulated PT2 needs validation.

12.5 Aftershocks and Source Models

The April 25 earthquake was followed by a series of aftershocks and those of the second large event formed a tight cluster 50–60 km east of the initial cluster (Fig. 12.6). Adhikari et al. (2015) used 553 aftershocks spanning over the first 45 days of the earthquake to suggest that the rupture was centered on the Kathmandu

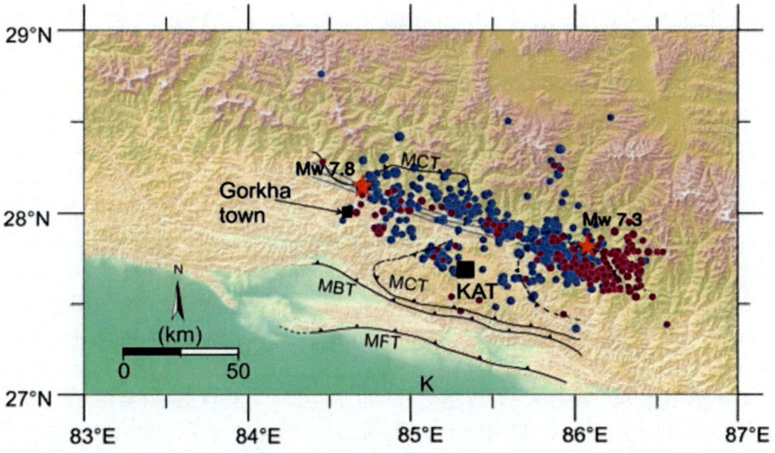

Fig. 12.6 Aftershocks of the Gorkha earthquake (total number 553) of magnitude ML ≥ 4 and depth <60 km, collected by the National Seismological Centre, Nepal, over a period of 45 days after the mainshock (Adhikari et al., 2015; modified after Parameswaran & Rajendran, 2017)

klippe, consistent with the mid-crustal ramp hypothesis that explains localization of earthquakes. Several source models have been put forward to explain the origin, rupture propagation, and aftershock pattern, including the initial USGS moment tensor solution that suggests a slip on ~11°-dipping fault with strike 295°.

In one of the early models, Avouac et al. (2015) imaged the rupture process by the back-projection of teleseismic P-waves from the Australian seismic network. They also determined a finite source model from the joint inversion of teleseismic waveforms in the 0.01–1 Hz frequency band and static surface displacements measured from SAR image offsets. Both these techniques generated similar results with a unilateral pulse-like rupture, a narrow strip of active slip, 20–30 km-wide, that propagated eastward at about 2.7–3.0 kms^{-1}. Both models suggested that the earthquake unzipped the downdip edge of the locked zone, with the rupture propagating eastwards for about 140 km, unzipping the lower edge of the locked portion of the fault, a process that was resumed during the May 12, M_w 7.3 aftershock.

Teleseismic moment inversions by others are consistent with the above results, suggesting a pure thrust faulting earthquake sourced at 15-km-depth on a plane with a strike of 293°, dip of 7°, and magnitude M_w 7.85 (e.g., Avouac et al., 2015). Based on these models, much of the net moment release (0.753×10^{21}Nm) occurred within the first 50 s of the ~70–80 s-long-rupture that extended for ~120 km in the southeast direction. The maximum dislocation was 4.18 m, sourced ~40 km southeast of the epicenter, and the average slip was ~0.1 m, distributed within 160×138 km^2 of the fault area. The slip vectors are directed up-dip, typical of thrust faults, and all the models suggest minimal slip in the Indo-Gangetic Plains. An improved model using InSAR and GPS data suggests a thrust motion on a shallow, and low dipping fault plane along a strike of 288.8° (Yang et al., 2021). This model also suggests that more than 90% of the seismic moment is released on the middle segment within a length of ~110 km, where the fault slip attains a maximum slip of 6.7 m. The slip pattern inferred from the source model suggests the presence of two significant slip deficit zones located on the northwest and southeast of the fault.

The precise, but sparsely-distributed GPS observations in the Central Nepal region provided another source of data to model the coseismic displacement. Tung and Masterlark (2016) suggests that the epicentral area was uplifted and moved southwestward with maximum vertical and horizontal displacements of 1.3 m and 1.9 m, respectively. The slip model shows a rhombus-shaped slip zone with three composite asperities, which are aligned along the strike direction at a depth of about 15 km and a rupture dominated by thrusting with minimal right-lateral strike-slip, with a maximum slip of 6.55 m, in agreement with seismological models (Avouac et al., 2015; Galetzka et al., 2015). A similar deformation pattern was observed from the line-of-sight displacements measured by the ALOS-2 (PALSAR-2) satellite (Lindsey et al., 2015). The line-of-sight (LOS) displacement map shows only limited displacements near the epicenter, but the maximum slip and surface displacement gradient were located at about 30 km northeast of Kathmandu.

12.6 Fault Plane Characteristics

The rupture is believed to have progressed much up-dip of the MHT and not downdip, either because the zone of aseismic slip inhibited its downward progression or because of a geometrical barrier created by the ramp. Duputel et al. (2016) conducted a reinvestigation of the Hi-CLIMB data and suggested that the horizontal extension of the LVZ corresponded with the along-dip distribution of coseismic slip and the aftershock epicentres. The imaging brought out the ramp-flat-ramp geometry of the MHT, the thrust fault flattening at 10–15 km depth, and the mid-crustal ramp connecting to a deeper and low-dipping thrust >25 km deep. Further, it has been argued that the 2015 earthquake must have ruptured a steeper part of the MHT. This geometry corresponds with the upper end of the mid-crustal ramp and is consistent with the mainshock hypocentre located on the up-dip edge of the crustal ramp, adding credence to this argument. Results presented by Wang et al. (2017), who inverted well-constrained medium-sized early aftershocks and modelled the teleseismic P-waves, indicate a clear double-ramp geometry of the MHT, with an almost flat décollement between them. The flat part dipping ~7° overlaps with the coseismic rupture of the mainshock. The inference is that the double-ramp fault geometry along the up-dip and down-dip directions must have played a significant role in inhibiting the rupture.

Studies that compared the coseismic rupture with the geological and structural features have shown that the maximum slip was placed immediately south of the ramp-flat transition (Hubbard et al., 2016; Mugnier et al., 2017). In the model by Hubbard and others two active blind ramps were found to bind the fault that slipped during the earthquake. One of these is a deep and active northern ramp, and the other is a middle ramp to the south. Mugnier et al. (2017) suggest that the deformation zone is bordered by two prominent N-S lineaments, and the rupture itself is bound on all sides by geological structures. Subevents characterized by multi-pulsed heterogeneous rupture process with time-dependent rupture velocity variation, is perhaps a reflection of lateral heterogeneities (Kumar et al., 2017). Whether or not these lateral discontinuities inhibited the eastward rupture or if they reflect the lateral ramps along the MHT, are to be validated from future investigations. An actively deforming duplex, propelled by the steeply dipping imbricate faults within it, could be another characteristic of the MHT in Nepal and other parts of the Himalaya (Mendoza et al., 2019). The complex geometrical transition of the MHT from planar, to duplex, and back to planar, is akin to an internally deforming tectonic 'doorstop' that essentially impedes subduction of the Indian plate. The resulting concentration of stress on highly vulnerable faults controls the nucleation and rupture of large earthquakes, such as the Gorkha event.

Distribution of static shear stress changes on the fault plane from a coseismic slip model suggest a dominant east-southeastward unilateral rupture process from the epicenter with an average rupture speed of ~3 km/s (Yin et al. (2017). Coseismic radiation shows clear frequency-dependent features and the lower frequency (0.05–0.3 Hz) radiation originates mainly from large coseismic slip regions with negative shear stress changes. In comparison, the higher frequency (0.3–0.6 Hz) radiation appears to originate from the down-dip part of the rupture. In their asperity model, such frequency-dependent radiation is related to the depth-varying frictional properties on the plate interface of the MHT in the Nepal segment.

Elliot et al. (2016) used geodetic, geologic, geomorphologic. and geophysical data to constrain the geometry of the MHT beneath Kathmandu. Based on structural analyses, together with interseismic and coseismic displacements, this model suggests a steep, shallow thrust fault flattening at a depth between 5 and 15 km and connecting to a mid-crustal, steeper thrust. The maintenance of the steep front of the high Himalayan range is supported by the mid-crustal ramp along the MHT, a transition zone that also coincides with the down-dip edge of the locked zone based on measurements of inter-seismic strain. In this model, the earthquake was sourced in the down-dip part of the locked portion of the MHT—at the interface of the frontal fault and a deeper, steeply dipping ramp of the MHT (Fig. 12.7). The position of the ramp is consistent with field observations of broadly folded foliations north of the Kathmandu Klippe, thought to be related to duplex development in the Lesser Himalaya, as proposed along a few geologic cross-sections across Nepal. The coseismic slip in their model is restricted to the deeper parts of the MHT, without any component of out-of-sequence thrusting on the MCT.

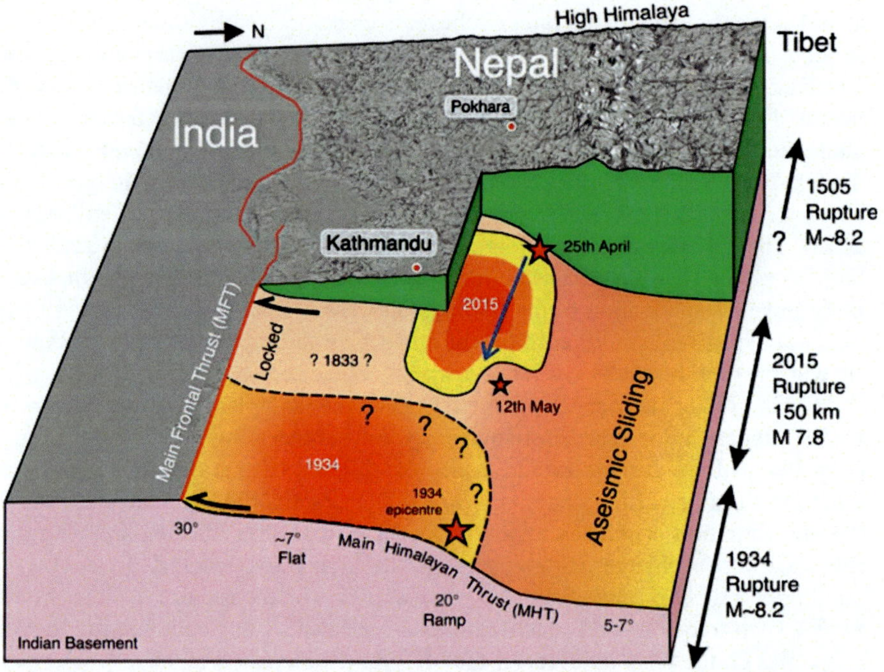

Fig. 12.7 Location of the 2015 earthquake and its aftershock on the resolved MHT geometry (with the upper plate removed to reveal the slip zone). The M_W 7.2 aftershock occurred at the eastern end of the main rupture. The rupture extents of previous earthquakes from 1934 (M_W 8.2), 1833 (M ~ 7.6) and 1505 (M_W 8.2) are poorly constrained (after Elliot et al., 2016)

12.7 Previous Earthquakes in Nepal

12.7.1 Archival Information

Reports of historically documented earthquakes are perhaps incomplete and there are only a few studies that have explored these archives (e.g., Chitrakar & Pandey, 1986; Iyengar et al., 1999; Pant, 2002; Rana, 2013). Bollinger et al. (2016) provide a detailed description of the various sources of information on the historical earthquakes in Nepal. Historical archives from Kathmandu (Nepal) indicate six major earthquakes (1223, 1255, 1344, 1408, 1833 and 1934 CE) that have impacted the valley region (see also Table 12.1). The earthquake reported to have occurred in 1223 has been documented based on chronicles in Sanskrit and Newari (Kathmandu vernacular) covering the period 1057–1389 CE. It is reported to have occurred on December 24, although Pant (2002) notes that the description is indecipherable due to defaced letters and words. The description on the next largest event (June 7, 1255) is quite vivid, specifying that temples and houses collapsed, killing one-third of the inhabitants, including the king, Abhaya Malladeva. The September 14, 1344 CE,

Table 12.1 Historical earthquakes in the Kathmandu area (Intensity MMI greater than VII in Kathmandu)

Year (CE)	Source-of information	Intensity (MMI)	Mag	Epicentre/localization	Main reference
1223	Historical	–	–	–	Pant (2002)
1255	Historical/trench	X		Surface rupture	Chitrakar and Pandey (1986), Wesnounsky et al. (2016)
*1344	Historical/trench	X	–	Surface rupture	Pant (2002), Bollinger et al. (2016), Mugnier et al. (2013)
1408	Historical/trench	X	–	Surface rupture in far western Himalaya	Chitrakar and Pandey (1986)
1681	Historical	≥VII		?	Chitrakar and Pandey (1986)
1767	Historical	≥VII		?	-do-
1810	Historical	IX		Eastern Nepal?	Chitrakar and Pandey (1986)
1833	Historical	IX–X	≥7.6	N28.00°, E86.00°	Oldham (1883), Bilham (1995)
1866	Historical	≥VII	≤7.6	Central or western Nepal (magnitude could be lower than what is shown)	Oldham (1883)
1934	Instrumental and historical	X	8.2	N27.55°, E87.09°	Chen and Molnar (1997), Pandey and Molnar (1988)

Notes Historical archives from Kathmandu (Nepal) indicate five major earthquakes (1255, 1408, 1344; 1866 and 1934 CE) that have impacted the valley region (Chitrakar & Pandey, 1986; Rana, 2013). The June 7, 1255 is considered to be a very damaging earthquake in the Kathmandu region, which reportedly killed the reigning king Abhaya Malla (Pant, 2002). According to Newari chronicles, a violent earthquake occurred on midday September 14, 1344 (Bollinger et al., 2016; Pant, 2002). The earthquake of 1408 CE cannot be authenticated and questions have been raised on whether this actually occurred (Bollinger et al., 2016). The Tibetan religious archives reveal another earthquake along the Nepal-Tibet border in 1505 CE (Ambraseys & Jackson, 2003), which does not find mention in the Nepalese or Indian sources (Rajendran et al., 2013; 2018); the 1505 CE had been assigned a magnitude (M_W) ranging from 8.2 to 8.6 (Ambraseys & Jackson, 2003; Bilham & Ambraseys, 2005)

earthquake was very destructive and the King Ari Malla died the next day after being fatally injured. Using the descriptions from the archives, Pant (2002) suggests that the destruction mostly occurred in Kathmandu, without much far field reports of destruction. According to the documents reviewed by Rana (1935), an earthquake is reported to have occurred in 1408 CE during the reign of Syamasimha, but there is no mention of its source.

The other documented great earthquake, based on historical source is reported to have occurred on June 6, 1505 CE. Historical documents during that period are scanty, and what is available makes no mention of any large event in Kathmandu. However, eyewitness reports in southwestern Tibet imply that the earthquake had affected a ~600 km-long stretch of the high Himalayan range (e.g., Ambraseys & Jackson, 2003). The size (~M_w 8.5) and location of the 1505 earthquake remain debated. There is a consensus that a medieval may have occurred along the central stretch of the Himalaya, and the year of the occurrence postulated either as 1344 or 1505 CE (e.g., Rajendran et al., 2015). Rana (1935) reports of another earthquake (M < 7.5), with an unconfirmed date, and is based on the 5-month-gap since the Great Comet of 1680 was spotted in Nepal.

12.7.2 Evidence from Trenching Excavations

Corroborative evidence for some of the historical earthquakes have been found in the trenches excavated at different locations by various researchers. From one of the early studies conducted in the central Himalayan region on the Indian side, Kumar et al. (2006) reported great earthquakes in medieval period. Based on this data, the last fault movement has been variously attributed to a great events between 1200 and 1400 CE, an interval that overlaps with many of the historically documented earthquakes. Mishra et al. (2016) used data from trenches on the eastern Himalaya to suggest that a mega-event in 1255 CE (M_w 9.0) had ruptured ~800 km of the Himalayan arc between 85.87° and 93.76°E longitudes. However, Pierce and Wesnousky (2016) have challenged this rupture attributed to the 1255 CE earthquake on the eastern segment and regard it as a misinterpretation of trench stratigraphy and incorrect explanation of radiocarbon age statistics. Other studies in the frontal Uttarakhand (India) Himalaya suggest that a great earthquake dated between 1259 and 1433 CE matches with a historical candidate of 1344 CE (Jayangondaperumal et al., 2017; Rajendran et al., 2015, 2019). It is also suggested that the overlapping segments from east to west, along the frontal thrust have been characterized by a successive pulse of great medieval earthquakes of the 11th, 13th, and fourteenth centuries. The paleo-liquefaction constraints from the northern Bihar Plains indicate that one of these medieval earthquakes (the 11th-century event) had caused liquefaction in the Bihar Plains (Rajendran et al., 2015, 2016a, b).

The central-eastern part of the Himalaya is known for two damaging earthquakes in recent history, namely, 1833 (M_w 7.6) and the 1934 (M_w 8.2) events. The source zones, rupture lengths, and magnitudes of the historically recent events of 1833 (M_w ~ 7.8) and 1934 are constrained by previous studies (e.g., Bilham, 1995; Chen & Molnar, 1977). The 1934 earthquake had been associated with an extensive liquefaction field in the Bihar Plains and the Kathmandu Valley, but the reports of the 1833 earthquake are not vocal about its liquefaction impact in this region. Possibly the heavy rains in Kathmandu on that day (the earthquake occurred in the late hours of August 26) must have washed away the sand-blow deposits and impaired the ground reports. This earthquake is considered to have originated in the northcentral Nepal, about 50 km north or northeast of Kathmandu (Bilham, 1995), and it resulted in moderate damage in the northern Bihar Plains.

The 2015 liquefaction features in the Kathmandu Valley offered an opportunity to focus on easy-to-identify liquefaction sites and probe for preserved stratigraphic records of earlier liquefaction events that may have impacted the region. Among them is an event around 1000 CE that matches with the evidence of a major earthquake from the Nepal frontal thrust (Rajendran, 2021). Liquefaction features generated by the preceding sequence of earthquakes (either 1833 or 1934) have been recognized in these locations along with older events. The spatial distribution of the 2015 liquefaction features may compare not only with the 1833 earthquake (in Nepal Himalaya) but also with the 1803 earthquake in the Garhwal Himalaya (e.g., Rajendran et al., 2018). These similarities may owe their origin to the analogous rupture characteristics and these are subjects of future investigations.

12.8 The Gorkha Earthquake and Its Implications

The 2015-type earthquakes may belong to the category of blind thrust events which leave minimal surface expression of their ruptures. Such events may have been more frequent, compared to the M +8 type earthquakes that occur over with longer temporal gaps. Thus, the 2015 earthquake may belong to a suite of large earthquakes whose ruptures get arrested mid-way on the MHT, like the 1803 or 1833 events (e.g., Mugnier et al., 2017). The bimodality of earthquakes in the Himalaya has been pointed out by many workers in the past (Ni & Barazangi, 1984; Pandey et al., 1995). Bimodality implies that blind earthquakes (up to M_W ~ 7.8) tend to cluster in the downdip part of the seismogenic zone, whereas infrequent great earthquakes (M_W 8+) propagate up to the Himalayan frontal thrust. Apparently the 2015 earthquake belongs to a category of blind earthquakes (M_W ~ 7.8) that typically rupture the MHT near the mid-crustal ramp, in contrast to the sporadic great earthquakes ($M_W > 8$) that rupture the entire width of the seismogenic zone (Zilio et al., 2019). The thermo-mechanical model presented by these authors argues that both frictional properties and non-planar geometry of the MHT influence the relative persistence of variations of seismic ruptures along the dip. They postulate that unzipping of downdip edge of the MHT and resultant transfer of static stress and critical loading of the up-dip parts, can lead to irregularities in the seismic cycle, including the bimodal seismic cycle. The variable recurrence of large ($M_W \leq 7.8$) and great (Mw > 8) earthquakes, which might originate from these complexities have implications for seismic hazard assessments. Perhaps an important outcome of the studies on the Gorkha earthquake include the generation of better constraints on the geometry of the MHT and their role in seismogenesis.

Seismic hazard associated with the central segment of the Himalaya is an issue of concern, especially in the backdrop of the forecast of an impending great earthquake (e.g., Bilham et al., 2001). Whether or not the 2015-type events would change the 'status quo' with regard to the predicted potential of the central gap to generate a great earthquake has been a point of discussion. The epicenter of the Gorkha earthquake was located on the eastern margin of the postulated central gap, leaving much of the western part unruptured. The great earthquakes in the central Himalaya are inferred to nucleate from moderate events near the base of the MHT. It has been suggested that identifying the preparation zones of such earthquakes may provide opportunities for forecasting the approach of future great earthquakes (Bilham et al., 2017). In this context one should recall the observation by Avouac et al. (2015) that the locked portion of the MHT, west of the 2015 rupture is a potential location for a future earthquake. The nearly 800 km-long stretch of the central segment between the 1833/2015 and the 1905 rupture is a well-identified seismic gap where the MHT is clearly locked, with a slip deficit of >10 m. All of these studies point to the longevity of quiescence on the segment west of the 2015 rupture, which should be of concern to the researchers and policy makers alike.

References

Adhikari, L. B., Gautam, U. P., Koirala, B. P., Bhattarai, M., Kandel, T., Gupta, R. M., Timsina, C., Maharjan, N., Maharjan, K., Dahal, T., & Hoste-Colomer, R. (2015). The aftershock sequence of the 2015 April 25 Gorkha-Nepal earthquake. *Geophysical Journal International, 203*(3), 2119–2124.

Ambraseys, N., & Jackson, D. (2003). A note on early earthquakes in northern India and southern Tibet. *Current Science, 84*, 570–582.

Armijo, R., Tapponnier, P., Mercier, J. L., & Han, T. L. (1986). Quaternary extension in southern Tibet: field observations and tectonic implications. *Journal of Geophysical Research, 91*, 13803–13872.

Avouac, J.-P., Bollinger, L., Lavé, L., Cattin, R., & Flouzat, M. (2001). Le cycle sismique en Himalaya. *Comptes rendus de l'Académie des Sciences, 333*, 513–529.

Avouac, J.-P., Meng, L., Wei, S., Wang, T., & Ampuero, J.-P. (2015) Lower edge of locked main Himalayan thrust unzipped by the 2015 Gorkha earthquake. *Nature Geoscience*, 6–11.

Basukala, B., Gutschow, N., & Kayastha, K. (2014). Towers in stone—Sikhara temples in Bhaktapur, Vatsala and Siddhilakshmi. In *Himal Books* (p. 159). Kathmandu, Nepal.

Bilham, R. (1995). Location and magnitude of the 1833 Nepal earthquake and its relation to the rupture zones of contiguous great Himalayan earthquakes. *Current Science, 69*, 101–127.

Bilham, R., & Ambraseys, N. (2005). Apparent Himalayan slip deficit from the summation of seismic moments for Himalayan earthquakes, 1500–2000. *Current Science, 80*, 1658–1663.

Bilham, R., Mencin, D., Bendick, R., & Bürgmann, R. (2017). Implications for elastic energy storage in the Himalaya from the Gorkha 2015 earthquake and other incomplete ruptures of the main Himalayan Thrust. *Quaternary International, 462*, 3–21.

Bilham, R., Gaur, V. K., Molnar, P. (2001). Himalayan seismic hazard. *Science, 293*, 1442–1444.

Bollinger, L., Tapponnier, P., Sapkota, S. N., & Klinger, Y. (2016). Slip deficit in central Nepal: Omen for a repeat of the 1344 CE earthquake? *Earth, Planets and Space, 68*, 1–12.

Bollinger, L., Avouac, J. P., Beyssac, O., Catlos, E. J., Harrison, T. M., & Grove, M., et al. (2004). Thermal structure and exhumation history of the lesser himalaya in central Nepal. *Tectonics, 23*, TC5015. https://doi.org/10.1029/2003TC001564.

Chen, W. P., & Molnar, P. (1977). Seismic moments of major earthquakes and the average rate of slip in central Asia. *Journal of Geophysical Research, 82*, 2945–2969.

Chitrakar, G. R., & Pandey, M. R. (1986). Historical earthquakes of Nepal. *Bulletin of Nepal Geological Society, 4*, 7–8.

Duputel, Z., Vergne, J., Rivera, L., Wittlinger, G., Farra, V., & Hetényi, G. (2016). The 2015 Gorkha earthquake: A large event illuminating the main himalayan thrust fault. *Geophysical Research Letters, 43*, 2517–2525. https://doi.org/10.1002/2016GL068083

Elliott, J. R., Jolivet, R., González, P. J., Avouac, J. P., Hollingsworth, J., Searle, M. P., Stevens , V. L. (2016). Himalayan megathrust geometry and relation to topography revealed by the Gorkha earthquake. *Nature Geoscience*. https://doi.org/10.1038/ngeo2623.

Fan, W., & Shearer, P. M. (2015). Detailed rupture imaging of the 25 April 2015 Nepal earthquake using teleseismic P waves. *Geophysical Research Letters, 42*(14), 5744–5752. https://doi.org/10.1002/2015GL064587

Galetzka, J., Melgar, D., Genrich, J. F., Geng, J., Owen, S., Lindsey, E. O., Xu, X., Bock, Y., Avouac, J. P., Adhikari, L. B., & Upreti, B. N. (2015). Slip pulse and resonance of the Kathmandu basin during the 2015 Gorkha earthquake. *Nepal Science, 349*(6252), 1091–1095.

Gautam, D. (2017). Unearthed lessons of 25 April 2015 Gorkha earthquake (M_w 7.8): Geotechnical earthquake engineering perspectives, Geomatics, *Natural Hazards and Risk*. https://doi.org/10.1080/19475705.2017.1337653.

Goda, K., Kiyota, T., Pokhrel, R., Chiaro, G., Katagiri, T., Sharma, K., & Wilkinson, S. (2015). The 2015 gorkha nepal earthquake: insights from earthquake damage survey. *Frontiers in Built Environment, 1*(8).

Grandin, R., Vallée, M., Satriano, C., Lacassin, R., Klinger, Y., Simoes, M., & Bollinger, L. (2015). Rupture process of the $M_w = 7.9$ 2015 Gorkha earthquake (Nepal): Insights into Himalayan megathrust segmentation. *Geophysical Research Letters, 42*(20), 8373–8382. https://doi.org/10.1002/2015GL066044.

Hodges, K. V., Wobus, C., Ruhl, K., Schildgen, T., & Whipple, K. (2004). Quaternary deformation, river steepening, and heavy precipitation at the front of the higher Himalayan ranges. *Earth and Planetary Science Letters, 220*, 379–389. https://doi.org/10.1016/S0012-821X(04)00063-9

Hubbard, J., Almeida, R., Foster, A., Sapkota, S. N., Bürgi, P., & Tapponnier, P. (2016). Structural segmentation controlled the 2015 M_w 7.8 Gorkha earthquake rupture in Nepal. *Geology, 44*(8), 639–642.

Iyengar, R. N., Sharma, D., & Siddiqui, J. M. (1999). Earthquake history of India in medieval times. *Indian Journal of History of Science, 34*, 181–237.

Jayangondaperumal, R., Daniels, R. L., & Niemi, T. M. (2017). A paleoseismic age model for large-magnitude earthquakes on fault segments of the Himalayan Frontal Thrust in the Central Seismic Gap of northern India. *Quaternary International*.

Kobayashi, T., Morishita, Y., & Yarai, H. (2015). Detailed crustal deformation and fault rupture of the 2015 Gorkha earthquake, Nepal, revealed from ScanSAR-based interferograms of ALOS-2. *Planets and Space 67*, 201. https://doi.org/10.1186/s40623-015-0359-z.

Kumar, G. P., Mahesh, P., Nagar, M., Mahender, E., Kumar, V., Mohan, K., & Kumar, R. M. (2017). Role of deep crustal fluids in the genesis of intraplate earthquakes in the Kachchh region, northwestern, India. *Geophysical Research Letters, 44*, 4054–4063. https://doi.org/10.1002/2017GL072936

Kumar, S., Wesnousky, S., Rockwell, T. K., Briggs, R. W., Thakur, V. C., Jayangondaperumal, R. (2006). Paleoseismic evidence of great surface rupture earthquakes along the Indian Himalaya. *Journal of Geophysical Research, 111*, B03304. https://doi.org/10.1029/2004JB003309.

Lavé, J., & Avouac, J. P. (2000). Active folding of fluvial terraces across the Siwaliks hills, himalayas of central Nepal. *Journal of Geophysical Research, 105*, 5735. https://doi.org/10.1029/1999JB900292

Lavé, J., & Avouac, J. P. (2001). Fluvial incision and tectonic uplift across the himalayas of central Nepal. *Journal of Geophysical Research, 106*(B11), 26561–26591. https://doi.org/10.1029/2001JB000359

Lindsey, E. O., Natsuaki, R., Xu, X., Shimada, M., Hashimoto, M., Melgar, D., & Sandwell, D. T. (2015). Line-of-sight displacement from ALOS-2 interferometry: M_w 7.8 GorkhaEarthquake and M_w 7.3 aftershock. *Geophysical Research Letters, 42*, 6655–6661.

Maharjan, M. (2017). Liquefaction in Kathmandu valley during 2015 Gorkha (Nepal) earthquake. In *16th world conference on earthquake engineering 16WCEE 2017*, Paper No. 3463.

Mendoza, M. M., Ghosh, A., Karplus, M. S., et al. (2019). Duplex in the main himalayan thrust illuminated by aftershocks of the 2015 M_w 7.8 Gorkha earthquake. *Nature Geoscience* , *12*, 1018–1022. https://doi.org/10.1038/s41561-019-0474-8

Mishra, R. L., Singh, I., Pandey, A., Rao, P. S., Sahoo, H. K., & Jayangondaperumal, R. (2016). Paleoseismic evidence of a giant medieval earthquake in the eastern Himalaya. *Geophysical Research Letters, 43*, 5707–5715.

Mugnier, J. L., Gajurel, A., Huyghe, P., et al. (2013). Structural interpretation of the great earthquakes of the last millennium in the central Himalaya. *Earth Science Reviews, 127*, 30–47.

Mugnier, J. L., Jouanne, F., Bhattarai, R., Cortes-Aranda, J., Gajurel, A., Leturmy, P., Robert, X., Upreti, B., & Vassallo, R. (2017). Segmentation of the Himalayan megathrust around the Gorkha earthquake (25 April 2015) in Nepal. *Journal of Asian Earth Sciences*

Murakami, H., & Kagami, H. (1991). Application of high-precision questionnaire intensity survey method to the modified Mercalli intensity scale. *Zishin, 44*, 271–281.

Nabelek, J., Hetényi, G., Vergne, J., Sapkota, S., Kafle, B., Jiang, M., Su, H., Chen, J., Huang, B. S., & Team, T. H. C. (2009). Underplating in the himalaya-tibet collision zone revealed by the Hi-CLIMB experiment. *Science, 325*(5946), 1371–1374.

Ni, J., & Barazangi, M. (1984). Seismotectonics of the Himalayan collision zone: Geometry of the underthrusting Indian plate beneath the Himalaya. *Journal of Geophysical Research, 80*, 1142–1163.

Oldham, T. (1883). A catalogue of Indian earthquakes from the earliest to the end of 1869. *Memoirs of the Geological Survey of India, 19*(3), 1–53.

Ohsumi, T., Mukai, Y., & Fujitani, H. (2016). Investigation of Damage in and Around Kathmandu Valley related to the 2015 Gorkha, Nepal earthquake and beyond. *Geotechnical and Geological Engineering, 34*, 1223–1245.

Pandey, M. R., & Molnar, P. (1988). The distribution of intensity of the Bihar-Nepal earthquake of 15 January 1934 and bounds on the extent of the rupture zone. *Journal of Nepal Geological Society, 5*, 22–44.

Pandey, M. R., Tandukar, R. P., Avouac, J. P., Lavé, J., & Massot, J. P. (1995). Interseimic strain accumulation on the Himalayan crustal ramp (Nepal). *Geophysical Research Letters, 22*, 751–754.

Pant, M. R. (2002). A step towards a historical seismicity of Nepal, Adarsa: A supplement to Purnima. *The Journal of the Samsodhana-Mandala* (Pundit Publications, Kathmandu), 2, 29–60.

Parameswaran, R. M., Rajendran, K. (2017). Structural context of the 2015 pair of Nepal earthquakes (M_w 7.8 and M_w 7.3): An analysis based on slip distribution, aftershock growth, and static stress changes. *International Journal of Earth Science (Geologische Rundschau), 106*, 1133–1146. https://doi.org/10.1007/s00531-016-1358-4.

Parameswaran, R. M., Natarajan, T., Rajendran, K., Rajendran, C. P., Mallick, R., Wood, M., & Lekhak, H. C. (2015). Seismotectonics of the April–May 2015 Nepal earthquakes: An assessment based on the aftershock patterns, surface effects and deformational characteristics. *Journal of Asian Earth Sciences, 111*, 161–174.

Pierce, I., & Wesnousky, S. G. (2016). On a flawed conclusion that the 1255 AD earthquake ruptured 800 km of the Himalayan frontal thrust east of Kathmandu. *Geophysical Research Letters, 43*(17), 9026–9029.

Rajendran, C. P., Rajendran, K., Sanwal, J., & Sandiford, M. (2013). Archaeological and historical database on the medieval earthquakes of the central Himalaya: Ambiguities and inferences. *Seismological Research Letters, 87*, 1098–1108.

Rajendran, C. P., John, B., Rajendran, K. (2015). Medieval pulse of great earthquakes in the central Himalaya: viewing past activities on the frontal thrust. *Journal of Geophysical Research Solid Earth*, 1623–1641.

Rajendran, C. P., John, B., Rajendran, K., & Sanwal, J. (2016a). Liquefaction record of the great 1934 earthquake predecessors from the north Bihar alluvial plains of India. *Journal of Seismology, 20*, 733–745. https://doi.org/10.1007/s10950-016-9554-z.

Rajendran, C. P., Sanwal, J., Morell, K., Sandiford, M., Kotlia, R. S., Hellstrom, J., & Rajendran, K. (2016b). Stalagmite growth perturbations from the Kumaun Himalaya as potential earthquake recorders. *Journal of Seismology, 20*, 579–594.

Rajendran, C. P., John, B., & Anandasabari, K., et al. (2018a). On the paleoseismic evidence of the 1803 earthquake rupture (or lack of it) along the frontal thrust of the Kumaun Himalaya. *Tectonophysics, 722*, 227–234.

Rajendran, C. P., Sanwal, J., John, B., Anandasabari, K., Rajendran, K., Kumar, P., Jaiswal, M., & Chopra, S. (2019). Footprints of an elusive mid-14th century earthquake in the central Himalaya: Consilience of evidence from Nepal and India. *Geological Journal, 54*, 2829–2846.

Rajendran, C. P. (2021). Constraints on previous earthquakes from the liquefaction sites in the Kathmandu Valley associated with the 2015 Gorkha earthquake and their regional implications. *Quaternary International, 585*, 44–54.

Rana, B. S. (1935). *Nepal Ko Maha Bhukampa (The Great Earthquake of Nepal)* (pp. 1–250). Jorganesh Press, Kathmandu (in Nepali).

Rana, B. S. (2013). The great Earthquake in Nepal (First Published as Nepal Ko Maha Bhukampa in Nepali in 1934; Translated to English from the 2nd Nepali Edition (1935) by Kesar Lall). Ratna Pustak Bhandar, Kathmandu, Nepal (136 p)

Schelling, D., & Arita, K. (1991). Thrust tectonics, crustal shortening and the structure of the far-Eastern Nepal himalaya. *Tectonics, 10*, 851–862.

Tung, S., & Masterlark, T. (2016). Coseismic slip distribution of the 2015 M_w 7.8 Gorkha, Nepal, earthquake from joint inversion of GPS and in SAR data for slip within a 3-D heterogeneous Domain. *Journal of Geophysical Research Solid Earth, 121*.

Wang, X., Wei, S., & Wu, W. (2017). Double-ramp on the main himalayan thrust revealed by broadband waveform modelling of the 2015 Gorkha earthquake sequence. *Earth and Planetary Science Letters, 473*, 83–93.

Wesnousky, S. G., Kumahara, Y., Chamlagain, D., Pierce, I., Karki, A., & Gautam, D. (2016). Geological observations on large earthquakes along the Himalayan frontal fault near Kathmandu, Nepal. *Earth and Planetary Science Letters, 457*, 366–375.

Wobus, C. W., Hodges, K. V., & Whipple, K. X. (2003). Has focused denudation sustained active thrusting at the Himalayan topographic front? *Geology, 31*, 861–864. https://doi.org/10.1130/G19730.1

Wobus, C., Heimsath, A., Whipple, K., & Hodges, K. (2005). Active out-of-sequence thrust faulting in the central Nepalese Himalaya. *Nature, 434*(7036), 1008–1011.

Wobus, C. W., Whipple, K. X., Hodges, K. V. (2006). Neotectonics of the central Nepalese Himalaya: constraints from geomorphology, detrital 40/39 Ar thermochronology, and thermal modeling. *Tectonics, 25*(4).

Yamada, M., Hayashida, Y., Mori, J., & Mooney, W. D. (2016). Building damage survey and microtremor measurements for the source region of the 2015 Gorkha, Nepal, earthquake. *Earth, Planets and Space, 68*, 117. https://doi.org/10.1186/s40623-016-0483-4.

Yang, Y., Chen, Q., Xu, Q., et al. (2021). Source model and Coulomb stress change of the 2015 M_w 7.8 Gorkha earthquake determined from improved inversion of geodetic surface deformation observations. *Journal of Geological, 93*, 333–351.

Yin, J. X., Yao, H. J., Yang, H. F., Liu, J., Qin, W. Z., Zhang, H. J. (2017). Frequency-dependent rupture process, stress change, and seismogenic mechanism of the 25 April 2015 Nepal Gorkha M_w 7.8 earthquake. *Science China Earth Sciences, 60*.

Zilio, D. L., van Dinther, Y., Gerya, T., et al. (2019). Bimodal seismicity in the Himalaya controlled by fault friction and geometry. *Nature Communication, 10*, 48.

Part III
Plate Boundary Earthquakes: Eastern Himalaya

Chapter 13
An Overview of the Tectonic Framework of the Eastern Himalaya

The Eastern Himalayan orogen is marked by a broad zone of complex deformation primarily caused by the northward indentation of the Indian plate, also reflected in the eastward escape of the Tibetan Plateau (e.g., Tapponnier et al., 1982) (Fig. 13.1). As the product of convergence between three major tectonic plates, namely, India, Eurasia and Sunda, the structural configuration of this collision boundary is different from that of the central part of the Himalayan arc. At the eastern terminus, the major structural elements take a sharp turn, and the regional strike changes its direction from NE-SW to NW–SE, forming the EHS, which is the most dominant structural entity in this collision zone. In the eastern Himalaya, the MFT links with the Sumatra subduction zone (e.g., Angelier & Baruah, 2009). Both the continental and oceanic parts of the plate boundary have hosted great earthquakes (e.g., 1950 Assam; 2004, Sumatra).

It is believed that the development of contractional structures in the Eastern Himalaya started at 5–10 Ma after the onset formation of such structures in the central Himalaya (e.g., Nandy, 2001). GPS velocity field shows a clockwise motion of crustal material around the eastern Himalayan syntaxis (EHS; Wang et al., 2001). To the west of the EHS, the deformation is dominated by east-west trending overthrust and north-south trending rifting (e.g., Armijo et al., 1989). Marked by clockwise rotation of the block, on the eastern part of the EHS, the regional deformation is dominated by strike-slip movement along southeast and north-south directions (e.g., England & Molnar, 1990). On the eastern side of the EHS, the Mishmi and Lohit thrusts abut against the Burmese Arc. In such a complex tectonic setting, modeling of relative velocities between India and Eurasia and between India and Burma/Myanmar plates has been a topic of great interest (Gupta et al., 2015 and references therein).

Zhang et al. (2004) suggested that at longitude 92°E, the northeastern part of Indian plate currently moves at a rate of 38 mm/yr, in the N25°E direction relative to stable Eurasia. This is faster than the velocity in the northwest part of India, where at the longitude 76°E, India moves at a rate of 35 mm/yr at N11°E relative to the stable Eurasia. The study by McQuarrie et al. (2008) suggested a minimum north–south

C. P. Rajendran and K. Rajendran, *Earthquakes of the Indian Subcontinent*, GeoPlanet: Earth and Planetary Sciences,
https://doi.org/10.1007/978-981-16-4748-2_13

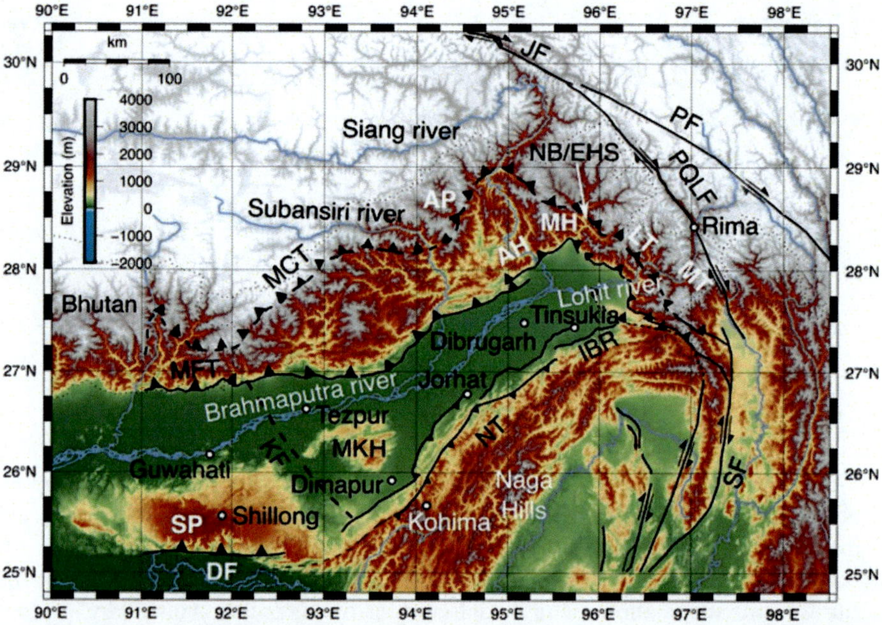

Fig. 13.1 Important structures in the Eastern Himalaya. NB/EHS: Namche Barwa/Eastern Himalaya Syntaxis; PF: Parlung fault; PQLF: Po-Qu-Lohit fault; JF: Jiali fault; SF: Sagaing fault; DF: Dauki fault; KF: Kopili fault; MT: Mishmi Thrust, LT: Lohit Thrust; NT: Naga Thrust; IBR: Indo-Burman Ranges; AH: Abhor Hills; AP: Arunachal Pradesh; MH: Mishmi Hills; MKH: Mikir Hills; SP: Shillong Plateau

shortening of 350 km since 22 Ma, yielding an average minimum shortening rate of 16 mm/yr. In a later study, a GPS shortening rate 15–20 mm/yr across the eastern Himalaya was obtained by Mukul et al. (2010). The model by Gupta et al. (2015), derived from GPS data along transects across the EHS, reveals a clockwise rotation of the rigid micro-plate comprising part of the Brahmaputra Valley, NE Himalaya and Northern Myanmar that rotate about a pole located at 14.5°N, 100.8°E at an angular rate of 1.75 ± 0.12°/Myr. Seismic imaging by Wang et al. (2019) shows steep subduction of the Indian plate to the west of the EHS and gentle subduction to the east. Thus, delamination and continental subduction are the dominant deep processes in the post-collisional stage in the eastern Himalayan collision zone.

Comprising the Mishmi Hills at the junction of the north-eastern Himalaya and the Indo-Burman Ranges, where the Himalayan arc makes the bend of about 90°, the northeast India region of the Himalaya arc exhibits intense structural deformation (e.g., Nandy, 2001; Yin et al., 2010a, b). An important tectonic element of the eastern Himalayan framework is the Shillong Plateau, the 400-km-long uplift involving the basement, considered to be a basement pop-up structure uplifted along the steep and seismically active crustal-scale reverse fault (Bilham & England, 2001; Rao & Kumar, 1997). The eastern Himalaya region exposes scattered outcrops of basement rocks without any modern foreland basin (Gansser, 1983; Yin et al., 2010a). In contrast to the central and NW Himalaya, the large-scale uplift involving the basement rocks and the lack of foreland basin are the two outstanding features of the eastern India collisional tectonics.

The Sikkim-Bhutan segment is regarded as a significant seismic belt, which has attracted much attention due to the state of stress accumulation and its status as a seismic gap (e.g., Li et al., 2020). Many workers have discussed the tectonics of the Arunachal segment and its western and eastern domains, separated by the Siang window (see Yin et a1., 2010a and references therein). Most importantly, it is noted that all the regional structural features such as the MCT, MFT, and the Indus-Tsangpo Suture make sharp U-turns around this window. In this chapter, we briefly outline the major characteristics of the broad tectonic zones in the eastern Himalayan collision zone: Indo-Burman Ranges, the Eastern Himalaya Syntaxis zone, the Shillong Plateau, the Upper Assam Valley, and the Bhutan segment (Fig. 13.1).

13.1 The Indo-Burman Ranges

Along the eastern margin of the Indian plate, the Indian lithospheric slab is subducted eastward beneath the Burmese plate forming the Indo-Burman Ranges (IBR) (Ni et al., 1989). The Indo-Burmese wedge has been recognized as the result of oblique subduction of the Bengal crust beneath the Burma sliver plate, and it presently propagates westward along the Shillong Plateau. Initially formed as an accretionary prism setting, the IBR gradually evolved to a sub-aerial fold-and-thrust belt during a highly oblique collision between Sundaland and the India Plate (Morley et al., 2020 and references therein). The width of the Indo-Burma fold belt exceeds 150 km in its central part and its N–S fold axes have resulted from the east–west convergence along the Burmese arc. The nearly 1100-km-long and north–south oriented range, including the Arakan-Yoma and Naga Hills, which is part of the IBR, links with the Himalaya at the eastern syntaxis zone. The Sagaing Fault, a large transform fault, defines the eastern limit of the Burmese lowlands, and it accommodates the oblique plate movement by right-lateral strike-slip and connects with the Andaman back arc (Curray, 1989).

More recent studies of the IBR using 2D seismic reflection data suggest that subduction ceased early in the Cenozoic, and instead, an inactive, dangling Indian Plate slab is presently undergoing lateral dextral translation. Cenozoic deformation in the IBR is reported as probably strongly strain partitioned, with strike-slip motion dominant in the inner belt and convergent deformation above a detachment. It remains debated whether the subduction is an ongoing process or not, but there is a general agreement that modern motions involve ~46 mm yr^{-1} of highly oblique motion of India with respect to Sundaland. About 21 mm yr^{-1} is accommodated by the Sagaing fault, and the remainder of the motion is to be absorbed as contractional and strike-slip deformation across the Indo-Burma Ranges (Morley et al., 2020). The January 3, 2016, Imphal (M_w 6.7) earthquake occurred within the complex interplate convergence zone between the India and Burma plates (Gahalaut et al.,

2016). An earlier earthquake on 6 August 1988 with analogous characteristics and magnitude M_w 7.2 nucleated at a greater depth (~90 km) at a location farther to the east (Person, 1988). These deep earthquakes (55–60 km) termed as intra-slab events are dominantly strike-slip events sourced within the bending part of the subducting Indian slab.

13.2 Eastern Himalayan Syntaxis

The major Himalayan structures take a sharp turn as the regional strike changes its direction from NE–SW to NW–SE, forming the Eastern Himalayan Syntaxis (EHS), expressed as a large antiform, named after the Namche Barwa peak (Wadia, 1931). This syntaxial zone includes the Mishmi Hills, at the junction of the north-eastern Himalaya and the Indo-Burma Ranges, where the Himalayan arc takes a sharp turn of about 90° and connects with the Indo-Burma Ranges. The major structural features within the NW–SE Mishmi Block are the NW-trending Mishmi and the Lohit thrusts. The Mishmi thrust, regarded as the equivalent of the frontal thrust of the Himalayan arc, and the Lohit thrust separates the trans-Himalayan plutonic rocks in the north from meta-volcanics and high-grade metamorphic rocks. Here, the NW–SE trending Mishmi (Arunachal) block defines the orographic linkage between the Himalaya collisional and the Burma subduction zones. The deformation style in the EHS transforms from dextral shear between India and SE Asia to collision along the Himalaya. As mentioned earlier, the plate boundary extends southward along the ~500 km-long Indo-Burman arc where the Indian lithosphere plunges eastwards, forming a subduction zone and a fold-and-thrust belt, with west-verging thrusts (Gahalaut & Gahalaut, 2007; Mukhopadhyay & Dasgupta, 1988; Ni et al., 1989).

The right-lateral strike-slip in southeast Tibet wraps around the EHS and connects with the Sagaing fault zone with the same sense of motion. A connection between the dextral Sagaing fault and the Jiali-Po-Qu faults would require an extensional right step on the northern terminus of the Sagaing Fault. The discontinuity of ophiolite belts in the syntaxis zone is regarded as the evidence for the plate motion accommodated by the strike-slip faults (Armijo et al., 1989). An alternate view is that the Cenozoic history of the EHS was dominated by thrusting and thickening to such an extent, that uplift and erosion have exposed mid-crustal rocks and removed the evidence of the Indus-Tsangpo suture here. Thus, the discontinuity of the ophiolite belt in the syntaxial zone may not necessarily originate from strike-slip motion, but it might be a consequence of crustal thickening, uplift, and erosion that would remove the suture zone rocks (Holt et al., 1991; Wang & Chu, 1988).

Convergence across the EHS has been poorly constrained until very recently. Devachandra et al. (2014) suggested that the detachment under the frontal EHS is locked, and that strain is accumulating at a rate corresponding to a slip rate of at least 20 ± 4 mm/yr. Consequently, a slip deficit of about 1.3 m might have accumulated since 1950, which is considered sufficient to generate at least one major earthquake, assuming that the entire strain was released in 1950. As there are no known earthquakes of comparable size during the past 200 years, the slip deficit of about 4 m holds potential for a great earthquake.

Along the EHS, the Indian Plate abuts against the Eurasia and the Burma (Myanmar) Plates, and approximately 50% of the resulting motion (~18 mm/year) is accommodated by strike-slip motion on the Sagaing Fault (Vigny et al., 2003). Towards the southeast of the EHS, the Mishmi and Lohit Thrusts (MT and LT) abut against the Northern Burmese Arc and to the northeast, right-lateral strike-slip motion occurs along Jiali, Po-Qu, and Parlung faults. The relative velocities between India and Eurasia plates and that between India and Burma (Myanmar) are likely to be very complex here, as discussed by several authors (see Devachandra et al., 2014 and references therein).

While the GPS models assume thin-skin tectonics (as elsewhere in the Himalaya), Yin et al. (2010a) argue that the development of major contractional structures in the eastern Himalaya started around 5–10 Ma, much after the onset of the equivalent structures in the central Himalaya. The crustal thickening was accomplished by thick-skinned thrusting involving the basement, rather than thin-skinned tectonics. Based on a detailed analysis of mesoscopic fold geometry, it has been argued that the traditional line-balancing methods could overestimate the total shortening by as much as 20%, and this scenario would call for a re-evaluation of the seismotectonics associated with the EHS. An assessment of these observations on the structural aspects of the EHS is important for elucidating the mechanism of the 1950 earthquake.

13.3 The Shillong Plateau and the Upper Assam Valley

The Shillong Plateau is a northward-tilting topographic feature, with its highest peaks reaching ~2 km along its southern rim. Underlain by the Proterozoic basement, the plateau is bounded by the Himalayan collisional zone in the north, the Indo-Burma subduction zone in the east and the Bengal basin in the south (Acharya & Mitra, 1986; Yin et al., 2010b). In the south, the plateau is bounded by the E–W oriented and north-dipping Dauki Fault, (Biswas & Grasemann, 2005; Evans, 1964), which separates the Precambrian basement of the Shillong Plateau from the Tertiary sediments of the southern Bengal basin. In the east, it is bound by the northwest-striking, right-slip Kopili fault and the Naga Hills thrust belt (Evans, 1964). In the north, the Oldham or Brahmaputra Valley faults binds the plateau (Bilham & England, 2001; Rajendran et al., 2004) and in the west, it is bound by the Jamuna fault (Gupta & Sen, 1988). Evans (1964) had originally considered the Dauki fault to be a major right-slip fault with >200 km of motion. However, Yin et al. (2010a) argue that the segment of the Dauki fault zone, ~12–14 km south of the Cherrapunji exhibits features of a thrust fault.

Internally, the Shillong Plateau exhibits a series of northeast, and occasionally north-trending linear topographic features, interpreted as extensional fissures related to the Cretaceous Gondwana breakup (e.g., Gupta & Sen, 1988). At ~ 35 km, the crust of the Shillong Plateau is slightly thinner than its surrounding regions (Mitra et al., 2018). These authors reported that the Moho, north of the Plateau, dips gently northward and reaches a depth of ~ 44 km at the Himalayan front. The crustal thickness increases by 12–13 km towards south across the Dauki fault due to the downfaulting of the plateau crust. It is now blanketed by a 13–14 km-thick sedimentary load of the Bengal Basin (Mitra et al., 2018). The strike-slip focal mechanisms dominate the plateau, with earthquakes occurring at a range of hypocentral depths, including in the uppermost mantle (Kayal & De, 1991). The late Cenozoic uplift of the plateau is considered as an important tectonic event, with its strong influence on the spatially varying exhumation history and stress distribution across the eastern Himalaya (Bilham & England, 2001; Clark & Bilham, 2008). The uplift of the plateau might have started between 4.9 and 5.2 Ma, coinciding with the northward deflection of the paleo-Brahmaputra River by the rising plateau, as reflected in the sedimentary record of the Bengal basin (Govin et al., 2018). The Mikir Hills, an extended part of the Plateau, are separated by the NW–SE trending Kopili fault, a 300 km-long structure that crosses the Assam Valley and reaches the MBT system (Nandy, 2001).

The Assam Valley, north of the Shillong Plateau, is a part of the Himalayan foreland basin covered by Tertiary sequences and recent alluvium. The valley is bounded by the Himalaya to the northwest, Mikir Hills to the southwest, Naga Hills to the southeast, and Mishmi Hills to the northeast and is being subjected to tectonic compression (Molnar & Stock, 2009). It has been suggested that the Brahmaputra River Valley might be a piggyback basin influenced by the Shillong Plateau uplift. Convergent tectonics in the region has severely affected the basement as well as the overlying sediments (Yin et al., 2010b). The depth to the basement inferred from geophysical surveys ranges from 3.5 to 7.2 km (Narula et al., 2000). Crystalline basement rocks remain exposed across the Brahmaputra Valley, due to the thin sediment cover, considered as a unique feature of this part of the Himalayan foreland basin (Gansser, 1983).

13.4 The Bhutan Segment

Bhutan and Arunachal Himalaya are two adjoining segments of the Himalayan arc, located respectively to the east and west of 92°E longitude, displaying different deformational characteristics, including GPS convergence rates. The convergence velocity between India and the Eurasian plate is estimated as ~31 mm/year to the east of this longitude, but the rate is significantly lower to its west in the Bhutan segment (Burgess et al., 2012). The rate of convergence is ~18 mm/year in eastern Nepal (Ader et al., 2012) and ~14–17 mm/year in Bhutan (Vernant et al., 2014). Towards the Bhutan Himalaya region, the seismic productivity is generally considered low, as compared to its adjoining segments. This lower seismicity is attributed to the purported stress shadow of the great 1897 Shillong earthquake (Gahalaut et al., 2011). These observations make the Bhutan segment worthy of specific mention, although this part of the Himalaya may not have experienced any great earthquakes in the recent history.

GPS-based studies by Li et al. (2020) suggests a higher convergence rate at 16.2–18.5 mm/yr across the Sikkim-Bhutan Himalaya, in agreement with the estimated slip rate of 18 mm/yr obtained by Mukul et al. (2018). Based on their inversion model by Li et al. (2020), the locking width in the Sikkim segment is 60–70 km, and in the western Bhutan, it increases to ~100 km, compared to the much-reduced locking width of ~70 km in the eastern Bhutan. These new estimates suggest parity in convergence rate with the rest of the Himalaya arc. The seismic activity in the Bhutan Himalaya, however, shows a significant 'low" compared to other segments of the Himalayan ranges. It is therefore possible that factors like the geometry of the subduction zone, rheology, and crustal thickening could influence the earthquake generation. Prominent transverse foreland structures abutting the Himalaya are considered as possible barriers that control the rupture (e.g., Zilio et al., 2020). For example, the Kopili fault zone, defined by seismicity, is one such transverse active structure in the area along with recently proposed Dhubri-Chungtang and Munger-Saharsa (Fig. 13.2). But the paleoseismic evidence for the 1714 CE from this region argues that the purported seismic gap in Bhutan has already been filled (Hetényi et al., 2016; Le Roux-Mallouf et al., 2016). These issues will be discussed further in the next two chapters. We believe the complexity of the Sikkim-Bhutan segment, that has remained less seismically productive, unlike the other segments of

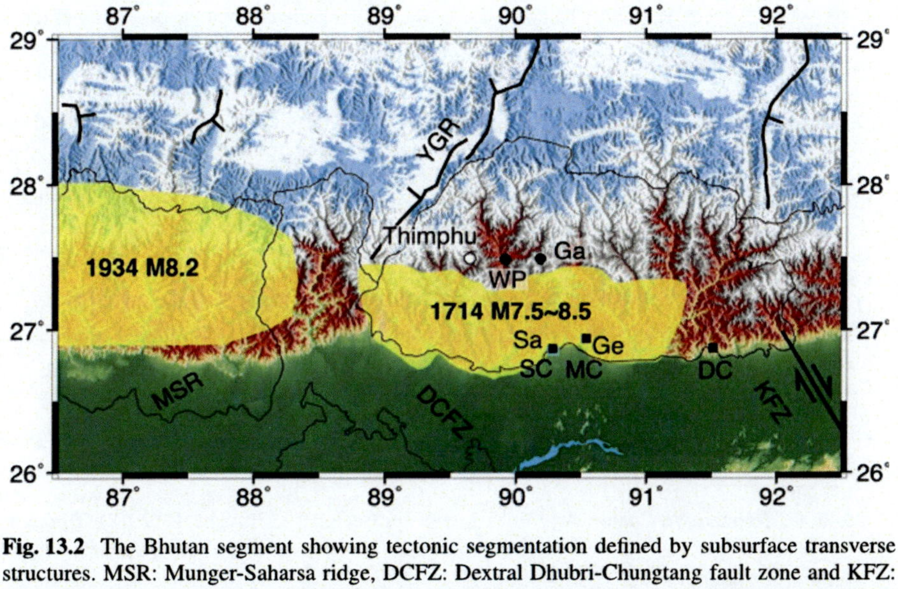

Fig. 13.2 The Bhutan segment showing tectonic segmentation defined by subsurface transverse structures. MSR: Munger-Saharsa ridge, DCFZ: Dextral Dhubri-Chungtang fault zone and KFZ: Kopili fault zone in the Sikkim-Bhutan Himalaya YGR: Yadong-Gulu graben. Ruptures of 1714 and 1934 earthquakes are shown. Black circle and squares are trench sites discussed in the text (modified after Li et al., 2020). WP: Wangdue Phodrang, Ga: Gangteng, Sa:Sapakhonghat, SC:Sarpang Chu,MC: Momo Creek, Ge: Gelephu; DC: Dungsam Chu,Ge: Gamon Chariali (see also Chap. 15)

the Himalaya, requires more elucidation, particularly in the context of the transverse structures and dominance of strike-slip earthquakes.

References

Acharya, S. K., & Mitra, N. D. (1986). Regional geology and tectonic setting of northeast India and adjoining region. *Memoirs - Geological Survey of India, 119*, 6–12.

Ader, T., Avouac, J.-P., Liu Zeng, J., Lyon-Caen, H., Bollinger, L., Galetzka, J., et al. (2012). Convergence rate across the Nepal Himalaya and interseismic coupling on the main Himalayan thrust: Implications for seismic hazard. Journal of Geophysical Research, *117*, B044403. https://doi.org/10.1029/2011JB009071.

Angelier, J., & Baruah, S. (2009). Seismotectonics in Northeast India: A stress analysis of focal mechanism solutions of earthquakes and its kinematic implications. *Geophysical Journal International, 178*, 303–326.

Armijo, R., Tapponnier, P., & Han, T. (1989). Late Cenozoic right-lateral strike slip faulting in Southern Tibet. Journal of Geophysical Research, *94*(B3), 2787–2838. https://doi.org/10.1029/JB094iB03p02787.

Bilham, R., & England, P. (2001). Plateau 'pop-up' in the great 1897 Assam earthquake. *Nature, 410*, 806–809.

Biswas, S., & Grasemann, B. (2005). Quantitative morphotectonics of the southern Shillong Plateau (Bangladesh/India). *Australian Journal of Earth Sciences, 97*, 82–03.

Burgess, W. P., Yin, A., Dubey, C. S., Shen, Z., Kelty, T. K. (2012) Holocene shortening across the Main Frontal Thrust zone in the eastern Himalaya. *Earth and Planetary Science Letters*, 357–358, 152–167; s 357–358; 152–167. https://doi.org/10.1016/j.epsl.2012.09.040.

Clark, M., & Bilham, R. (2008). Miocene rise of the Shillong Plateau and the beginning of the end for the Eastern Himalaya. *Earth and Planetary Science Letters, 269*, 337–351.

Curray, J. R. (1989). The Sunda Arc: A model for oblique plate convergence. *Netherlands Journal of Sea Research, 24*(2–3), 131–140.

Devachandra, M., Kundu, B., Catherine, J., Kumar, A., & Gahalaut, V. K. (2014). Global Positioning System (GPS) measurements of crustal deformation across the frontal eastern Himalayan syntaxis and seismic-hazard assessment. *Bulletin of the Seismological Society of America, 104*, 1518–1524.

England, P., & Molnar, P. (1990). Right-lateral shear and rotation as the explanation for strike-slip faulting in eastern Tibet. *Nature, 344*, 141–142.

Evans, P. (1964). The tectonic framework of Assam. *Geological Society of India Journal, 5*, 80–96.

Gansser, A. (1983). Geology of the Bhutan Himalaya (181 p.). Boston: Birkhäuser Verlag.

Gahalaut, V. K., Rajput, S., & Kundu, B. (2011) Low seismicity in the Bhutan Himalaya and the stress shadow of the 1897 Shillong Plateau earthquake. *Physics of the Earth and Planetary Interiors, 186*, 97–102. https://doi.org/10.1016/j.pepi.2011.04.009.

Gahalaut, V. K., & Gahalaut, K. (2007). Burma plate motion. *Journal of Geophysical Research, 112*, B10402. https://doi.org/10.1029/2007JB004928

Gahalaut, V. K., Martin, S. S., Srinagesh, D. et al (2016). Seismological, geodetic, macroseismic and historical context of the 2016 Mw 6.7 Tamenglong (Manipur) India earthquake. *Tectonophysics, 688*, 36–48.

Govin, G., Najman, Y., Copley, A., Millar, I., van der Beek, P., & Huyghe, P. (2018). Timing and mechanism of the rise of the Shillong Plateau in the Himalayan foreland. *Geology, 46*, 279–282. https://doi.org/10.1130/G39864.1

Gupta, T.D., Riguzzi, F., Dasgupta, S., Mukhopadhyay, B., Roy, S., & Sharma, S. (2015). Kinematics and strain rates of the Eastern Himalayan Syntaxis from new GPS campaigns in Northeast India. *Tectonophysics, 655*, 15–26 (2015)

Gupta, R. P., & Sen, A. K. (1988). Imprints of Ninety-East Ridge in the Shillong Plateau. *Indian Shield: Tectonophysics, 154*, 335–341.

Hetényi, G., Le Roux-Mallouf, R., Berthet, T., Cattin, R., Cauzzi, C., Phuntsho, K., & Grolimund, R. (2016). Joint approach combining damage and paleoseismology observations constrains the 1714 A.D. Bhutan earthquake at magnitude 8 ± 0.5. *Geophysical Research Letters, 43*, 10695–10702. https://doi.org/10.1002/2016GL071033

Holt, W. E., Ni, J. F., Wallace, T. C., & Haines, A. J. (1991). The active tectonics of the eastern Himalayan syntaxis and surrounding regions. *Journal of Geophysical Research, 96*(B9), 14595–14632.

Kayal, J. R., & De, R. (1991). Microseismicity and tectonics in northeast India. *Bulletin of the Seismological Society of America, 81*, 131–138.

Le Roux-Mallouf, R., Ferry, M., Ritz, J-F., Berthet, T., Cattin, R., & Drukpa, D. (2016). First paleoseismic evidence for great surface-rupturing earthquakes in the Bhutan Himalayas. *Journal of Geophysical Research: Solid Earth, 121*. 10.1002/ 2015JB012733.

Li, S., Tao, T., Gao, F., Qu, X., Zhu, Y., & Huang, J. (2020). Interseismic coupling beneath the Sikkim–Bhutan Himalaya constrained by GPS measurements and its implication for strain segmentation and seismic Activity. *Remote Sens, 12*, 2202. https://doi.org/10.3390/rs12142202.

McQuarrie, N., Robinson, D., Long, S., Tobgay, T., Grujic, D., Gehrels, G., & Ducea, M. (2008). Preliminary stratigraphic and structural architecture of Bhutan: Implications for the along strike architecture of the Himalayan system. *Earth and Planetary Science Letters, 272*, 105–117.

Mitra, S., Priestley, K. F., Borah, K., & Gaur, V. K. (2018). Crustal structure and evolution of the Eastern Himalayan plate boundary system, Northeast India. *Journal of Geophysical Research: Solid Earth, 123*, 621–640.

Molnar, P., & Stock, J. M. (2009). Slowing of India's convergence with Eurasia since 20 Ma and its implications for Tibetan mantle dynamics. *Tectonics, 28*, TC3001.

Morley, C. K., Naing, T. T., Searle, M., & Robinson, S. A. (2020). Structural and tectonic development of the Indo-Burma ranges. *Earth-Science Reviews, 200*, 102992.

Mukul, M., Jade, S., Ansari, K., Matin, A., & Joshi, V. (2018). Structural insights from geodetic Global Positioning System measurements in the Darjiling-Sikkim Himalaya. *Journal Structural Geology, 114*, 346–356

Mukhopadhyay, M., & Dasgupta, S. (1988). Deep structure and tectonics of the Burmese arc: Constraints from earthquake and gravity data. *Tectonophysics, 149*, 299–322.

Mukul, M., Jade, S., Bhattacharya, A. K., & Bhusan, K. (2010). Crustal shortening in convergent orogens: Insights from global positioning system (GPS) measurements in Northeast India. *Journal of the Geological Society of India, 75*, 302–312.

Nandy, D. R. (2001). *Geodynamics of Northeastern India and adjoining region* (209 pp.) acb Publ., Kolkata, India.

Narula, P., Acharyya, S. K., & Banerjee, J. (2000). Seismotectonic Atlas of India and its environs. *Geological Survey of India, SEISAT-15*, 16 and 17.

Ni, J. F., Guzman-Speziale, M., Bevis, M., Holt, W. E., Wallace, T. C., & Seager, W. R. (1989). Accretionary tectonics of Burma and the three-dimensional geometry of the Burma subduction zone. *Geology, 17*(1), 68–71.

Person, W. J. (Ed.) (1988). Seismological notes- January-February. Bulletin of the Seismological Society of America, *78*, 2115–2119.

Rajendran, C. P., Rajendran, K., Duarah, B. P., Baruah, S., & Earnest, A. (2004). Interpreting the style of faulting and paleoseismicity associated with the 1897 Shillong, northeast India, earthquake: Implications for regional tectonism. *Tectonics, 23*, TC 4009. 1029/2003/TC001605

Rao, P. C., & Kumar, R. M. (1997). Uplift and tectonics of the Shillong plateau, northeast India. *Journal of Physics of the Earth, 45*, 167–176.

Tapponnier, P., Pelzer, G., Ledain, A. Y., Armijo, R., & Cobbold, P. (1982). Propagating extrusion tectonics in Asia: New insights from simple experiments with plasticine. *Geology, 10*, 611–616.

Vernant, P., Bilham, R., Szeliga, W., Drupka, D., Kalita, S., Bhattacharyya, A. K., Gaur, V. K., Pelgay, P., Cattin, R., & Berthet, T. (2014). Clockwise rotation of the Brahmaputra Valley relative to India: Tectonic convergence in the eastern Himalaya, Naga Hills, and Shillong Plateau. *Journal of Geophysical Research: Solid Earth, 119*, 6558–6571.

Vigny, C., Socquet, A., Rangin, C., Chamot-Rooke, N., Pubellier, M., Bouin, M. N., Bertrand, G., & Becker, M. (2003). Present-day crustal deformation around Sagaing fault, Myanmar. *Journal of Geophysical Research, 108*(B11), 2533.

Wadia, D. N. (1931). The syntaxis of the northwest Himalaya: Its rocks, tectonics and orogeny. *Records of the. Geological Survey of India, 65*(2), 189–220.

Wang, Q., Zhang, P. Z., Freymueller, J. T., Bilham, R., Larson, K. M., Lai, X., You, X., Niu, Z., Wu, J., Li, Y., Liu, J., Yang, Z., & Chen, Q. (2001). Present day crustal deformation in China constrained by global positioning measurements. *Science, 294*, 574–577.

Wang, C.-Y., Mooney, W. D., Zhu, L., Wang, X., Lou, H., You, H., et al. (2019). Deep structure of the eastern Himalayan collision zone: Evidence for underthrusting and delamination in the postcollisional stage. *Tectonics, 38*, 3614–3628.

Wang, E., & Chu, J. J. (1988). Collision tectonics in the Cenozoic orogenic zone bordering China. *India, and Burma, Tectonophysics, 147*, 71–84.

Yin, A., Dubey, C. S., Kelty, T. K., Webb, A. A. G., Harrison, T. M., & Chou, C. Y. (2010a). Geologic correlation of the Himalayan orogeny and Indian craton: Part 2. Structural geology, geochronology and tectonic evolution of the Eastern Himalaya. *Geological Society of America Bulletin, 122*(3–4), 360–395.

Yin, A., Dubey, C. S., Webb, A. A. G., Kelty, T. K., Grove, M., Gehrels, G. E., & Burgess, W. P. (2010b). Geologic correlation of the Himalayan orogen and Indian craton: Part I. Structural geology, U-Pb zircon geochronology, and tectonic evolution of the Shillong Plateau and its neighbouring regions in NE India. *Geological Society of America Bulletin, 122*, 336–359.

Zhang, P., Shen, Z., Wang, M., Gan, W., Burgmann, R., Molnar, P., et al. (2004). Continuous deformation of the Tibetan Plateau from global positioning system data. *Geology, 33*, 809–812. https://doi.org/10.1130/G20554.1.

Chapter 14
Shillong 1897

14.1 Introduction

The great 12 June 1897, northeast India earthquake, sourced beneath the Shillong Plateau has been described as one of the most damaging tremblors in India (Fig. 14.1). Formerly part of the undivided province of Assam, Shillong has now become part of the Meghalaya State and currently it is its capital. The earthquake mainly impacted Shillong and the surrounding areas, including the alluvial plains of the Brahmaputra River. This event opened new vistas in observational seismology and resulted in one of the finest monographs on any of the 19th-century earthquakes globally. Later studies of this earthquake owed much to this report prepared by R. D. Oldham (Oldham, 1899), who was working with the Geological Survey of India at that time.

The 1897 earthquake attained global attention, not only due to its large size and impact, but also because of its occurrence at a time when modern seismographs had just come into wider use and instrumentally recorded data became available. It is considered a landmark event in seismology as it led to several ground-breaking initiatives, including the use of seismic waves to map the interior of the earth. Oldham used teleseismic waves of this earthquake to distinguish between the three types of seismic waves and eventually applied them to map the Earth's core. Gutenberg (1956) calculated its magnitude at ~8.7, using the records from Milne's instruments. This event is among a few 19th-century earthquakes assigned with an instrumentally recorded magnitude (Richter, 1958). However, this estimate has been subsequently revised as M_w 8.2, after re-appraising the isoseismals (Ambraseys & Bilham, 2003).

Oldham's report not only gives the details of destruction and ground failures from the earthquake, but also provides information on ground accelerations, which reportedly exceeded the Earth's gravitational acceleration. The report has also presented details of measurements from the triangulation survey of the area, initially conducted in 1860 (Fig. 14.2). It is one of the earliest documents presenting observations based on a repeat leveling survey. Oldham's account formed the basis for most of the later studies, including the intensity/magnitude calculations. More than a century after

C. P. Rajendran and K. Rajendran, *Earthquakes of the Indian Subcontinent*,
GeoPlanet: Earth and Planetary Sciences,
https://doi.org/10.1007/978-981-16-4748-2_14

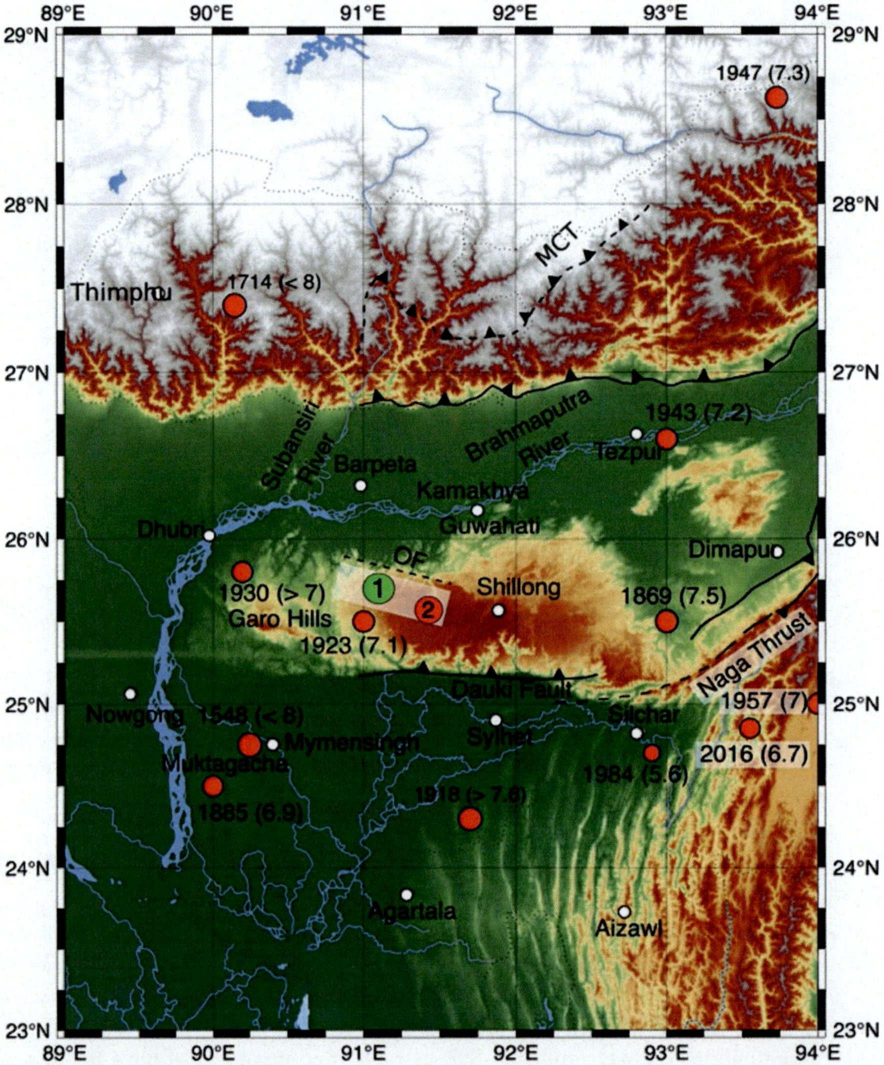

Fig. 14.1 Distribution of earthquakes near the Shillong Plateau (SP). Earthquake locations marked as 1 and 2 are from Ambraseys and Bilham (2003) and England and Bilham, (2015), respectively. The geodetically constrained rupture zone of the 1897 earthquake (the Oldham fault; OF) is shown as a white transparent rectangle. Its eastern and northern edges are constrained by triangulation across the SP and in the Brahmaputra Valley, respectively, and its western edge is constrained by 11 m of slip up-to-the east on the Chedrang fault, which moved in response to subsurface reverse faulting beneath it on the Oldham fault. The northern edge of the rupture lies at ~6 km depth and its southern edge near the base of Earth's crust at 31 km. A dashed line shows the projection of the buried Oldham fault (OF) to the surface, where it is marked by a line of increased stream gradients and incision

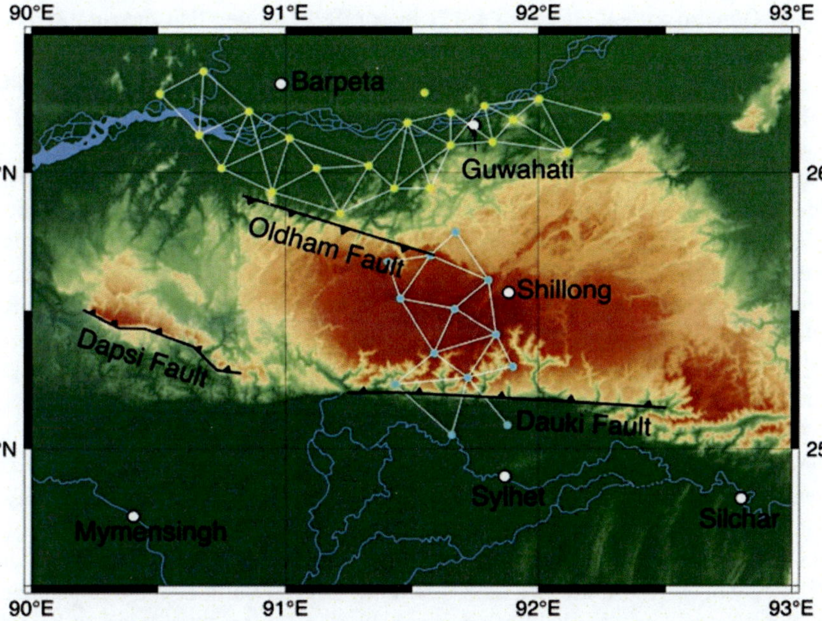

Fig. 14.2 Setting of the triangulation network of the Survey of India in the southern Brahmaputra Valley and the Shillong Plateau (adapted from England and Bilham, 2015)

Oldham published his report, more information on the earthquake effects has come to light from a detailed narrative in the form of letters written by La Touche, a geologist working in the Geological Survey of India at the time of the earthquake. The transcripts of these letters, which provide day-by-day accounts of post-earthquake survey is an excellent micro-detailed documentation on the effects of this great earthquake (La Touche, 1897; Bilham, 2008).

Although Oldham's monograph established the status of the 1897 earthquake as the best documented among the 19th-century events, many issues concerning its mechanism remained controversial. One poorly understood aspect was the location and geometry of the causative fault and its relation to the Shillong Plateau (also abbreviated as SP) and the Himalaya plate boundary. Oldham (1899) initially believed that the earthquake was caused by the movement along the thrust plane located below the plateau with a length of 320 km and width of 80 km, and at a depth ranging from 8 to 14 km. It was assumed later that the earthquake occurred on a north-dipping fault that crops up on the southern boundary of the SP, following the traditional understanding of the great Himalayan earthquakes (e.g., Seeber & Armbruster, 1981). It was proposed that the rupture was about 550 km long (east–west) and 300 km wide (north–south), on a shallow thrust-type dipping detachment under the Himalayan arc that extends towards the SP in the up-dip direction. Molnar and Pandey (1989) and Molnar (1990) suggested that the rupture was about 200 km long, with its southern limit near the Dauki fault, but the northern limit of the rupture remained unspecified.

Using Oldham's data, Gahalaut & Chander (1992) argued that the rupture occurred on a gently dipping thrust fault confined to regions beneath the SP.

The Dauki fault system that extends along the southern margin of the Shillong (Meghalaya) Plateau is a north-dipping structure, including a system of E-W faults and folds (Biswas & Grasemann, 2005; Evans, 1964; Nandy, 2001). In the north, the plateau margin is marked by the Brahmaputra River, and towards the east, there occurs the NW–SE trending Kopili fault. In the east, the plateau is bound the Naga Hills thrust belt (Evans, 1964), and to the west, the boundary is marked by the Jamuna fault (Gupta & Sen, 1988). The SP is believed to have been a popped up landform between two active faults on its southern and northern borders: (e.g., Rao & Kumar, 1997; Bilham & England, 2001).

Bilham and England (2001) used the pre- and post-earthquake survey data from one of the lines first measured in 1859 (Strahan, 1891) and re-measured in 1936/37 (Wilson, 1939), to model the causative fault. Deviating from the previous models, they proposed for the first time that the earthquake occurred on a south-dipping reverse fault, named as the Oldham fault $\pm$ 5 km south of 26°N and spatially coinciding with the northern topographic front of the plateau. An assessment by Rajendran et al. (2004) had presented a different view. While agreeing with the south-dipping geometry of the fault, the Brahmaputra fault, located north of the topographic front of the plateau beneath the Brahmaputra alluvium, was considered to be the host structure.

In a subsequent note, Rajendran et al. (2006) commented on the lack of any active tectonic expressions in the surface geology, landforms, gravity anomaly, or microseismicity associated with the Oldham fault. The issue of the stability of the 1859–1937 survey monuments that were placed on the alluvial setting of the Assam Valley, had also been a matter of concern, as pointed out by Bomford, an officer with the Survey of India (Bomford, 1939; p. 32). In a later article, England and Bilham (2015) have reiterated their original hypothesis after further evaluating the survey data, covering the southern part of the Assam Valley and a N-S transect in the SP (Fig. 14.2). Burrard's (1898) report that lists angle changes in the geodetic network on the hanging wall of the Oldham fault provided additional database for further strengthening of the model constraints. We will return to the discussion of these fault models towards the end of this chapter.

The repeat period of the 1897-type earthquakes is another vexing issue as the entire region is dense with seismic sources capable of generating high-magnitude earthquakes. Thus, the liquefaction records are most likely to be overprinted by the succession of moderate events. Rastogi et al. (1993) were the first to report a paleoliquefaction feature from this region and used it as an earthquake proxy. The liquefaction chronology developed by Sukhija et al. (1999a, b) suggested a regional recurrence period of the order of 500 years for large earthquakes in the Brahmaputra Valley. Differing considerably from the ages derived from the liquefaction chronology, Bilham and England (2001) argued for a source-specific recurrence interval of 3000–8000 years, based on model-specific fault-slip data. Rajendran et al. (2004) suggested that the 1897-type of earthquakes may occur over intervals of about 1200 years, an interpretation based on historical information and independently obtained paleoseismological data.

As already mentioned, earlier workers had grouped the 1897 earthquake with the great Himalayan events, considering the SP as an immediate outlier of the Himalaya. However, it is now being referred to as an "intraplate" earthquake (e.g., Ambraseys & Bilham, 2003). Most papers consider its origin in the tectonic environment of the SP, proximal to the India-Eurasia plate boundary. This chapter provides a broad overview of the coseismic observations, various interpretations of the seismotectonic setting of the 1897 earthquake, and observations on the past earthquakes that have affected this region.

14.2 Felt Reports and Intensity

The 1897 earthquake generated widespread landslips and severe coseismic liquefaction, leading to sand venting in areas far beyond the plateau margins. It caused maximum destruction in Shillong and Guwahati, where most of the government buildings were located. It is reported to have destroyed most of the masonry buildings, and almost all stone works in and around Shillong, including many bridges, stone houses, and churches (Fig. 14.3). The earthquake triggered landslides mostly reported from the Garo Hills and from distant locations like Cherrapunji in the south and Sikkim-Darjeeling Hills in the north. One of the major hazards in the Brahmaputra floodplain and the neighboring areas was related to soil liquefaction (Fig. 14.4). Fissures and sandblows due to liquefaction occurred over a wide area of Lower Assam, Meghalaya, West Bengal, and northern Bangladesh. Large-scale liquefaction-related subsidence was also reported from distant sites such as a locality called Muktagacha, Mymensingh District, on the southern side of the Dauki fault, in Bangladesh.

In Oldham's isoseismal map, published in 1899, the highest intensity was equated with X on Rossi-Forel scale, and a "hat-shaped" patch defined the corresponding

Fig. 14.3 Collapsed bridge outside Shillong, thrown off from its abutments by the earthquake; photographed by La Touche and reproduced by Oldham (1899). *Source* Bilham (2008)

Fig. 14.4 Sandblow formed during the 1897earthquake photographed by La Touche. Image reproduced from Bilham (2008)

region (Fig. 14.5). This was based on an approximation between the Rossi-Forel (RF) and Mercalli (M) scale to describe the local shaking effects. Gutenberg (1956) had located the epicenter in the middle of this patch. The southern boundary of this zone is somewhat linear and was reported to roughly coincides with the topographic front of the plateau. However, the somewhat circuitous shape of its northern boundary extends beyond the SP-the shape of the high- intensity zone could have resulted from the selective amplification of seismic energy in the alluvial valley (Fig. 14.5).

In a later report focusing on the secondary effects such as ground deformations and fractures, Oldham (1920) provided more details on land-level changes and cumulative damage from aftershocks. The new observations prompted him to report that the intensities were VIII, closer to and reaching X and even XII, at some places throughout the greater part of the western region of the Plateau. Using Oldham's descriptions, Seeber and Armbruster (1981) drew another contour to show the limit of intensity of VII to IX, which extends ~ 550 km in an E-W direction. This contour line delimits the intense liquefaction in the alluvial plains, and according to them the 1897 rupture probably did not extend beyond 93ºE.

Recognizing the limitations of Oldham's assessment of intensities based on the Rossi-Forel-Mercalli approximation, Ambraseys and Bilham (2003) pointed out that the European intensity scales cannot be blindly applied for the assessment of acceleration-related damage to indigenous structures in Assam. Thus, based on a re-evaluation of the felt effects, it was surmised that Oldham's intensities were probably exaggerated by 1.5 to 3 isoseismal units and thus, the magnitude was downgraded as M_W 8.2 from M_W 8.7. Apart from Oldham's detailed report, various sources including the 1898–99 report of the public works department, as well as the transcripts of the

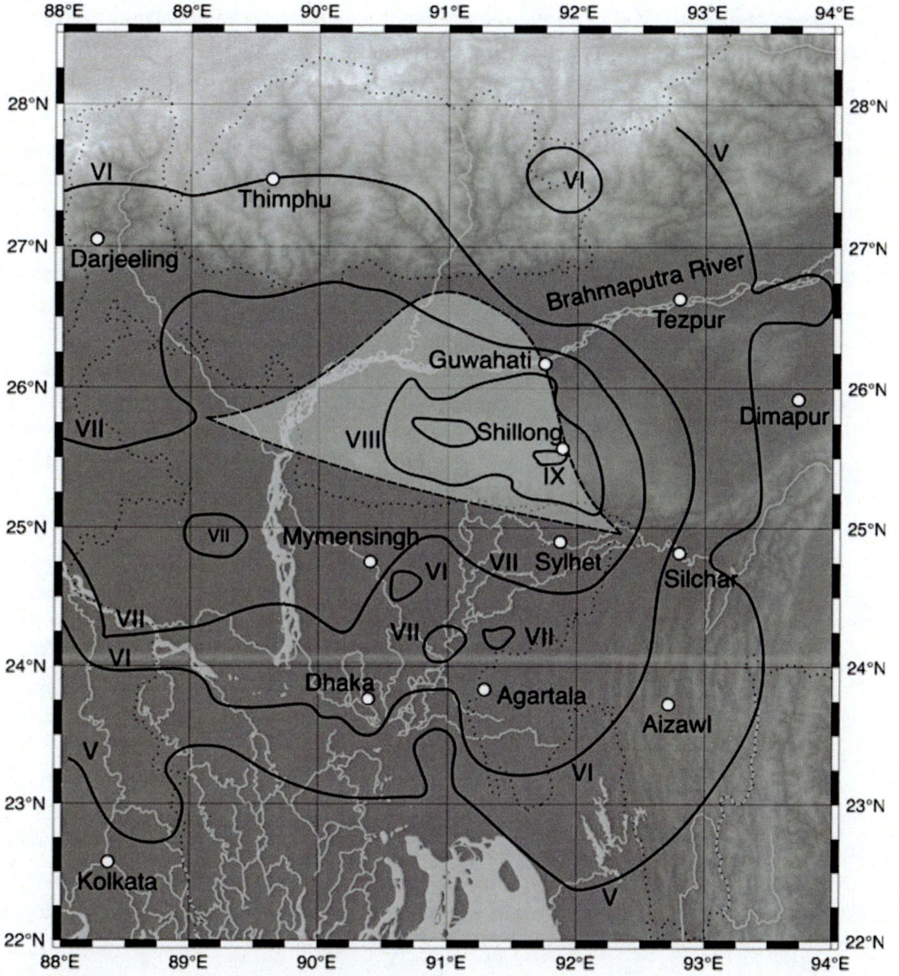

Fig. 14.5 The hat-shaped patch of Oldham (1899) overlapped on the isoseismal map of Ambraseys and Bilham (2003)

letters of La Touche (1897) to his wife, describing the effects of the earthquake became useful for the re-evaluation of the felt reports. The revised MSK isoseismal map by Ambraseys and Bilham (2003) is presented in Fig. 14.5.

14.3 Geomorphologic and Structural Setting

The SP is a gently northward-tilting topographic feature, with a Precambrian Indian plate basement. With the length of about 250 km and width of 80 km, and at an elevation of about 1,500 m above the plains, this is the only raised topography in the Himalayan foreland, with well-expressed structural boundaries defined by geology and geomorphology. The Brahmaputra River flows along the northern and western peripheries of the plateau and the Sylhet Trough (basin) binds its southern side. The geographical placement of the plateau is such that it influences the trajectory of the Indian Summer Monsoon and receives an average of 11.4 m of monsoonal rains, the second highest in the world. The bedrock of the plateau consists of Archaean/Proterozoic gneiss and schist, compositionally like the peninsular shield rocks (Evans, 1964; Nandy, 2001). These basement rocks, along with the intrusive bodies have been dated to be Neoproterozoic to Early Paleozoic ages: 900 and 450 Ma (Yin et al., 2010 and references therein).

Although the whole of the plateau is composed of a similar suite of crystalline rocks, its northern and southern boundaries are morphologically quite distinctive (Fig. 14.6). Spectacular gorges, incised river valleys, and waterfalls mark its southern boundary, compared to its northern boundary which shows a relatively smoother, stair-case-type topography. The crustal thickness of the SP is ~35 km, thinner than its surrounding regions (Mitra et al., 2005). To the north of the plateau, the Moho reportedly dips gently northward and reaches a depth of ~ 44 km at the Himalayan front (Kumar et al., 2004). It is likely that the extensive exposures of the basement rocks across the Brahmaputra Valley requires the presence of an underlying north-dipping structure that could have uplifted both the Brahmaputra Valley and the SP (Yin et al., 2010). The geological and morphometric studies conducted in this area have confirmed the ongoing active crustal deformation along the northern, southern, and western margins of the plateau (Mishra, 2019).

The mafic complexes, which are part of the Rajmahal-Sylhet basalt province, have been considered to be associated with Kerguelen volcanism of 120 to 110 Ma (e.g., Kent et al., 2002). The raising of the SP began around 5.3 million years in the Pliocene Epoch (Najman et al., 2016). GPS data indicate block rotations under the Brahmaputra Valley relative to India, with implications for faster convergence between the SP and Bangladesh across the Dauki fault or slowing in the convergence

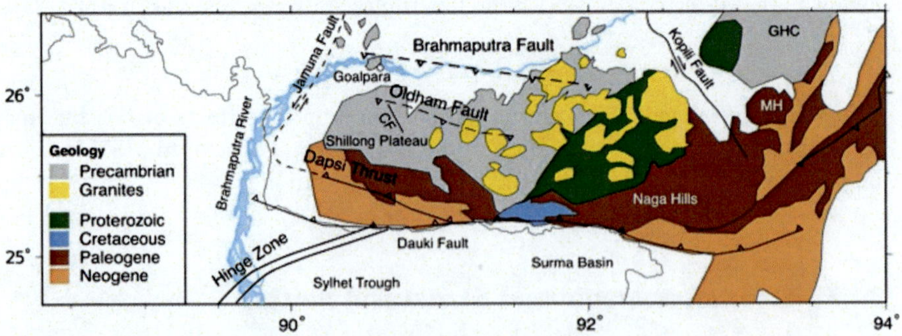

Fig. 14.6 Lithology and important structural features of the Shillong Plateau and its vicinity. MH: Mikir Hills (modified from Mallick et al., 2020 and Yin et al., 2010)

rate across the Bhutan Himalaya (Vernant et al., 2014). Thus, it is possible that the plateau may have a role in the regional tectonism in response to stresses caused by the bending of the Indian lithosphere below the Himalaya (Govin et al., 2018).

During the last decade, geophysical studies have provided new data on the faults associated with the SP. Gokarn et al. (2008) carried out broadband magneto-telluric (MT) surveys covering 26 sites over their 200-km long linear profile running from north of 25°N to 27°N latitude, and almost 92° E parallel, from the southern boundary of the SP towards the Himalayan front, along the Lower Brahmaputra Valley. However, their geoelectric section could not delineate the Oldham fault believed to mark the northern margin of SP. The results indicate the presence of two low resistivity features with a rather weak resistivity contrast of a factor of 5 (100 Ω-m against the surrounding resistivity of 500 Ω-m), extending over four sites immediately south of 26°N, admittedly, based on limited station coverage. In this study, the Dauki fault was identified as a NE–SW striking thrust zone with a dip angle of about 30°, along which the low resistivity layer of Bengal sediments and the underlying oceanic crust subducted to the northwest. The model shows that about 50 km length of these sequences has been subducted beneath the SP presently, and they are traced up to a depth of about 40 km. Another thrust zone, sub-parallel to the Dauki thrust is also observed in the lower Brahmaputra Valley, corresponding to the Brahmaputra fault, interpreted to be an intra-cratonic thrust within the Indian plate. Thus, it is concluded that a large fraction of the seismicity over the SP is associated with the NE–SW striking Dauki thrust. Further, it is suggested that the SP and the adjoining sedimentary layers act as a supracrustal block, not directly participating in the subduction process.

Additional information on the subsurface structure of the region around the SP is available from the receiver function analysis based on 24 broadband seismographs (Mitra et al., 2018). Although these studies have provided more clarity on the nature of the Dauki fault, their results do not provide any confirmation on the faults that bind the northern boundary of the plateau. The results suggest that immediately south of the SP, beneath the northern Bengal Basin, the plateau crust is down-faulted with a Moho offset of 12–13 km. The down faulting, and associated uplift of the plateau, is mediated by reverse motion on the south bounding Dauki fault, with a $\sim$1 km high fault scarp. The estimate of the vertical uplift of the plateau matches well with the sedimentary records from the Surma Basin (Najman et al., 2016) and the GPS-derived convergence rate across the southern margin of the plateau (Vernant et al., 2014).

A new tectonic model based on GPS data, exhumation rates, and gravity by Mallick et al. (2020) proposes subduction initiation below the Bangladesh basin. The idea of a transitional-oceanic crust beneath the Surma Basin that is negatively buoyant that facilitates subduction is central to this model wherein the slip on the Dauki thrust caused by the sinking of the footwall pulled the plateau to the south and caused it to rise. The force required to raise the plateau is attributed to the sinking of the negatively buoyant transitional-oceanic lithosphere along a passive margin where subduction has been initiated. The erosion rates within the metamorphic core of the plateau is suggested to be very close to zero and therefore, the growth of topography is entirely attributed to tectonic uplift.

14.4 Coseismic Surface Features

14.4.1 Morphological Changes in the Brahmaputra Valley

Widespread liquefaction and landslides that followed the earthquake added to the overall severity of destruction. The earthquake destroyed settlements and small towns in the western part of the plateau, and caused heavy damage in the surrounding districts, chiefly due to the extensive liquefaction of the ground. As reported by Oldham (1899), "beds of rivers, tanks, even wells were ridged up or filled by the outpouring sand from fissures, thereby generally disturbing the drainage system of the land and causing extensive flooding." Massive influx of sediment contributed to the sudden rise of the water levels of the rivers, ranging from 0.5 to 3 m, although water level subsided soon. There is a detailed documentation of the coseismic water level changes in the Brahmaputra River, which gave indications of a general raising of the riverbed. The earthquake triggered morphological changes around Guwahati (Gauhati) and Goalpara, situated on the banks of the Brahmaputra River (Fig. 14.7). Several surface features, like fresh ponds, raised riverbeds, and swamps were formed within the valley. On pages 335 to 336, the report quotes an official, from Kamrup

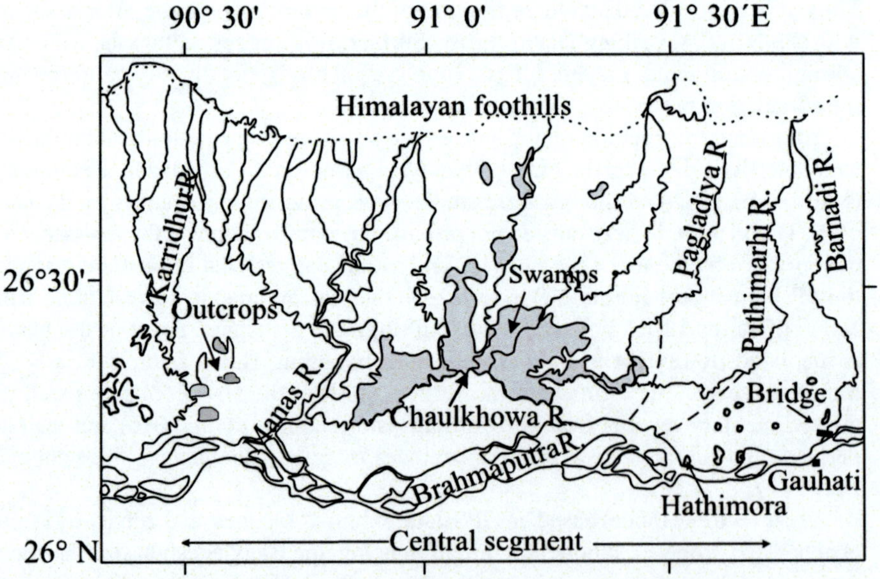

Fig. 14.7 The channel system of the Brahmaputra basin, based on pre-earthquake maps of the area (Allen, 1905). Shaded area represents the zone of post-earthquake swamps and marshes based on the 1911 Survey of India maps. The change in the courses of the Pagladia and Puthimarhi Rivers was a post-earthquake phenomenon. Dashed lines represent the original courses of these rivers on a pre-earthquake landscape. The stone bridge indicated at the downstream of Barnadi River collapsed during the 1897 earthquake (after Rajendran et al., 2004)

(an administrative district in Assam), "there could not be much doubt that the river channels have been raised, and that the lands near the riverbanks have generally subsided to a depth varying from 3 to 7 feet or more, while the tracts of the subdivision towards the north nearer the Bhutan hills remain higher as before".

Oldham (1899) reported that the Brahmaputra bed rose over a stretch of 300 km although this forcing up was varied—more extensive in some locations than others. Although the outpouring of water from sand venting was a factor, the damming of the river by debris was pointed out as a major cause of the level changes, and the reports do not suggest fault movement as a factor for level changes, Elevation changes were more pervasive in the region between 90.5° and 91.5°E, as evident from a comparison of the 1893 map of the region (Allen, 1905) with another one prepared during 1911–13. Extensive growth of a swampy area, restricted to the northern bank of the Chaulkhowa River, has been interpreted as a post-earthquake phenomenon. Another manifestation of morphological change is observed in the eastward migration of the south-flowing rivers of Pagladiya and Puthimarhi, both of which have abandoned their original courses and joined the Brahmaputra River (Das et al., 1996; Fig. 14.7). Elevated beds of 'older alluvium' at 4–5 m (occasionally ~ 8 m) above the present riverbed have been cited as manifestations of past vertical movements within the river valley (e.g., Goswami, 1985).

Although Oldham did not report evidence of any through-going surface features that could have helped identify the causative fault, he makes some pertinent observations, which may have tectonic implications. He estimated that the bed of Brahmaputra was raised over a length of 300 km, while the levees and the floodplains subsided. To explain this divergence in the level changes between riverbeds and floodplains, Oldham (1899; p.89) writes that "waves traversing the alluvium would throw each part into alternate compression and extension as the surface was bent into concave or convex curve." From a hazard perspective, the raised riverbeds and other related changes in the landscape accentuated the intensity and areal spread of floods that occurred after the earthquake. The elevated riverbeds also resulted in lateral shifting of many river channels. As discussed earlier, Pagaldiya (Fig. 14.7), a strong north bank tributary of the Brahmaputra shifted as a new channel abandoning the original course permanently (Saikia, 2019 and references therein).

14.4.2 Secondary Faulting in the Plateau

Oldham (1899) initially proposed that the earthquake might have occurred on a shallow, north-dipping thrust fault underneath the Shillong Plateau. Based on the high accelerations and spatial distribution of aftershocks, it was proposed that the rupture might have extended from the southern topographic front of the plateau to Guwahati and Goalpara in the north. However, Oldham (1926) retracted from this earlier view and favored a mechanism controlled by a deep-seated laccolithic intrusion. Although the report of Oldham made no references to what could be termed as primary faults along the northern topographic front of the plateau, it contained descriptions of what are now being termed as secondary ground ruptures.

Among the notable feature was a vertical displacement up to 10 m along a 19-km long N-S Chedrang (Dudhnai) fault, a tear fault on the western part of the Shillong Plateau (Fig. 14.8). Similarly, a vertical displacement was reported along a four-km-long Samin fault (25°41': 93°43') and a two-km-long WNW Mandalgiri fault (25°39': 90° 33') (Nandy, 2001). These authors have proposed that all these faults form part of a N-S striking, what they termed as 'Dudhnai' fault, and considered this to be the primary source of the 1897 earthquake.

Named after the river through the structure passes, the 20–24-km long NNW Chedrang fault with a vertical throw varying from 0.60 m to 10.60 m with an upthrow to the east must have either slipped coseismically or coinciding with the immediate major aftershocks (Ambrasseys and Bilham, 2003). Oldham's map of the structural feature extending from the headwater of the Chedrang River through Dalbot (25°50':90°44'), Dilma (25°52':90°43') and Jira (25°55':90°41') has been further refined by England and Bilham (2015) with the help of Google Earth image. The Chedrang fault is associated with some tell-tale indications of faulting, such as sag pond near the southern bend of the river named Wolding Wari, fault scarps and waterfalls and abandoned river courses (Fig. 14.8). This fault, considered by Bilham and England (2001) as a secondary fault extending to ~ 9 km is suggested to have moved due to the stresses induced by the reverse movement on the buried main Oldham fault, whose rupture may not have reached any closer than 5 km from the surface. In a more recent work, Subedi and Hetényi (2021) propose that the 1897 earthquake took place at the junction of the Chedrang and Oldham faults and propagated eastwards. This location is arrived at from early instrumental data and 3D velocity models.

14.5 Instrumental Data

Ambraseys and Bilham (2003) have reviewed the instrumental data and the various magnitude estimates of the 1897 earthquake. The earthquake occurred at 17 h 15 m, local time (11 h 09 m GMT) and was recorded by twelve pre-modern seismographs in Europe (Germany, Italy, France, and the UK) at epicentral distances of 64° and 72°. The techniques available at the end of the nineteenth century were clearly insufficient to obtain precise instrumental locations, the more so because the stations were distant. The macro-seismic location of the event was estimated at 26°N, 91°E by various workers (the mean geodetic location being 25.7°N. 91.1°E). England and Bilham (2015) placed the epicenter further south, based on the extent of rupture area (see Fig. 14.1). Surface wave magnitude (Ms) was estimated from the trace amplitudes recorded by six calibrated instruments operating in Italy, one in Russia, and two in the UK (Ambraseys & Bilham, 2003 and references therein). The newly proposed magnitude estimate based on Abe's method is M_S 8.0 ± 0.15, a value close to 8.2 as estimated by Kanamori and Abe (1979), and to 8.0 (Abe, 1994) and to the unified magnitude M = 8.0 (Gutenberg, 1956). A higher value of 8.7 that is widely quoted

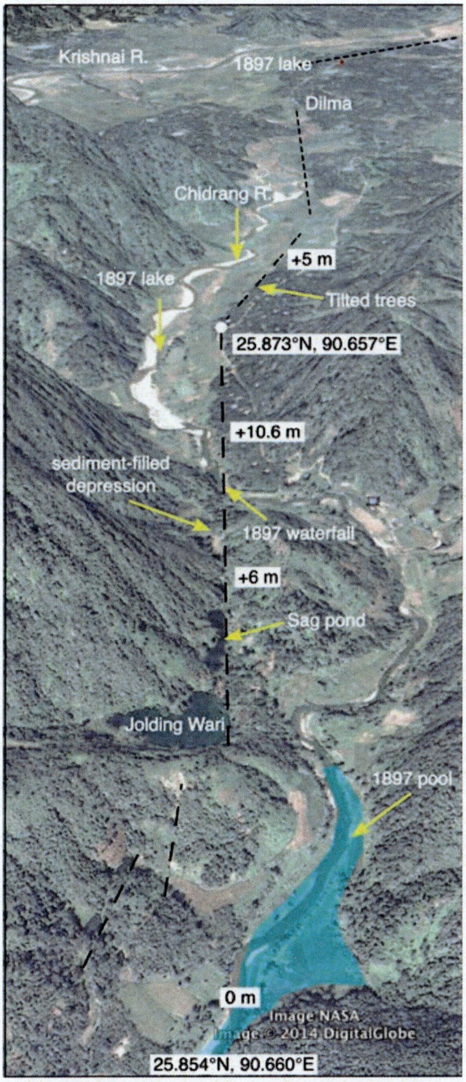

Fig. 14.8 View of the Chedrang fault constructed from the narrative description given by Oldham (1899) and plotted on Google Earth map. Oblique N40W view of Chedrang fault is showing features described by Oldham (1899). Indicated slip is up to the NE on vertical slickensides. The foreground pool (now filled with sediments) occurred at a 30o bend in the fault where the projected path of the fault showed no offset. Oldham ascribed the presence of the lake to the back tilting of the bed of the stream. The precise path of the fault through the sediments in the distant valley remains conjectural (after England & Bilham, 2015)

in the literature, was obtained by Richter (1958) by converting Gutenberg's unified magnitude "m" of 8.0 as M_S and using the empirical relation ($M_S = 1.59$ m-4.0) applicable to earthquakes in California based on surface waves with periods close to 20 s. The re-evaluated instrumental magnitude of $Ms = 8.0 \pm 0.1$ (Ambraseys, 2000) and the geodetic moment estimate of $M_W = 8.1 \pm 0.1$ (Bilham & England, 2001), further revised as $8.15 < Mw < 8.35$ (England & Bilham, 2015) are the currently followed estimates.

14.6 Predecessors of the 1897 Earthquake

Various workers have addressed the question on the predecessors of the 1819 earthquake. From a sedimentary section located in the northern part of the Chedrang fault. Rastogi et al (1993) reported sand dikes and slump structures attributable to a previous earthquake dated as 1220 ± 100 yr BP. They also reported about two terraces along the northern extension of the River Krishnai at a height ~ 10 m above the river, possibly uplifted during previous earthquakes. In a subsequent work detailing the liquefaction chronology from the excavated sections along the Krishnai, Sukhija et al. (1999a and b) reported the evidence for three previous earthquakes: 500 ± 150, 1100 ± 150 and $> 1500 \pm 150$ yr BP and suggests a return period of 400–600 years for large earthquakes in the Shillong Plateau. Bilham and England (2001) propose a source-specific recurrence interval of 3000–8000 years, based on inferred fault slip data using the triangulation data. Combining the paleoseismologic evidence and the historical evidence, Rajendran et al. (2004) suggested that the penultimate earthquake had occurred ~1200 yr BP. Later work by England and Bilham (2015) aver that, great earthquakes in the region are separated by several thousand years, estimated from GPS velocities that show a shortening rate of ~ 5 mm/yr across the plateau, while magnitude 7 events could occur every 100 years. The following section highlights the constraints on the past earthquakes in the region.

14.6.1 Historical Data

Historical records on Assam are quite informative about its rich cultural heritage, in particular its ancient temples – some of them are in a continued existence lasting more than1500 years. Their documented history of restoration and/or reconstruction bears testimony to the ground-shaking effects from great earthquakes that must have impacted this area (see Table 14.1). The ruins of almost all the stone temples discovered around Dhubri, Goalpara, Barpeta, Guwahati, Nowgong, and Tezpur (see Fig. 14.1 for locations) belong to *circa* ninth-tenth century CE point to widespread and contemporaneous destruction (Chaudhury, 1964). It has also been reported that there was a wave of reconstruction activity around 835–860 CE, as testified by a copper plate inscription near Guwahati that mentions of re-erection of a "lofty white

Table 14.1 Earthquake Chronology in the Brahmaputra Valley from the Historical Data

Year of earthquake	Location/effects	Intensity/size
ninth century (between. 825 and 835 CE)	Destruction of temples and palaces from Dhubri in the west through Guwahati to Tezpur in the east (Banerji, 1923; Chaudhury, 1964). Location: near Guwahati-southern Brahmaputra Valley	Very large earthquake; magnitude $\geq$ 8.0; location and timing of the earthquake are based on the present study
1663 CE	Earthquake felt at Guwahati (Gauhati); probably a local earthquake; no damages reported (Gait, 1906)	Probably a mild tremor
1697 CE	Earthquake strongly felt at Sadiya (27°48 N; 94° 38'E), about 300 km east of Guwahati. Massive damage; appreciable aftershock sequence; ground fissuring and liquefaction at the town of Sadiya (Iyengar et al., 1999). Either this earthquake occurred at the 1950 source zone, or it could be an independent source in the upper Brahmaputra Valley or Himalaya	Very large earthquake; magnitude $\geq$ 8.0;
1930 CE	Earthquake at Dhubri (25°57'N; 90°00'E); liquefaction and ground failure within the alluvial tract of north Bengal and NW Bangladesh and western Brahmaputra Valley (Gee, 1934)	Moderate earthquake; M ~6.5
1870–1880 CE	Series of tremors occurred near Guwahati, followed by a relatively quiet period, starting from 1880, until the 1897 event (Keatings, 1877, 1878; Bapat et al., 1983)	Minor earthquakes; none of them damaging
June 12, 1897	Heavy destruction in the valley (Tezpur in the east to Goalpara in the west) and many parts of the Shillong Plateau including monuments and stone bridge near Hajo	Very large earthquake; magnitude $\geq$ 8.0

Note The medieval records include moderate earthquakes of 1548 CE (Garhgaon, SE of Sibsagar; 26°45'N; 94°50'E; about 300 km east of Guwahati) and 1596 CE (Gajala; modern location uncertain), minor earthquakes of 1642 and 1649 CE that occurred in the Upper Assam (Iyengar et al., 1999). Large earthquakes reportedly also occurred in the Higher Himalaya during 1713 and 1806 near Upper Assam at 27.5°N; 93.0E and 28.5 N; 92.0°E, respectively (Ambraseys & Jackson, 2003) temple, which had fallen down" (e.g., Barua, 1966). References to earthquake-related damage are also found in ancient books such as *Kalikapurana* and *Yoginitantra* (late ninth and sixteenth century CE, respectively) (referred by Iyengar et al., 1999). For example, these texts mention of the destruction of the temple of Kamakhya (7- eighth century CE) situated on the top of Nilanchal Hills, located about 8 km from the town of Guwahati.

At least in some instances where the structures had not totally collapsed, the foundations or other surviving elements were used for their restoration. For example, it has been documented the circa sixth century CE temple located at Da-Parbatia near Tezpur was reconstructed over the ruins of a previous structure. Barua (1966) has noted that when the younger structure of *circa* eighteenth century CE collapsed during 1897, the stone doorframe that survived the previous event was exposed. A single structure following distinct styles of construction, belonging to different vintages, reveal its antiquity and its history of multiple episodes of destruction that it has endured. Such ancient ruins are found at many locations along the Brahmaputra Valley (Banerji, 1923; Chaudhury, 1964), pointing to the regional impact of destruction and the regional impact the 1897 earthquake.

14.6.2 Archeoseismological Data

Places of archaeological importance and sites of prehistoric settlements are the usual targets for paleoseismological investigations due to the possibility of finding cultural artefacts and organic remains that can be dated with clarity. The temple of Kamakhya (seventh-eighth century CE) on the top of Nilanchal Hills, northeast of Shillong, is considered as one such, with an archive of information. For example, ancient books such as Kalikapurana (ninth century CE) and Yoginitantra (sixteenth century CE) mention earthquake-related damage to this temple (referred by Iyengar et al., 1999). Excavations in this site exposed evidence of two levels of destructions, the younger one from 1897, and an older one, possibly from the ~ 1200 CE event, possibly the one described in the ancient books (Rajendran et al., 2004, Fig. 14.9). The age of the pre-1897 earthquake was determined using samples of pottery from the pre-1897 destruction level and the bricks from the bottommost layer. The thermoluminescence (TL) dates of 1194$\pm$120 yr and the 1464$\pm$150 yr from these samples were interpreted as minimum and maximum age limits for the pre-1897 earthquake. Reports of damage to other local monuments of the same vintage and epigraphical evidence also mention of destruction to structures around the ninth century CE, coinciding probably with the earthquake inferred from the buried ruins at the Kamakhya temple.

The damage to *Sil Sáko,* the ancient bridge built over Barnadi River, can also be used as a constraint on the previous earthquake/s. Although a dry channel presently, the 42-m-long stone bridge across the Barnadi existed sometime during 5th to eleventh century CE (Hannay (1851). The exact date of its construction is not known, the design and style of architecture suggest 5th to eleventh century CE, according to Hannay (1851), who also provided a sketch (Fig. 14.10). Captain Dalton, who exam-

Fig. 14.9 Photograph showing a pit excavated at the base of the southern compound wall of the Kamakhya temple, with only the bottom part is represented in the photograph (b) Log showing complete stratigraphy with different levels of temple foundations and corresponding thermoluminescence (TL) dates. The upper bricks layer represents ruins from the 1897 earthquake (after Rajendran et al., 2004)

ined this bridge in 1851 (as quoted in Hannay, 1851) describes the bridge as: "… *solid masonry, built without lime or mortar…. the superstructure being a platform with a slight curve 140 feet long and 8 feet in breadth composed of slabs of stone six feet nine inches long and ten inches thick, numbering five in the whole breadth, resting on an under-structure of sixteen pillars, three in a row, equally divided by three large solid buttresses; with a half buttress projecting from a circular mass of masonry forming the abutments at each end of the road, there being in the whole, 21 passages for the water*".

The bridge was used in 1205 CE by an invading cavalry to cross over to reach Bhutan, an indication that the bridge was strong and usable at that time. Historians note that the *Sil Háko* bridge is the only structure within this kingdom that match with the descriptions by Captain Dalton and conclude that what collapsed in 1897 is *Sil Sáko,* which the cavalry used in 1205–1206 (Gait, 1906; Barua, 1966; p. 141). From this example of a long-standing bridge across the Barnadi, it is concluded that an 1897-type earthquake might not have occurred in this region at least for the 690 years (Rajendran & Rajendran, 2011; Rajendran et al., 2004). Hannay (1851) mentions about an imperfectly done repair work on the bridge in the intervening

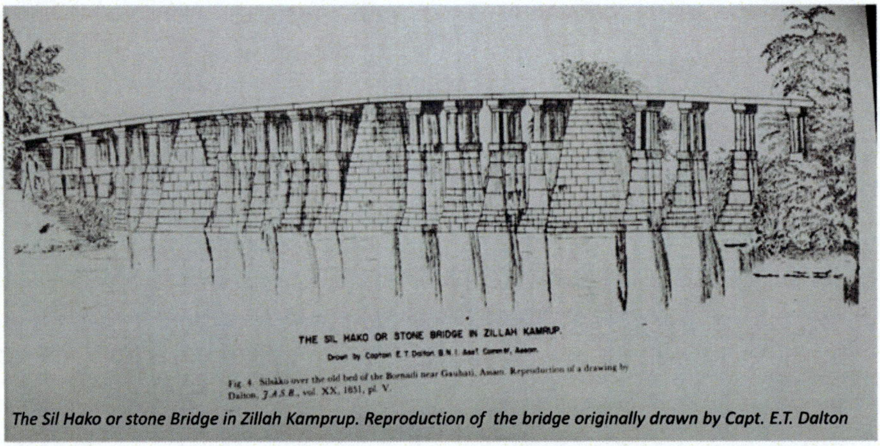

Fig. 14.10 Sil Sako: an ancient stone bridge over river Barnadi, which was an important transit route in ancient and medieval times. The cavalcade led by Muhammed Bin Bukhtiyar, are believed to have crossed the bridge during 1205–06 CE. This historical bridge was destroyed completely in the 1897 Assam earthquake (Hannay, 1851)

time. It is not known if this repair work was done consequent to some peripheral damage caused by any intervening earthquake. However, if indeed the bridge had survived in the form as described by Hannay till 1851, its total collapse can only be attributed to be an unprecedented ground acceleration experienced after 1851, quite possibly during the 1897 earthquake.

14.6.3 Evidence from Liquefaction Features

The spectacular liquefaction features generated within the Brahmaputra Valley documented in detail by Oldham (1899) served as proxies for site selection for trenching excavations. As mentioned earlier, Sukhija et al. (1999a and b) had suggested evidence for at least two older earthquakes during 1450–1650 and 700–1050 CE. Rajendran et al. (2004) bracketed the liquefaction features in this area as younger features (1450–1650 CE) characterized by ~ 10 cm wide vents, and older (700–1050 CE) with vents that are $\geq$ 50 cm, confirming that multiple events have occurred here prior to 1897.

The paleoliquefaction investigations conducted by Rajendran et al. (2004) along the Chedrang River near Dilma (Fig. 14.11), indicate multiple events- dated at 645–980 CE (1250$\pm$80 yr B.P.) and 535–530 BCE (2170$\pm$140 yr B.P.). Averaging with the radiocarbon dates published by Sukhija et al. (1999a), they suggested that the penultimate earthquake must have occurred around 1208$\pm$40 yr B.P. This age falls in the range of 1220$\pm$100 yr B.P. obtained from a tree trunk embedded in a sandblow excavated near Dilma (Rastogi et al., 1993).

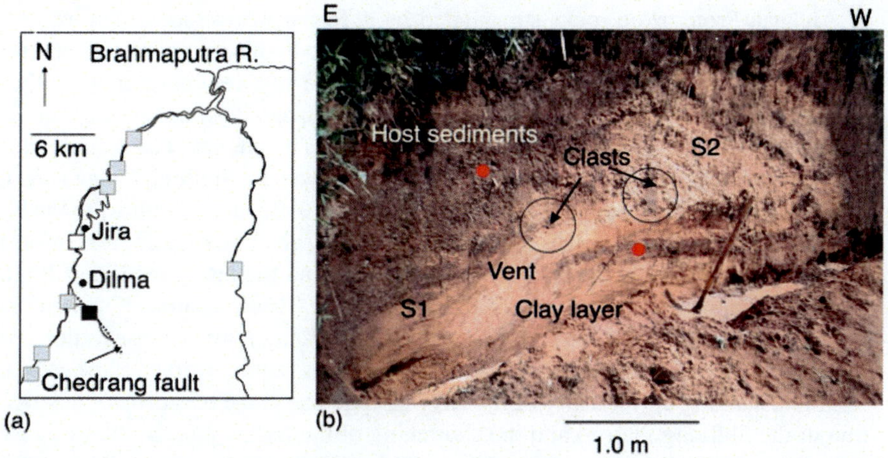

Fig. 14.11 **a** Locations of paleoliquefaction sites around the Chedrang River. Sites investigated (open circles: Rastogi et al. (1993); shaded circles: Sukhija et al. (1999a) and solid squares: Rajendran et al., 2004). **b** A view (facing north) showing the paleoliquefaction feature exposed at Dilma-1 located on the western bank of the Chedrang River. S1 is the older sill, and S2 is the younger sill. Samples collected from the ruptured clay layer and the host sediments (locations shown by red dots) yielded calibrated ages of 2170 ± 140 years B.P and 1250 ± 80 years B.P., respectively. The section shown here is 1 m below the surface

While there are some age constraints for the faults associated with the 1897 source, the history of the Dauki fault is still quite unclear. Despite its significant morphological and geophysical expressions, this southern boundary fault of the plateau is not known to have generated any major earthquakes during historical or recent times (Chen & Molnar, 1990). The ~ 300 km length of the Dauki, together with the Dapsi Thrust, has not ruptured least during the last 500 years (Rajendran et al., 2004; Steckler et al., 2008). While previous studies have not shown any specific evidence to suggest that the Dauki fault is active, the more recent trenching investigation near the Dauki fault on the Bangladesh side exposed evidence for past earthquakes. A possible earthquake (M < 8), dated between 1500 and 1630 CE, can be correlated with the 1548 earthquake in Bangladesh (Morino et al., 2011).

14.7 Seismotectonics of the 1897 Earthquake: An Ongoing Debate

The geological evolution of the Eastern Himalayan region is influenced initially by the dilatational tectonics of the Triassic as the Indian plate moved away from the Gondwanaland, to be followed by the compressional tectonics in the Tertiary Period, consequent to the Indo-Eurasian and Indo-Burman collisions (Dasgupta & Biswas, 2000). The placement of the Shillong Plateau (along with Mikir Hills), consisting of Archean-Proterozoic rocks surrounded by a Tertiary Basin as an outlier in the NE Himalayan plate boundary, region added much complexity to the regional seismotectonic processes. The 1897 earthquake was the first reminder of the hitherto underestimated active tectonic potential of the Plateau and its boundary regions. The earthquake occurred when the instrumentation was in its infancy. The seismic monitoring of the region indicated a cluster of seismicity over the Shillong Plateau and the adjoining Lower Brahmaputra Valley (e.g., Khattri 1992a, b), in contrast to the high seismicity of the Indo-Burman ranges. Much of the subsurface geological information of the region is obtained from gravity studies and the interpretation of limited geological exposures. Therefore, the proposals of the 'Brahmaputra', 'Oldham', and 'Dauki' faults, although backed variously by a different set of observations and models, their nature and the spatial and depth extents require further clarification.

Understanding the tectonics of the 1897 earthquake, in the backdrop of the evolution of the Shillong Plateau and the shortening of the Indian plate at the foot of the Himalaya is still a live issue. At the heart of this debate is the question of the location and geometry of the 1897 rupture, which began during the lifetime of R.D. Oldham. As mentioned earlier, the plateau uplift is constrained by two border faults on the north and south. Modern workers were initially inclined towards Oldham's initial proposal of a north dipping thrust fault, a structure considered as an extension of the Himalayan frontal dècollement (e.g., Seeber & Armbruster, 1981; Chen & Molnar, 1990; Gahalaut and Chander, 1992).

Oldham's initial proposal of a north dipping thrust as a causative structure of the 1897 earthquake would have to structurally extend onto the Dauki fault, which was considered as an extension of the Himalayan frontal dècollement. Oldham's initial supposition of the Dauki fault as the causative structure has been followed by various modern workers (Gahalaut and Chander, 1992; Mukhopadhyay et at., 1993; Morino et al., 2014). Gahalaut and Chander (1992) had offered a geodetic solution consisting of 5 m of oblique slip on the Dauki fault, but this was based on Oldham's observations of subsidence in the Brahmaputra Valley and not on the triangulation data. A recent analysis of the geodetic data shows that strain decreases from north of the Plateau to the south, ruling out slip on the Dauki fault during the earthquake (England & Bilham, 2015).

The east–west trending 300 km-long Dauki fault was originally regarded as a major right-lateral strike-slip fault with >200 km of motion (Evans, 1964). The drop of ~90 mGal Bouguer gravity suggested vertical movement of the order of 13 km, based on which Hiller & Elahi (1984) proposed a dominant thrust-type of movement. Yin et al. (2010) also regard it as a thrust contact with a clear morphologic expression that displays two levels of incised terrace surfaces in the hanging wall against a flat flood plain in the footwall. The positive Bouguer gravity anomaly (Nayak et al., 2008) over the Shillong Plateau (~50 mGal) indicates uncompensated topography with no crustal root to support the topography. In the latest model on regional tectonics, Mallick et al. (2020) used gravity and GPS observations to suggest subduction initiation of the Surma Basin along the pre-existent weak fabric of the Dauki thrust. With all these scenarios in the backdrop, a big-picture question in this context is, why hasn't the décollement broken through the sedimentary pile and created a wedge? That is one way of accommodating the India-Shillong convergence without having to lift an entire landform made of crystalline rocks.

The prevailing understanding about the geometry of the causative fault changed when Bilham and England (2001) used the triangulation data to propose a steep (50°) SSE-dipping reverse fault, close to the northern topographic front of the Shillong Plateau, which they named as the Oldham fault. In their model, a pair of reverse faults of opposing vergence has raised the Shillong Plateau as a pop-up structure (Fig. 14.12). The refined geodetic solution published by England and Bilham (2015) requires an average slip of 25 ± 5 m on a fault that dips south at ~ 40° beneath the northern topographic front of the plateau with a rupture length of 79 km. The Chedrang fault located west of the Oldham fault is suggested to have sympathetically slipped >10 m of down-to-the-west. So, the main rupture needed to acquire a length of 95 km to reach close to the Chedrang fault, and the estimated magnitude range arrived at is 8.15 < Mw < 8.35.

An alternate faulting scenario during the 1897 earthquake was presented by Rajendran et al. (2004). According to this view, the south-dipping Brahmaputra fault, located farther north of the plateau front in the alluvial plains could have hosted the

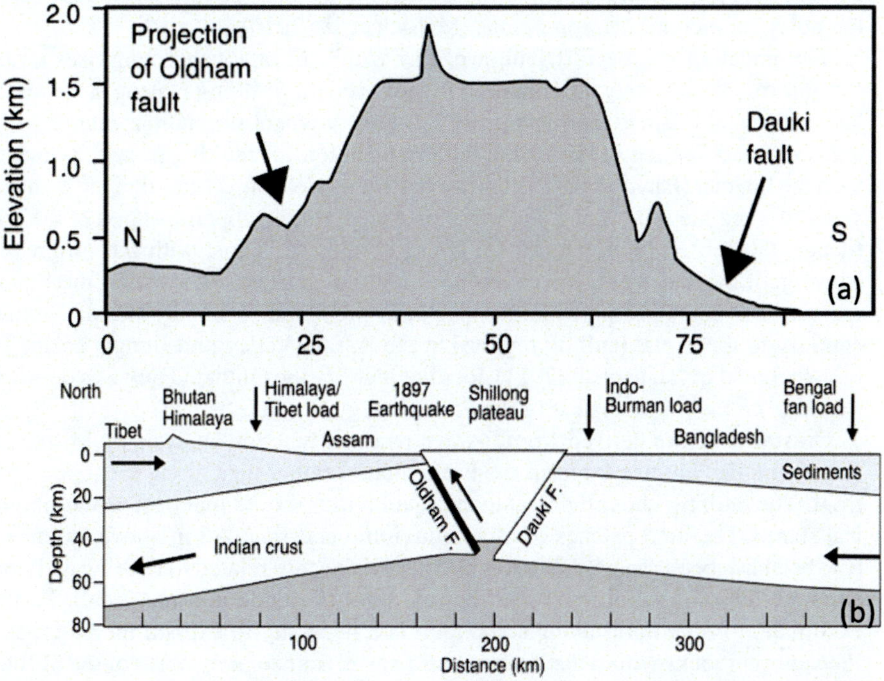

Fig. 14.12 **a** Topographic section through the central Shillong Plateau showing the projected surface location of the Oldham fault. **b** N/S section from Tibet to the Bay of Bengal showing schematic geometry of plateau pop-up. The dip of the Dauki fault is conjectural and it would intersect the 1897 rupture where it dips less than 40° (after Bilham & England, 2001)

1897 earthquake. Differing with this view, in a subsequent work, England and Bilham (2015) pointed out the low strain observed in the Brahmaputra Valley, in contrast to the high strain rate on the plateau region. They state that, "had a reverse faulting of large slip occurred there it would have resulted in several meters of uplift the bed of the Brahmaputra precisely where Oldham describes a depression." However, these arguments must be considered in the backdrop of the reported morphological changes that were mostly confined to the northern bank of the river. As discussed earlier in this chapter, whether these changes were of tectonic or non-tectonic origin is open to debate, especially because of the lack of geodetic control points in the northern part of the alluvial plains.

These divergent views need to be reconciled in the overall tectonic context and subsequent work using GPS geodesy. The paucity of surface expressions from a great earthquake on the plateau (Srinivasan, 2003; Rajendran et al., 2006; Morino et al., 2011) and lack of any gravity anomalies do not favor the proposal of "Oldham fault' (Rajendran et al. 2004, 2006). Admittedly, both the gravity and seismic data for this region are inadequate and not of high quality. The current GPS data too are inadequate to help resolve the debate since the horizontal velocity field mainly shows inter-block convergence and inter-seismic locking on the block bounding faults; the vertical data are overwhelmed by the annual monsoonal loading signal and appear too noisy for tectonic interpretations (Mallick et al., 2020).

The modeling of stress distribution pattern within the region that comprises various tectonic regimes, namely the Himalaya in the north, the Shillong Plateau in the middle and the Bengal Basin in the south using 2-D finite element simulations predict a non-uniform stress regime in the region with implication for faulting pattern associated with the Plateau (Islam et al., 2011). The best-fit model conjectures that the compressive stress regime is dominant in the region except for the uppermost part of the crust (depth <30 km), where extensional stress dominates. Thus, with the progressive convergent displacement, the extensional failures get restricted within the Plateau region, and the upper part of the plateau is predicted to fail by normal faulting contrary to the thrust faulting regimes in the Assam Valley and Bengal basin. The best-fit model of Islam et al. (2011) does not identify the Oldham Fault as a dominant structure on the northern side of the Plateau (Fig. 14.13).

Crustal structure derived from P-wave receiver function analysis by Mitra et al. (2018) implies faulting beneath the Brahmaputra rather than along the topographic front. The faulting under the Brahmaputra alluvium should manifest as morphological scarps. Elevated patches of older alluvium occurring 4–8 m above the present riverbed have been reported, but it is not clear if they are related to faulting activities, although these possibilities cannot be ruled out (Rajendran et al., 2004). Further, exhumation in the north along some fault is a necessity to explain the outcrops of Precambrian rocks within the Brahmaputra basin, immediately north of the Shillong Plateau (Fig. 14.7), where multiple *en-echelon* faults could have provided a step-like topography.

While the previous papers commented mostly about the northern boundary of the Shillong Plateau as the main structural entity, a recent model of the regional plate motion and tectonics notes the necessity of combining both the northern and southern boundaries of the plateau to explain the available geophysical and geological data

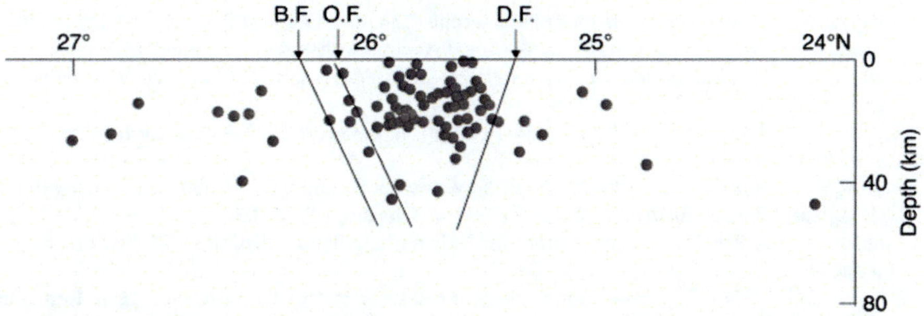

Fig. 14.13 N-S cross-section of the Shillong Plateau showing all the proposed faults (D.F: Dauki Fault; O.F: Oldham Fault; B.F: Brahmaputra Fault) in a pop-up model (after Islam et al., 2011; distribution of earthquakes from Nayak et al. (2008). The best-fit model from 2-D numerical modeling for the stress distribution of the plateau region by Islam et al. (2011) predicts that the Oldham fault does not contribute to the regional deformation, although the results of the model do not exactly match the epicentral distribution

(Mallick et al., 2020). These authors argue that the observed inter-plate convergence and plateau uplift in this region is driven by subduction initiation of the Indian lithosphere south of the plateau. The vertical motion of the plateau, apparent in its surface topography as well as the shape of its Moho, is merely a consequence of the Indian lithosphere attempting to penetrate the mantle, which drives shortening and uplift of the upper plate along relatively narrow shear zones on its southern and northern margins. On the southern side of the Shillong Plateau, the deformation architecture is represented by the Dauki and Dapsi thrusts, while on the northern side, with a step-like topography, the complex erosional and depositional behavior of the Brahmaputra River to a large extent obfuscates manifestations of active structure(s). The seismogenesis associated with the Shillong Plateau continues to be enigmatic and needs to be further investigated by employing various investigative tools, including seismic reflection surveys that will facilitate mapping the subsurface structures.

References

Abe, K. (1994). Instrumental magnitudes of historical earthquakes, 1892 to 1898. *Bulletin of the Seismological Society of America, 84*, 415–425.

Allen, B. C. (1905). Assam District Gazetteers (Vols. 3, 4, and 5). Govt. of Assam, Gauhati, India.

Ambraseys, N. (2000). Reappraisal of north Indian earthquakes at the turn of the 20th century. *Current Science, 79*, 101–114.

Ambraseys, N., & Bilham, R. (2003). Re-evaluated intensities for the great Assam earthquake of 12 June 1897, Shillong, India. *Bulletin of the Seismological Society of America, 93*(2), 655–673.

Ambraseys, N., & Jackson, D. (2003). A note on early earthquakes in northern India and southern Tibet. *Current Science, 84*, 570–582.

Bapat, A., Kulkarni, R.C., & Guha, S. K. (1983). Catalogue of Earthquakes in India and Neighbourhood from historical Period up to 1979, 211 pp. Indian Soc. of Earthquake Technol Roorkee.

Banerji, R. D (1923) Annual report, Archaeological Soc. of India, pp. 80 – 81, New Delhi.

Barua, R. K. L. (1966) Early History of Kamarupa (234 pp.), Lawyers Book Stall, Gauhati, India.

Bilham, R. (2008). Tom La Touche and the Great Assam Earthquake of 12 June 1897: Letters from the Epicenter. *Seismological Research Letters - SEISMOL RES LETT., 79*, 426–437. https://doi.org/10.1785/gssrl.79.3.426

Bilham, R., & England, P. (2001). Plateau 'pop-up' in the great 1897 Assam earthquake. *Nature, 410*, 806–809.

Biswas, S., & Grasemann, B. (2005). Quantitative morphotectonics of the southern Shillong Plateau (Bangladesh/India). *Australian Journal of Earth Sciences, 97*, 82–03.

Bomford, G. (1939). The readjustment of the Indian triangulation, Prof. Pap. 28, 100 pp., Surv. of India, Dehra Dun.

Burrard, S.G. (1898). Report on trigonometrical results of earthquake in Assam, Tech. Rep., Surv. of India, Dehra Dun, India.

Chaudhury, P. D. (1964). Archaeology in Assam, 77 pp., Department of Archaeology, Gauhati, India.

Chen, W.-P., & Molnar, P. (1990). Source parameters of earthquakes and intraplate deformation beneath the Shillong Plateau and the northern Indoburman ranges. *Journal of Geophysical Research, 95*, 12,527–12,551.

Das, P. K., Duarah, B. P., & Goswami, D. C. (1996). Channel migration study of the Pagladiya River, Assam using remote sensing and field survey data. *Journal of Geosciences, 1*, 38–45.

Dasgupta, A. B., & Biswas, A. K. (2000). *Geology of Assam*, Geological Society of India, Bangalore, India (pp. 170).

England, P. C., & Bilham, R. (2015). The Shillong Plateau and the great 1897 Assam Earthquake: Tectonics (Vol. 34, pp. 1792–1812). https:// doi .org /10 .1002/2015TC003902.

Evans, P. (1964). The tectonic framework of Assam. *Geological Society of India Journal, 5*, 80–96.

Gahalaut, V. K., & Chander, R. (1992). A rupture model for the great earthquake of 1897, northeast India. *Tectonophysics, 204*, 163–174.

Gait, E. A. (1906). A history of Assam. Thacker, Spink & Co. 407p.

Gee, E. R. (1934). The Dhubri earthquake of July 8, 1930, Mem Geol Surv Ind 65, part 1, 106 pp.

Gokarn, S. G., Gupta, G., Walia, D., Sanabam, S. S., Hazarika, N. (2008). Deep geoelectric structure over the Lower Brahmaputra valley and Shillong Plateau, NE India using magnetotellurics. *Geophysical Journal International, 173*, 92–104

Goswami, D. C. (1985). Brahmaputra River, Assam, India: Physiography, basin denudation and channel aggradation. *Water Resources Research, 21*, 959–979.

Govin, G., Najman, Y., Copley, A., Millar, I., van der Beek, P., Huyghe, P., et al. (2018). Timing and mechanism of the rise of the Shillong Plateau in theHimalayan foreland. *Geology, 46*, 279–282. https://doi.org/10.1130/G39864.1

Gupta, R. P., & Sen, A. K. (1988). Imprints of Ninety-East Ridge in the Shillong Plateau. *Indian Shield: Tectonophysics, 154*, 335–341.

Gutenberg, B. (1956). Great earthquakes between the period 1896–1903. *Eos, Transactions of the American Geophysical Union, 37*, 608.

Hannay, S. T. (1851). Brief notice of the Sil-sako or stone bridge in Zillah Kamrup. *Journal of the Asiatic Society of Bengal, 20*, 291–294.

Hiller, K., & Elahi, M. (1984). Structural development and hydro-carbon entrapment in the Surma Basin, Bangladesh (northwest Indo–Burman Fold belt). In *Proceedings of 4th Offshore Southeast Asia Conference*, Singapore (pp. 6.50–6.63).

Islam, M. S., Shinjo, R., & Kayal, J. R. (2011). Pop-up tectonics of the Shillong Plateau in northeastern India: Insight from numerical simulations. *Gondwana Research*. https://doi.org/10.1016/j.gr.2010.11.007

Iyengar, R. N., Sharma, D., & Siddiqui, J. M. (1999). Earthquake history of India in medieval times. *Indian Journal of History of Science, 34*, 181–237.

Kanamori, H., & Abe, K. (1979). *Journal of Geophysical Research, 84*, 6131–6139.

Keatings, R. H. (1877). Record of the occurrences of earthquakes in Assam during the year 1874. *The Journal of the Asiatic Society of Bengal, 46*, 294–309.

Keatings, R. H. (1878). Record of the occurrences of earthquakes in Assam during 1877. *The Journal of the Asiatic Society of Bengal, 47*, 4–11.

Kent, R. W., Pringle, M. S., Müller, R. D., Saunders, A. D., & Ghose, N. C. (2002). 40Ar/39Ar geochronology of the Raj-mahal basalts, India, and their relationship to the Kerguelen Plateau. *Journal of Petrology, 43*, 1141–1153.

Khattri, K. N. (1992a). Seismic hazard in Indian region. *Current Science, 62*(1–2), 109–116.

Khattri, K. N. (1992b). Seismological investigations in north eastern region of India. *Memoirs - Geological Society of India, 23*, 275–302.

Kumar, R. M., Raju, P. S., Devi, E. U., Saul, J., & Ramesh, D. S. (2004). Crustal structure variations in northeast India from converted phases. *Geophysical Research Letters, 31*, L17605. https://doi.org/10.1029/2004GL020576.

La Touche Rev, J. D. (1897). The late earthquake in India. *Nature, 56*, 444–445.

Mallick, R., Hubbard, J. A., Lindsey, E. O., Bradley, K. E., Moore, J. D. P., et al. (2020). Subduction initiation and the rise of the Shillong Plateau. *Earth and Planetary Science Letters, 543*. https://doi.org/10.1016/j.epsl.2020.116351

Mishra, M. N. (2019). Active tectonic deformation of the Shillong plateau, India: Inferences from river profiles and stream-gradients. *Journal of Asian Earth Sciences, 181*, 103904

Mitra, S., Priestley, K., Bhattqacharyya, A. K., & Gaur, V. K. (2005). Crustal structure and earthquake focal depths beneath northeastern India and southern Tibet. *Geophysical Journal International, 160*, 227–248.

Mitra, S., Priestley, K. F., Borah, K., & Gaur, V. K. (2018). Crustal structure and evolution of the Eastern Himalayan plate boundary system, Northeast India. *Journal of Geophysical Research: Solid Earth, 123*, 621–640.

Molnar, P., & Pandey, M. R. (1989). Rupture zones of great earthquakes in the Himalayan region. *Proceedings of the Indian Academy of Sciences - Earth & Planetary Sciences, 98*(1), 61–70.

Molnar, P. (1990). A review of the seismicity and the rates of active underthrusting and deformation at the Himalaya. *Journal Himalayan Geology, 1*, 131–154.

Morino, M., Kamal, A. S. M. M., Muslim, D., Ali, R. Md. E., Kamal, M. A., Rahman, Md.Z., & Kaneko, F. (2011). Seismic event of the Dauki fault in 16th Century confirmed by trench investigation at Gabrakhari Village, Haluaghat, Mymensingh, Bangladesh. *Journal of Asian Earth Sciences, 42*, 492–498.

Morino, M., Kamal, A., Maksud, S. M. et al. (2014). A paleo-seismological study of the Dauki fault at Jaflong, Sylhet, Bangladesh: Historical seismic events and an attempted rupture segmentation model. *Journal of Asian Earth Sciences, 91*, 218–226. 10.1016/j.jseaes.2014.06.002.

Muhkopadhyay, S., Chander, R., & Khattri, K. N. (1993). Fine structure of seismotectonics in western Shillong Massif, north east India. *Proceedings of the Indian Academy of Sciences - Earth & Planetary Sciences, 102*, 383–398.

Najman, Y., Bracciali, L., Parrish, R. R., Chisty, E., & Copley, A. (2016). Evolving strain partitioning in the Eastern Himalaya: The growth of the Shillong Plateau. *Earth and Planetary Science Letters, 433*, 1–9.

Nandy, D. R. (2001). Geodynamics of Northeastern India and Adjoining Region, 209 pp., acb Publ., Kolkata, India.

Nayak, G. K., Rao, V. K., Rambabu, H. V., & Kayal, J. R. (2008). Pop-up tectonics of the Shillong Plateau in the great 1897 earthquake (Ms 8.7): Insights from the gravity in conjunction with the recent seismological results. *Tectonics, 27*, 1–8.

Oldham, R. D. (1899). Report of the great earthquake of 12thJune 1897.Mem. Geol. Surv. India reprinted: 1981, Geological Survey India, Calcutta, 379p.

Oldham, R. D. (1920). The Cutch earthquake of 16th June 1819 with a revision of the Great earthquake of the 12th June 1897, Mem. Geol. Survey of India (Vol. 46, pp. 80–147). Calcutta.

Oldham, R. D. (1926). The Cutch (Kachh) earthquake of the 16th June, 1819 with a revision of the great earthquake of the 12th June, 1897. *India Geological Survey Memoir, 46*, 71–147.

Rajendran, K., & Rajendran, C. P. (2011). Revisiting some significant earthquake sources in the Himalaya: Perspectives on past seismicity. *Tectonophysics, 504*, 75–88.

Rajendran, C. P., Rajendran, K., Duarah, B. P., Baruah, S., & Earnest, A. (2004). Interpreting the style of faulting and paleoseismicity associated with the 1897 Shillong, northeast India, earthquake: Implications for regional tectonism. *Tectonics, 23*, TC 4009. 1029/2003/TC001605

Rajendran, C. P., Rajendran, K., Duarah, B. P., Baruah, S., & Earnest, A. (2006). Reply to comment by R. Bilham on "Interpreting the style of faulting and paleoseismicity associated with the 1897 Shillong, northeast India, earthquake: Implications for regional tectonism." *Tectonics, 25*, TC2002.

Rao, P. C., & Kumar, R. M. (1997). Uplift and Tectonics of the Shillong Plateau, Northeast India. *Journal of Physics of the Earth, 45*, 167–176.

Rastogi, B. K., Chadha, R. K., & Rajagopalan, G. (1993). Palaeoseismicity studies in Meghalaya. *Current Science, 64*, 933–935.

Richter, C. F. (1958). Elementary Seismology. W.H. Freeman Company. Reprinted: S. Chand and Company, New Delhi, 768 p.

Saikia, A. (2019). *The unquiet river* (583p.). Oxford University Press.

Seeber, L., & Armbruster, J. (1981). Great detachment earthquakes along the Himalayan arc and long-term forecasting. In: Simpson, D. W., Richards, P. G. (Eds.), *Earthquake prediction-an international review*. Maurice Ewing Series. American Geophysical Union (Vol. 4, pp. 259–277).

Srinivasan, V. (2003). Deciphering differential uplift in Shillong Plateau using remote sensing. *Journal of the Geological Society of India, 612*, 773–777.

Steckler, M. S., Akhter, S. H., & Seeber, L. (2008). Collision of the Ganges-Brahmaputra Delta with the Burma Arc: Implications for earthquake hazard. *Earth and Planetary Science Letters, 273*, 367–378. https://doi.org/10.1016/j.epsl.2008.07.009

Strahan, G. (1891). Great trigonometrical survey of India, Assam Valley triangulation, Tech. Rep., Survey of India, Calcutta, India.

Subedi, S. and Hetényi, G. (2021) Precise locating of the great 1897 Shillong Plateau earthquake using teleseismic and regional seismic phase data. The seismic record *1*, 135–144. https://doi.org/10.1785/0320210031

Sukhija, B. S., Rao, M. N., Reddy, D. V., Nagabhushanam, P., Hussain, S., Chadha, R. K., & Gupta, H. K. (1999a). Paleoliquefaction evidence and periodicity of large prehistoric earthquakes in Shillong Plateau, Indian Earth Planet. *Science Letter, 167*, 269–282.

Sukhija, B. S., Rao, M. N., Reddy, D. V., Nagabhushanam, P., Hussain, S., Chadha, R. K., & Gupta, H. K. (1999b). Timing and return period of major paleoseismic events in the Shillong Plateau, India. *Tectonophysics, 308*, 53–65.

Vernant, P., Bilham, R., Szeliga, W., Drupka, D., Kalita, S., Bhattacharyya, A. K., Gaur, V. K., Pelgay, P., Cattin, R., & Berthet, T. (2014). Clockwise rotation of the Brahmaputra Valley relative to India: Tectonic convergence in the eastern Himalaya, Naga Hills, and Shillong Plateau. *Journal of Geophysical Research: Solid Earth, 119*, 6558–6571.

Wilson, C. A. K. (1939). Triangulation of the Assam valley series, geodetic report 1938, survey of India, Dehra Dun, India, pp 10-22.

Yin, A., Dubey, C. S., Webb, A. A. G., Kelty, T. K., Grove, M., Gehrels, G. E., & Burgess, W. P. (2010). Geologic correlation of the Himalayan orogen and Indian craton: Part I. Structural geology, U-Pb zircon geochronology, and tectonic evolution of the Shillong Plateau and its neighbouring regions in NE India. *Geological Society of America Bulletin, 122*, 336–359.

Chapter 15
Upper Assam 1950 Earthquake

15.1 Introduction

The August 15, 1950, M_w 8.6 earthquake is the largest among those that occurred in the Himalaya during the last 200 years. Popularly known as the Assam earthquake (and as Chayu or Medog earthquake), this event had originated in the rugged mountainous areas between the Himalaya and the Hengudan (southwest China), with its located in the China-Tibet region north of a village known as Rima (Fig. 15.1). The largest among the continental earthquakes, this was felt over an area of 3 million km^2 in India, Myanmar (Burma), Bangladesh (then East Pakistan), Tibet and southern China. As for the Indian side, it caused extensive damage in many towns of Upper Assam and regions across the eastern Himalaya. The area most affected is an arc bordered on the west by the Ranganadi River (west of Subansiri River) and on the east by the Dhing River (Gee, 1952). This area includes the Abor Hills, the Mishmi Hills, and a part of the Assam Plains. The seismotectonic significance of this earthquake has been recognized in the early years of modern seismology. As with the other great Himalayan earthquakes of the last century, several aspects of the 1950 earthquake continue to be debated and more recent studies, using modern-day techniques are providing fresh insights. Identification of the host structure, nature of morphological manifestations and the earthquake history of the region have been discussed in the recent published papers.

Initial analyses of the available seismograms had suggested a strike-slip mechanism for the 1950 earthquake, with predominantly dextral motion on a NW–SE-oriented steep fault (Ben-Menahem et al., 1974). A later study by Chen and Molnar (1977) suggested thrust faulting on a gently northward-dipping plane, comparable to other earthquakes in the Himalayan arc. It was also expected, on geometrical considerations, that the slip from an earthquake sourced on the EHS would be strike-slip, whereas if it were based on the MHT, the faulting would be dominated by thrust-type mechanism (Devachandra et al., 2014). Post-earthquake surveys did not

C. P. Rajendran and K. Rajendran, *Earthquakes of the Indian Subcontinent*, GeoPlanet: Earth and Planetary Sciences,
https://doi.org/10.1007/978-981-16-4748-2_15

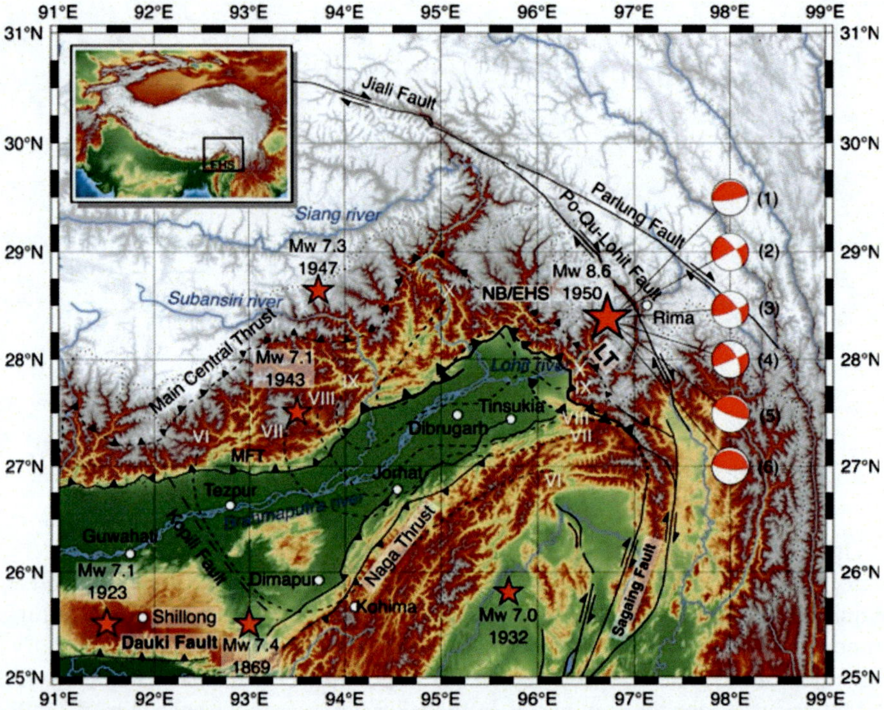

Fig. 15.1 Major structures and earthquakes in the eastern Himalaya: NB: Namche Barwa. EHS: the Eastern Himalayan Syntaxis. LT: Lohit Thrust. The focal mechanism solutions for the 1950 earthquake are from, 1: Chen and Molnar (1977); 2: Tandon and Dube (1973); 3: Sylman et al. (1974); 4: Ben-Menahem et al. (1974); 5: Devachandra et al. (2014) and 6: Coudurier-Curveur et al. (2020). Dashed lines represent isoseismals for the 1950 earthquake (after Poddar, 1950)

record any primary surface rupture and thus this earthquake also fell into the category of the great Himalayan earthquakes whose primary ruptures do not reach the surface, as discussed in the context of other earthquakes. This was the prevailing understanding until Coudurier-Curveur et al. (2020) reported a 200 km-long surface rupture along the EHS—a proposal made from extensive field-based observations and high-resolution satellite data. It was suggested that the earthquake had ruptured both the MFT (locally represented by the Abor Hills) and the Mishmi Thrust (MT), producing clear tectonic scarps cutting across the Quaternary alluvial terraces.

Paleoseismological investigations in the Brahmaputra Valley have been useful to assess the earthquake history for providing a window into the potential predecessor/s of the 1950 earthquake (Reddy et al., 2009; Rajendran & Rajendran, 2011). The historical and paleoseismic records imply that the Assam earthquake of 1950 may not have any predecessors mimicking in its magnitude, at least during the last 1000 years or even longer (Rajendran & Rajendran, 2021). Unlike the other parts of the Himalaya, the challenges of field studies in the Upper Assam Valley need also to be underscored, as this region is being constantly impacted by floods and landslides that could either potentially obfuscate or smother the original sedimentary structures/features. Despite these challenges, the studies that have gained speed in the recent years have added to the understanding of coseismic surface deformation in the region and the mechanism of the earthquake (Priyanka et al., 2017; Coudurier-Curveur et. al., 2020; Singh et al., 2021). These studies have also raised new questions. This chapter provides a review of the current "knowns and unknowns" about the 1950 and the previous earthquakes in the region.

15.2 Felt Reports of the 1950 Earthquake and Its Impact on the Landscape

The Assam earthquake occurred in the late evening of August 15, 1950, at about half past seven, Indian Standard Time, on the third anniversary of India's independence. It was felt at half-past seven local time when the celebrations were closing. Across the political border, in the Indian and Tibetan villages in the mountainous regions, people were getting ready to retire to their homes when the earth shook violently, accompanied by a thundering noise. Lasting for about four minutes, the earthquake was felt over an area of 3 million km^2, spread over India, Myanmar (Burma), Bangladesh (then East Pakistan), Tibet and southern China. As mentioned in the beginning, it caused extensive damage to many towns of the Upper Assam, and there were about 1500 casualties on the Indian side (Poddar, 1950; Ramachandra Rao, 1953). It is most likely that an equal number or more had died in the hilly the terrain on the Chinese or the Burmese side. Highest felt intensities were reported from a narrow quadrant to the SW, possibly because of the amplification of shaking effects within the alluvium (Poddar, 1950; Tandon, 1954) (see Fig. 15.1).

Much of the real-time information about the effects of this earthquake comes from the accounts of Frank Kingdon-Ward, a botanist, plant collector, explorer, and writer, who was camping at Rima in the border area of southern Tibet on that day. The commentary goes like this: "*I was seated writing my diary near the entrance to our tent. Suddenly after the faintest tremor (felt by my wife and not by me), there came an appalling noise, and the earth began to shake violently.... We were immediately thrown to the ground, the lantern too was knocked over and went out instantly*" (Kingdon-Ward, 1951). In the words of a resident from the Upper Assam: "the earth heaved and rolled with a sickly undulating movement.... At the end there was a succession of loud booming noises variously described as resembling anti-aircraft fire and express trains" (Gee, 1952). The earthquake attracted much global attention because of its Himalayan source and its magnitude and soon found its way into a classic seismology text, 'Elementary Seismology' (Richter, 1958). This book gives a detailed account of how the earthquake had severely impacted the landscape of Upper Assam, and scarred the hill slopes, by triggering spatially extensive incidents of landslide as well as its seismological aspects known at that time.

Occurring, as it did in the height of monsoon, when the "ground was saturated and the rivers swollen to their maximum volume", the damage was greatest in the Abhor and Mishmi region (Gee, 1952). Here the hills were sheared from top to bottom, their sides crashing down into the valleys below, thus generating temporary dams of rocks, sediment, and vegetation, and eventually to burst one by one or simultaneously. In some instances, some of these temporary lakes endured for a longer period and some rarest ones could still be remaining intact (e.g., Sangestar Tso, also known as Madhuri Lake located in Tawang, Arunachal Pradesh). The floods following the rupturing of temporary dams carried vast quantities of silt and debris raised the river channels including that of the Brahmaputra itself, and often causing changing of river courses. The floods impacted the forests, leaving dead tree stumps standing in upright positions, embedded in the sand sheets along the river (Ramesh & Gadagkar, 1990). The influx of massive sediment load that raised the riverbed along with tens of thousands of uprooted trees and bodies of wild animals and cattle is reported to have affected the river's navigability (Gee, 1952; Saikia, 2019).

The earthquake-induced landslides, floods, and the erosion of the riverbanks not only impacted the forests, but it also destroyed vegetation and wildlife (Gee, 1952). The loss of cattle is estimated to be 50,000–100,000 and depletion of river-borne fish, stifled by the increased amount of mud and oxygen depletion. The earthquake could have mobilized about 47 km^3 of material, resulting in the injection of a considerable sediment load into the Brahmaputra, that took many years to flush through its lower courses (Mathur, 1953). Some recent studies have highlighted how this earthquake drastically changed the sediment budget in the Ganga–Brahmaputra basins (Akter, 2016). The sediments transported through the Brahmaputra resulted in a huge load in the Meghna estuary at the delta front in Bangladesh; the movement of the finer fractions happening within a few years, and the coarser ones must have taken nearly five decades to reach the deltaic region. The earthquake thus contributed to accelerated land build-up and delta progradation in the downstream part of the drainage.

There was extensive damage to roads, railways, bridges, and buildings located in the plains and the estimated acceleration on alluvium was 0.5 g (Ben-Menahem et al., 1974). The ground shaking was most extensive in Sadiya and the nearby towns of the Upper Assam. Mathur (1953) had reported widespread occurrence of fissures and widespread liquefaction in many localities in the alluvial plain of the Lohit River. Soil liquefaction damaged large areas of cultivable land, which later got transformed into grasslands, and finally ended up as fallow land, leading to downstream migration of rural population. Considerable damage was caused to the town of Dibrugarh and nearby villages and tea estates with their factories. The Assam Railway suffered much, its "rail tracks being torn up and twisted into snake-like patterns" (Gee, 1952; see Fig. 15.2).

Fig. 15.2 Damage scenes near Dibrugarh in Assam during the 1950 earthquake; photographed by Poddar (1950): (i) twisted rail track; (ii) close-up of an engine shed; (iii) subsidence of a building due to lateral spread; adopted from Bilham (2019)

15.3 Tectonic Setting of the Eastern Himalaya

The continued north–south convergence along the Himalayan arc and the east–west convergence along the Burmese arc since the beginning of the Tertiary Period have culminated in a complex tectonic framework along the north-eastern convergent boundary (Nandy, 2001; Figs. 15.1 and 15.3). Here, the E–W regional strike of the Himalaya abruptly changes along the eastern syntaxial bend, defining the zone of collision centering on Namcha Barwa peak (Fig. 15.3). This is also where the Indian Plate abuts against the Eurasia and the Burma (Myanmar) Plates, and approximately 50% of the resulting motion (~18 mm/year) is accommodated by strike-slip motion on the Sagaing fault (Vigny et al., 2003). The region exhibits intense structural deformation manifestly represented by well-defined structural and topographic trends (Searle et al., 2007).

To the west of the EHS is the Arunachal Himalaya, extending to the Bhutan segment, and to the south, is the Indo-Burmese arc region where the subducted Indian plate slides past the Burma plate. The deformation style in the EHS changes from dextral shear between India and SE Asia to collision along the Himalaya. The Mishmi Thrust at the cusp of the EHS is the equivalent of the Main Frontal Thrust (MFT) of the Himalayan arc. The Lohit Thrust separates the trans-Himalayan plutonic rocks in the north from metavolcanics and high-grade metamorphic rocks (Fig. 15.3). Balanced cross sections across the NW-oriented Lohit Thrust (Ningthoujam et al., 2015) indicate that the deformation here has led to large scale erosion and that

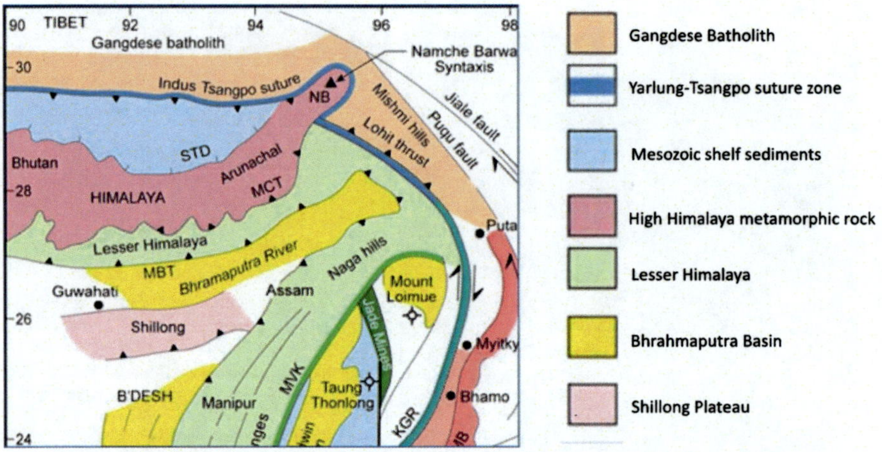

Fig. 15.3 Geology of eastern India and neighboring regions showing the major suture zones, faults, and terrane boundaries (modified after Searle et al., 2007)

in comparison with other parts of Himalaya, crustal shortening in the easternmost Arunachal is maximum, with a shortening strain of 83.28%.

The GPS-based deformation models suggest significant differences in convergence rates between the segments of the Himalayan arc located to the east and west of 92°E longitude. While the convergence velocity between India and Eurasian plate is estimated as ~31 mm/year to the east of this longitude, the rate is significantly lower to its west (Burgess et al., 2012). For example, the rate of convergence is ~18 mm/year in eastern Nepal (Ader et al., 2012) and ~14–17 mm/year in Bhutan (Vernant et al., 2014). Devachandra et al. (2014) argue that the strain corresponding to a slip rate of at least 20 ± 4 mm/yr is accumulating under the locked frontal EHS, accounting for a deficit of about 1.3 m, since 1950. Gupta et al. (2015) suggest a convergence of 32.4 mm/year between the Mishmi and Lohit Thrusts, postulating a recurrence period of ~200 years for the 1950-type of earthquakes. With no report of a similar-size event during the past 200 years, the inferred slip deficit is ~4 m is enough to generate a great earthquake.

15.4 Epicentral Location and Fault Models

The epicentral location of the 1950 earthquake at 28.6ºN and 96.5ºE and hypocentral depth of ~14 km was derived using the data from the Bureau Central International de Séismologie, Strasbourg (Tillotson, 1951). Initially, the India Meteorological Department located the earthquake at 29°N and 97°E (Tandon, 1954), and the magnitude of 8.6–8.7 (in Richter instrumental scale) was assigned using two seismograms, from Pasadena and Strasburg (Richter, 1958). The epicenter was located ~40 km west of

Rima, a village near the India-Tibet border, which was reported to have experienced the maximum severity. The relocated epicenter proposed by Chen and Molnar (1977) at 28.36°N 96.45°E, is shown in Fig. 15.1. The aftershocks, first located by Tandon (1954), and later relocated by Chen and Molnar (1977), were distributed over a wide zone.

Diverse mechanisms have been suggested for the 1950 earthquake (e.g., Tandon, 1954; Ben-Menahem et al., 1974; Chen & Molnar, 1977; Molnar & Deng, 1984; Coudurier-Curveur et al., 2020). The solution by Ben-Menahem et al. (1974) which has been widely referred to, had suggested strike-slip faulting on a fault plane trending NNW-SSE, spatially associated with the EHS with a steep north-easterly dip. While most early studies agreed on strike-slip faulting, Chen and Molnar (1977), relying on the available first-motion data, suggested thrust faulting on an easterly striking, gently north-dipping fault at the eastern end of the Himalayan plate boundary. In a later paper, Molnar (1990) used the distribution of relocated aftershocks to suggest a wide zone of strike-slip and shallow thrust faulting occurring on multiple planes.

Devachandra et al. (2014) used GPS data to estimate of azimuth of slip on the detachment and the dip of the underneath detachment, to derive a possible focal mechanism solution. The thrust solution appears like that of Chen and Molnar (1977), suggesting that faulting along the EHS is analogous to the faulting pattern observed elsewhere in the Himalaya arc. Previous solutions that mostly related the earthquake to the EHS have been challenged by Coudurier-Curveur et al. (2020), who suggest ruptures on multiple planes involving both the MFT (Abor Hills) and the Mishmi Thrust (MT). The existence of tectonic scarps on these thrusts that cut through the Quaternary alluvial terraces at a high angle, is suggested as supporting evidence. The oblique slip on two planes—the shallow-dipping part of the Himalaya's Main Frontal Thrust and the steeply dipping Mishmi Thrust, orthogonal to each other, and the 200 km-long surface rupture along the Mishmi and Abor Hills are the highlights of the new model. The two planes represent the shallow-dipping Main Frontal Thrust (13–15°) and the significantly steeper Mishmi Thrust (25–28°). Larger amounts of co-seismic slip and uplifts occurred on the Mishmi Thrust (>30 m), as compared to the slip of >7 m on the MFT. From the spatial distribution characteristics of landslide scars and relocated aftershock distributions, the rupture area was inferred as 330 km × 90 km, across the East Himalayan Syntaxis. How the two orthogonal planes with variable dips would have ruptured simultaneously without any time delay remain ambiguous, but the proposed slip and seismic moment are compatible with the composite main shock mechanism.

15.5 Earthquake History of the Region

Information on past earthquakes in the Assam Valley and the neighboring regions has been extracted mostly from the written documents such as the indigenous *Ahom* records and the colonial writings (e.g., Iyengar et al., 1999). The book "A history of Assam", prepared by Edward Gait (1906) is another source of information. Paleoliquefaction investigations around the Brahmaputra Valley and nearby areas have provided geological evidence for earthquakes, although the chronology needs further validation (Rajendran & Rajendran, 2011; Reddy et al., 2009).

Significant among the historically known events is the Garhgaon earthquake of 1548 CE that presumably occurred near the Naga Hills (Fig. 15.4). Based on trenching excavations in Gabrakhan Village near Mymensingh, Bangladesh, Morino et al. (2009) reported a medieval offset dated at 1500–1630 CE, possibly linked to the 1548 CE earthquake. If that observation holds up, it is likely that this event may have occurred outside the Assam Valley, but distantly impacted the region. The 1596 CE Gajala earthquake has been documented by Gait (1906), but with very little information on its epicentral details. Among the better-known events, the important one is the 1697 Sadiya earthquake, felt strongly in and around Sadiya Town, ~150 km southwest of the 1950 source (Iyengar et al., 1999). Descriptions collated from various

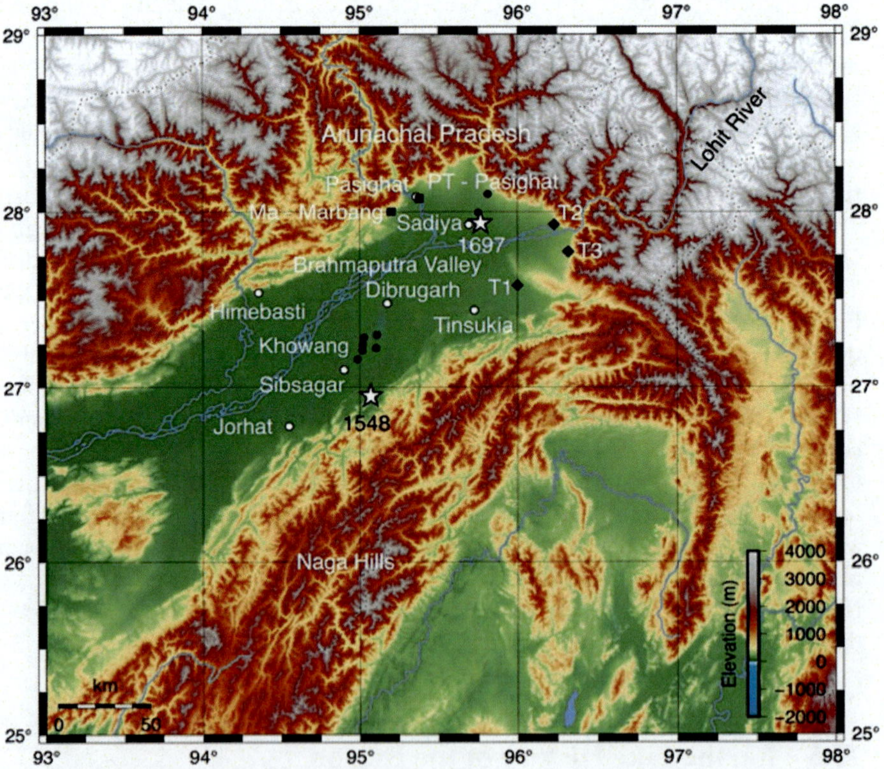

Fig. 15.4 Locations of historical earthquakes and selected sites of paleoseismological investigations in the northeast region (Arunachal Himalaya, Brahmaputra Valley and adjoining regions): PT: Pasighat; Ma: Marbang (Priyanka et al., 2017); T1 and T2 (Rajendran & Rajendran, 2011); T3 (Kamlang Nagar; Singh et al., 2021). Solid black circles mark the sites excavated by Reddy et al. (2009); sedimentary section from a site at Khowang is shown in Fig. 15.5)

sources point to a major earthquake, followed by significant aftershocks, landslides, ground fissuring and liquefaction suggesting intensity not less than X. Another event discussed in various documents is the Tingkhang earthquake. Ambraseys and Jackson (2003) regard this as a significant earthquake.

The paleoseismic data from a trench in the central Bhutan Himalaya (Le Roux-Mallouf et al., 2016) indicates an earthquake in the age ranges of 1642 and 1836 CE. Believed to have occurred in the spring of 1713, with a likely magnitude of Ms 7. The year, location and magnitude of this earthquake have been revised by Hetényi et al. (2016) and it is suggested to be a frontal earthquake on the Bhutan Himalaya that occurred in the year of 1714. In an excavation study at a location called Dungsam Chu (26.79194°N, 91.51085°E) on the frontal thrust of Bhutan, further east of the site of Hetényi et al., (2016; Fig. 13.2 for trench locations), Zhao et al. (2021) from another trench site (Dungsam Chu, see Fig. 13.2) reiterate the faulting episode from the 1714 CE earthquake. However, they also add a caveat that "due to the lack of an undisturbed layer sealing the fault, we could not determine the minimum age of the E1 event", i.e., the event 1—the youngest event exposed in the section. However, much of the damage from the 1714 earthquake has been reported from the monasteries like Gangteng near Thimphu (central Bhutan; see Fig. 13.2). The temporal context of this earthquake is further discussed at the end of this chapter.

Morphological evidence of faulting has been used as proxies in the explorations of previous earthquakes. For example, terrace uplifts on the frontal area of the Mishmi Hills and the ^{10}Be exposure ages of 173 ± 27 yr, 1188 ± 180 yr, 1636 ± 486 yr, 1710 ± 513 yr and 2734 ± 405 yr) from the terrace deposits are assumed to represent episodes of uplift due to multiple earthquakes including the 1950 earthquake, the last event in the sequence (Coudurier-Curveur et al., 2020). However, it must be mentioned here that the two differing 10Be surface exposure dates, namely, ~173 years and ~1188 years, are obtained from T1 surface. The presence of a scarp that shows a maximum cumulative height ~3 m, identified near Pasighat, located on the frontal part of the MFT, is regarded as potential evidence for the coseismic rupture Priyanka et al. (2017). From the excavations at Kamlang Nagar along the Mishmi Thrust in the Eastern Himalayan Syntaxis (Fig. 15.4), Singh et al. (2021) infer coseismic near surface offsets from the 1950 earthquake. From this site, a dip-slip displacement of 24.6 ± 4.6 m on a 25 ± 5°E dipping fault has been reported. Previous excavations in Himebasti Village had revealed evidence for an earthquake-induced surface rupture from an event that occurred between 1650 and 1890 CE, possibly the 17th earthquake of 1697 CE Sadiya earthquake (Pandey et al., 2019; see Fig. 15.4 for locations).

Finding well-preserved liquefaction features and obtaining materials to date whatever features are exposed, is a challenging task in the highly dynamic riverine environment of the Brahmaputra and its adjacent areas. Although constrained by such limitations, paleoseismological investigations have provided some information on the nature of geological signatures left by great earthquakes like the 1950 event. Reddy et al. (2009) have presented an inventory of sandblow features documented from the Brahmaputra Valley as the example shown in Fig. 15.5. They have also presented evidence for one medieval earthquake, dated in the age range of 1262–1635 CE.

Fig. 15.5 Liquefaction feature exposed on the left bank of Burhi Dihing River at Khowang. The solid circle represents the organic material collected as an upper bound with its ^{14}C age. About 1.2 m high sand dyke (Dy) is seen with connectivity to the source sand and subsequent deposition over it (Figure courtesy: D.V. Reddy)

15.6 Unresolved Questions

The region around Eastern Himalaya had hosted two great earthquakes during the last 200 years, namely the 1897 Shillong (M_w 8.2) and the 1950 Upper Assam (M_w 8.6). The potential for future great earthquakes in the central segment as well as the north-eastern segments has been recognized. Advances in the understanding of the seismotectonics of the eastern Himalaya have been slow compared to other parts of the Himalaya. The last two decades have witnessed much progress in providing newer insights into the earthquake processes in the complex tectonic regime of the NE India.

The recently gathered data also bring up some additional questions on the earthquake rupture processes in the eastern Himalaya. The rupture mechanism of the 1950 earthquake involving two orthogonal planes dipping differently needs further elucidation. One question is whether it was a single continuous rupture that involved multiple faults. If so, is it tenable that these orthogonal faults are connected at depth? It may be recalled that such perplexing rupture complexities are generally associated with strike-slip faults. For example, the 1992 Landers (M_w 7.3), 1999 Hector Mine (M_w 7.1), 2010 Haiti (M_w 7.0), 2012 Indian Ocean (M_w 8.6 and 8.2) and the 2019 Ridgecrest, California, earthquake sequences (M_w 6.4, 5.4 and 7.1) have ruptured on complex fault systems and in some cases involving intersecting faults producing multiple earthquakes (Li et al., 2020). It is noted that the rupture processes of these earthquake sequences follow a cascading nature, rather than a single continuous rupture along multiple faults. The seismotectonic scenario does also begs yet another question whether the complex rupture geometry assigned to the 1950 earthquake is repeatable or it is just a rare one-time phenomenon. Both the Arunachal MFT (the Abor Hills) and the Mishmi Thrust must have been activated independently multiple times in the past rather than in tandem, as reported in the case of the 1950 earthquake (Rajendran & Rajendran, 2021). The faulting kinematics of single continuous rupture involving two orthogonal fault planes, as invoked for the 1950 earthquake, must be an extremely complex phenomenon and a rarity that needs more validation.

The morphology of the Himalayan foreland fold-and-thrust belt owes a lot to the surface processes influenced by climate. For example, the eastward narrowing of the fold-and-thrust belt (despite the increase in the eastward convergence) is considered as an outcome of higher rainfall rates (Hirschmiller et al., 2014). As the active erosion causes rapid retreat of the mountain-front, the traces of the MFT and other active thrusts become fuzzy and difficult to trace. The task of identifying surface ruptures along the MFT becomes challenging, and as noted by Coudurier-Curveur et al. (2020), the application of classical paleoseismological techniques in such a terrain can be ambiguous, at best. It is suggested that co-seismic slip (>30 m) and uplift (>7 m) observed on the Mishmi Thrust, where the thrust cuts the youngest deposits at the level of a permanent river channel, may lead to erroneous estimates on the timings of past earthquakes. These complexities must be accounted for in any earthquake recurrence models for the region.

Although proxy signatures of past earthquakes were exposed in trenches, establishing their chronological constraints has been problematic, due to the lack of exposures of multiple features and non-availability and/or reworking of dateable material. In the high liquefaction susceptibility alluvial plains that are vulnerable to earthquakes from near and far sources, it is difficult to discriminate between sandblows caused by multiple sources (Rajendran & Rajendran, 2021). For example, the moderate January 5, 2017, earthquake (M_w 5.6; 24.0°N 91.9°E; focal depth of 31 ± 6.0 km) caused liquefaction features including lateral spreads and sandblows (Debbarma et al., 2017). Developed at distances of ~25 away from the epicentre, these authors consider them as anomalous for an earthquake of its size and focal depth. Similarly, the April 26, 2021 (M_w 6.0) magnitude event near Tezpur (Table 1.1) also generated liquefaction (eyewitness reports) testifying to the high liquefaction potential in these regions. The paleoliquefaction investigation around Kopili fault, located between Tezpur and Guwahati (Fig. 15.1), towards the lower Brahmaputra Valley reported events during 250 ± 25 yr. BP, between 400 to 770 yr. BP and 900 ± 50 yr. BP (Kumar et al., 2016). The available liquefaction data, with all its limitations and uncertainties, however, point to a major medieval earthquake likely to have occurred in the Upper Assam between 1260 and 1635 CE, and this might as well be the well-documented 1697 Sadiya earthquake. As for the 1950 source, Coudurier-Curveur et al. (2020) suggest bi-millennial return period.

Field evidence from Bhutan and Arunachal obtained by two independent groups suggest two great medieval earthquakes (1697 and 1714). Hetényi et al. (2016) suggest that the 1714 CE earthquake ruptured the megathrust in Bhutan, most likely during a M 7.5–8.5 event, ruling out the possibility that a seismic gap exists in the Bhutan segment. Does this mean that the western Arunachal Pradesh (east of Bhutan and west of 93.5°E) is the only segment in the eastern Himalaya with no known big event? But if the 1697 CE earthquake had ruptured the western Arunachal Pradesh (Pandey et al., 2021), no seismic gap exists in the Arunachal segment.

If indeed both 1697 Sadiya and 1714 Bhutan were great earthquakes, one would expect to see reports of devastation from both events along with some geological indications. However, devastation from the 1714 earthquake does not find any mention in the regional historical records from Assam. The major damage from this earthquake is reported from Gangteng monastery, located closer to Thimphu, central Bhutan (see Fig. 13.2). For an earthquake of magnitude M ~ 8, it is surprising that no major devastation is historically reported from the middle and lower parts of the Brahmaputra Valley that parallel the Bhutan central segment. It should, however, be mentioned here that Erteleva et al. (2014) mention about the 1714 earthquake among the historically reported earthquakes from Assam that "shattered temples in Guwahati", without indicating the source of that information or any specific details of the impacted temples. Further clarity on the events of 1697 and 1714 in terms of their epicenters and their magnitudes will contribute to the understanding of the paleoseismic context of the predecessors of the 1950 earthquake (Rajendran & Rajendran, 2021). It is reasonable to conclude from the available geological records that no earthquakes that mimic the magnitude and style of rupturing of the 1950 earthquake occurred at least in the last 1000 years or may be more. The northeastern Himalayan front and the syntaxial bend need to be further investigated considering the intricacies of the regional seismotectonics.

A recent trenching investigation by Singh et al. (2021) across a 10 m-high-scarp at a site called Kamlang Nagar along the Mishmi Thrust in the Eastern Himalayan Syntaxis (T3 in Fig. 15.4), reveals an offset, ~4 m out of the estimated 24.6 ± 4.6 m, that is attributed to the 1950 earthquake. They surmise that only a multiple earthquake scenario can explain the 10 m- high scarp, after combining geological evidence from the present trench and the earlier one on the MFT at Marbang, Arunachal Himalaya (Jayangondaperumal et al., 2011). They also propose an earlier earthquake between 590 BCE–66 CE. This study on the Mishmi Thrust and the previous one at Pasighat on the HFT (Priyanka et al., 2017) add further credence to the presumption that the 1950 event may have ruptured along the two orthogonal fault systems, as originally suggested by Coudurier-Curveur et al. (2020). But what is remarkable is the fact that the Marbang site, located ~12 km southwest of Pasighat, shows no evidence for the 1950 offset. Such contradictory observations need to be reconciled to throw light on the surface rupture mechanics of the 1950 earthquake.

References

Ader, T., Avouac, J.-P., Liu Zeng, J., Lyon-Caen, H., Bollinger, L., Galetzka, J., Genrich, J., Thomas, M., Chanard, K., Sapkota, S. N., Raujari, S., Shrestha, P., Ding, L., Flouzat, M. (2012). Convergence rate across the Nepal Himalaya and interseismic coupling on the main Himalayan thrust: Implications for seismic hazard. *Journal of Geophysical Research, 117*, B044403. https://doi.org/10.1029/2011JB009071.

Ambraseys, N., & Jackson, D. (2003). A note on early earthquakes in northern India and southern Tibet. *Current Science, 84*, 570–582.

Akter, M. (2016). Conceptualizing environmental governance on the GBM basin. *Bandung J of Global South, 3*, 25. https://doi.org/10.1186/s40728-015-0025-7.

Ben-Menahem, A., Aboodi, E., & Schild, R. (1974). The source of the great Assam earthquake—An interplate wedge motion. *Physics of the Earth and Planetary Interiors, 9*(4), 265–289.

Bilham, R. (2019). Himalayan earthquakes: A review of historical seismicity and early 21st century slip potential. In Treloar, P. J., & Searle, M. P. (Eds.), *Himalayan tectonics: A modern synthesis* (Vol. 483, pp. 423–482). London: Geological Society. Special Publications.

Burgess, W. P., Yin, A., Dubey, C. S., Shen, Z., Kelty, T. K (2012) Holocene shortening across the main frontal thrust zone in the eastern Himalaya. *Earth and Planetary Science Letters*, 357–358, 152–167. https://doi.org/10.1016/j.epsl.2012.09.040.

Chen, W. P., & Molnar, P. (1977). Seismic moments of major earthquakes and the average rate of slip in central Asia. *Journal of Geophysical Research, 82*, 2945–2969.

Coudurier-Curveur, A., Tapponnier, P., Okal, E., Vander Woerd, J., Kali, E., Choudhury, S., Baruah, S., Etchebes, M., & Karakas, Ç. (2020). A composite rupture model for the great 1950 Assam earthquake across the cusp of the East Himalaya syntaxis. *Earth and Planetary Science Letters*. https://doi.org/10.1016/j.epsl.2019.115928.

Debbarma, J, Martin, S., Suresh, G., Ahsan, A, Gahalaut, V. K. (2017). Preliminary observations from the 3 January 2017, Mw 5.6 Manu, Tripura (India) earthquake. *Journal of Asian Earth Sciences, 148*, 173–180.

Devachandra, M., Kundu, B., Catherine, J., Kumar, A., & Gahalaut, V. K. (2014). Global positioning system (GPS) measurements of crustal deformation across the frontal eastern Himalayan syntaxis and seismic-hazard assessment. *Bulletin of the Seismological Society of America, 104*, 1518–1524.

Erteleva, O., Aptikaev, F., Baruah, S., et al. (2014). Seismic treatment for a maximal credible earthquake in Guwahati city area of northeast India region. *Natural Hazards, 70*, 733–753.

Gait, E. A. (1906). *A history of Assam* (407 p.). Kolkata: Thacker, Spink & Co.

Gee, E. P. (1952). The Assam earthquake of 1950. *Journal of the Bombay Natural History Society, 50*(3), 629–635.

Gupta, T. D., Riguzzi, F., Dasgupta, S., Mukhopadhyay, B., Roy, S., & Sharma, S. (2015). Kinematics and strain rates of the eastern Himalayan Syntaxis from new GPS campaigns in Northeast India. *Tectonophysics, 655*, 15–26.

Hetényi, G., Le Roux-Mallouf, R., Berthet, T., Cattin, R., Cauzzi, C., Phuntsho, K., & Grolimund, R. (2016). Joint approach combining damage and paleoseismology observations constrains the 1714 A.D. Bhutan earthquake at magnitude 8 ± 0.5. *Geophysical Research Letters, 43*, 10695–10702. https://doi.org/10.1002/2016GL071033

Hirschmiller, J., Grujic, D., Bookhagen, B., Coutand, I., Huyghe, P., Mugnier, J. L., & Ojha, T. (2014). What controls the growth of the Himalayan foreland fold-and-thrust belt? *Geology, 42*(3), 247–250.

Iyengar, R. N., Sharma, D., & Siddiqui, J. M. (1999). Earthquake history of India in medieval times. *Indian Journal of History of Science, 34*, 181–237.

Kingdon-Ward, F. (1951). Notes on the Assam earthquake. *Nature, 167*, 130–131.

Kumar, A., Singh, S. K., Mitra, S., Priestley, K. F., & Shankar, D. (2016). The 2015 April 25 Gorkha (Nepal) earthquake and its aftershocks: Implications for lateral heterogeneity on the main Himalayan thrust. *Geophysical Journal International, 208*(2), 992–1008.

Le Roux-Mallouf, R., Ferry, M., Ritz, J.-F., Berthet, T., Cattin, R., & Drukpa, D. (2016). First paleoseismic evidence for great surface-rupturing earthquakes in the Bhutan Himalayas. *Journal of Geophysical Research: Solid Earth, 121*. https://doi.org/10.1002/2015JB012733.

Li, C., Zhang, G., Shan, X., Zhao, D., Li, Y., Huang, Z., Jia, R., Li, J., & Nie, J. (2020). Surface rupture kinematics and coseismic slip distribution during the 2019 Mw7.1 Ridgecrest, California earthquake sequence revealed by SAR and Optical images. *Remote Sensing, 12*, 3883.https://doi.org/10.3390/rs12233883.

Mathur, L. P. (1953). The Assam earthquake of 15 August 1950. In Rao, M. B. (Ed.), *A compilation of papers on the Assam earthquake of August 15, 1950* (pp. 56–60). Publication No. 1, Geological Survey of India.

Molnar, P. (1990). A review of the seismicity and the rates of active underthrusting and deformation at the Himalaya. *Journal of Himalayan Geology, 1*, 131–154.

Molnar, P., & Deng, Q. (1984). Faulting associated with large earthquakes and the average rate of deformation in Central and Eastern Asia. *Journal of Geophysical Research, 89*(B7), 6203–6227.

Morino, M., Kamal, A. S. M. M., Muslim, D., Ali, R. M. E., Kamal, M. A., Kaneko, F. (2009). Activity of Dauki Fault during the 1897 Ms 8.0 Great Assam earthquake confirmed by trench investigation at Gabrakahri Village, Haluaghat, Mymensingh, Bangladesh. *Journal of the Geological Society of Japan, 115*, 11–12.

Nandy, D. R. (2001). *Geodynamics of northeastern India and adjoining region* (209 p.). Kolkata: ACB Publication.

Ningthoujam, P. S., Dubey, C. S., Lolee, L. K., Shukla, D. P., Naorem, S. S., & Singh, S. K. (2015). Tectonic studies and crustal shortening across Easternmost Arunachal Himalaya. *Journal of Asian Earth Sciences, 111*, 339–349.

Pandey, A., Jayangondaperumal, R., Hetényi, G., Priyanka, R. S., Singh, I., Mishra, R. L., Srivastava, H.B. (2019). Primary surface faulting of 1697-1950 great earthquakes along the eastern Himalayan Frontal Thrust. In *Conference Paper*.

Pandey, P., Jayangondaperumal, R., Hetényi, G., Priyanka, R. S., & Singh, I. (2021). Establishing primary surface rupture evidence and magnitude of the 1697 CE Sadiya earthquake at the Eastern Himalayan Frontal thrust, India. *Scientific reports, 11*(1), 1–14.

Poddar, M. C. (1950). The Assam earthquake of 15th August 1950. *Indian Miner., 4*, 167–176.

Priyanka, R. S., Jayangondaperumal, R., Pandey, A., Mishra, R. L., Singh, I., Bhushan, R., Srivastava, P., Ramachandran, S., Shah, C., Kedia, S., Sharma, A. K. (2017). Primary surface rupture of the 1950 Tibet-Assam great earthquake along the eastern Himalayan front, India. *Scientific Reports, 7*. https://doi.org/10.1038/s41598-017-05644-y.

Rajendran, K., & Rajendran, C. P. (2011). Revisiting some significant earthquake sources in the Himalaya: Perspectives on past seismicity. *Tectonophysics, 504*, 75–88.

Rajendran, C. P., Rajendran, K. (2021). Paleoseismic context of the 1950 earthquake: Implications for seismic gaps in the Eastern Himalaya. *Physics and Chemistry of the Earth, Parts A/B/C*. https://doi.org/10.1016/j.pce.2021.103055.

Ramachandra Rao, M. B. (1953). *A compilation of papers on the Assam Earthquake of August 15, 1950* (112 p.). Publication No. 1. Calcutta: The Central Board of Geophysics.

Ramesh, N. R., Gadagkar, N. S. (1990). Effect of 1950 earthquake on geomorphology in a part of Brahmaputra valley near Teju, Arunachal Pradesh. *Journal of the Geological Society of India, 35*, 87–90.

Reddy, D. V., Nagabhushanam, P., Kumar, D., Sukhija, B. S., Thomas, P. J., Pandey, A., Sahoo, R. N., Ravi Prasad, G. V., & Datta, K. (2009). The great 1950 Assam earthquake revisited: Field evidence of liquefaction and search for paleoseismic events. *Tectonophysics, 474*, 463–472.

Richter, C. F. (1958). *Elementary seismology* (768 p.). W.H. Freeman Company. Reprinted: New Delhi: S. Chand and Company.

Saikia, A. (2019). *The unquiet river* (583 p.). Oxford: Oxford University Press.

Searle, M. P., Noble, S. R., Cottle, J. M., Waters, D. J., Mitchell, A. H. G., Hlaing, T.,Horstwood, M. S. A. (2007). Tectonic evolution of the Mogok metamorphic belt, Burma (Myanmar) constrained by U–Th–Pb dating of metamorphic and magmatic rocks. *Tectonics, 26*, TC3014. https://doi.org/10.1029/2006TC002083.

Singh, I., Pandey, A., Mishra, R. L., Priyanka, R. S. Brice, A. Jayangondaperumal, R., & Srivastava, V. (2021). Evidence of the 1950 great Assam earthquake surface break along the Mishmi thrust at Namche Barwa Himalayan Syntaxis. *Geophysical Research Letters, 48*, e2020GL090893. https://doi.org/10.1029/2020GL090893.

Sylman, Y., Shimshonj, M., & Ben-Menahem, A. (1974). Theoretical amplitudes of body waves from a dislocation source in the earth, III. P, S, and surface reflections. *Physics of the Earth and Planetary Interiors, 8*, 130–147.

Tandon, A. N. (1954). Study of the great Assam earthquake of August 1950 and its aftershocks. *Indian Journal of Meteorology and Geophysics, 5*, 95–137.

Tandon, A. N., & Dube, R. K. (1973). A study of the crustal structure beneath the Himalayas from body waves. *Pure and Applied Geophysics, 111*, 2207–2215.

Tillotson, E. (1951). The great Assam earthquake of August 15, 1950. *Nature*, 128–130.

Vernant, P., Bilham, R., Szeliga, W., Drupka, D., Kalita, S., Bhattacharyya, A. K., Gaur, V. K., Pelgay, P., Cattin, R., & Berthet, T. (2014). Clockwise rotation of the Brahmaputra Valley relative to India: Tectonic convergence in the eastern Himalaya, Naga Hills, and Shillong Plateau. *Journal of Geophysical Research: Solid Earth, 119*, 6558–6571.

Vigny, C., Socquet, A., Rangin, C., Chamot-Rooke, N., Pubellier, M., Bouin, M.N., Bertrand, G., & Becker, M. (2003) Present-day crustal deformation around Sagaing fault, Myanmar. *Journal of Geophysical Research, 108*(B11), 2533. https://doi.org/10.1029/2002JB001999.

Zhao, Y., Grujic, D., Baruah, S., et al. (2021). Paleoseismological findings at a new trench indicate the 1714 M8.1 earthquake ruptured the main frontal thrust over all the Bhutan Himalaya. *Frontiers in Earth Science*. https://doi.org/10.3389/feart.2021.689457.

Chapter 16
Epilogue

Globally, earthquakes and related hazards lead the list of natural disasters in terms of damage and human loss, and they affect very large areas, causing death and destruction on a massive scale. The cumulative fatality globally since 1800 due to earthquakes is reported to be 138,000. The extent of fatality depends not just on the magnitude of the earthquakes, but also on the time of their occurrences, proximity to human settlements, construction practices locally followed, and also the ground conditions. India, one of the most populous countries with its unique geology and seismotectonic setting, is highly susceptible to earthquake-related hazards. The Indian subcontinent, characterized by several seismotectonic units, namely, the Himalayan collision zone in the north, Indo-Burmese arc in the northeast, rift zones in the Peninsular Indian shield, and the Andaman Sumatra trench in the Bay of Bengal, is one of most earthquake prone regions in the world. According to the vulnerability atlas of India prepared by Building Materials and Technology Promotion Council, Government of India (BMTPC 2019), almost 60% of the total landcover in India is susceptible to seismic hazards. This situation is further made untenable by the rapid population growth and the ever-expanding and largely under-regulated urbanization, wherein buildings are constructed without adequate earthquake safety features as enumerated in Indian Standard Building Codes. Since the last mega-Himalayan earthquake of M_w~8.6 on the 15th of August 1950 in the Upper Assam, the population of the country more than tripled. Larger numbers are concentrated in large cities that show an yearly growth rate of more than 20% (The Registrar General and Census Commissioner, India: www.censusindia.gov.in).

The damage and casualty statistics for the 1819 and the 2001 earthquakes (M_w 7.7) provide appropriate comparative examples. The 2001 earthquake killed 20,085 people and destroyed about 3,39,000 buildings, whereas the 1819 earthquake toppled around 7,000 houses and killed approximately 1,500 people. Such large variance in the impact of two earthquakes of the same magnitude and sourced in similar terrain conditions reflects the dependency of damage potential on the growth of urbanization and population. In a hypothetical scenario, if the M 7.5–8.0 earthquake were to

C. P. Rajendran and K. Rajendran, *Earthquakes of the Indian Subcontinent*,
GeoPlanet: Earth and Planetary Sciences,
https://doi.org/10.1007/978-981-16-4748-2_16

occur in the census year of 1991 in the Kangra region of Himachal Pradesh, the loss of life is projected to be 65,000 as compared to 20,000 lives lost in 1905—yet another example of scaling up of damage levels due to the exponential impact of urbanization and population growth (Arya, 2000). No such recent projections are available, but these are not exaggerated figures, as one can learn from one of the most destructive contemporary earthquakes. The 2005 Kashmir (Muzaffarabad) caused more than 80,000 casualties and severe destruction to 1000 s of built structures. The images of destruction from the 2005 Kashmir earthquake would somewhat mimic the projections for a future Kangra-type earthquake along the Himalayan arc.

In terms of earthquake hazard preparedness, even the technologically advanced countries undergo massive devastation, as exemplified by the 2011 Japan earthquake and the consequent tsunami that killed about 21,000 people and caused massive destruction, including wrecking of a nuclear plant. The damaged Fukushima plant will take at least decades to decommission. A swath of land around the plant will remain uninhabitable for years to come, and thousands of residents have been displaced. The wastewater disposal is another example of the complex, long-term effects of the 2011 tsunami inundation. The case of 2011 damage may be attributed to underestimating the maximum expected earthquake, despite warnings to the contrary derived from academic research. Another set of global examples shows that even moderate earthquakes have become increasingly deadly in many parts of the world, more of a consequence of population growth and expansion of the built environment. A case in point is the 2003 Bam earthquake of magnitude M_w 6.6 that struck the Kerman Province of southeastern Iran which caused a large death toll of ~26,200. The damage to buildings, mostly exacerbated by the local practices that used mud bricks was too severe for its moderate magnitude.

The urgency of earthquake hazard assessment in India began to be felt intensely after the 1993 Killari (Latur) earthquake. For a magnitude M_w 6.3 earthquake, the damage it inflicted in the region was incommensurately high. The global as well as country-wide data is clear on the fact that the earthquake hazard vulnerability is on the increase. The task of reducing the social and economic cost can be achieved only through a systematic assessment of seismic hazard and introduction of better construction practices. Land zonation and other earthquake regulations, based on scientific data on seismic source characteristics and ground motion variability will aid these activities (see Nath & Thingbaijam, 2012 and references therein).

The data shows that the seismic source regions in the Peninsular India structurally coincide with failed-rift settings such as in the Kachchh-Saurashtra, Narmada-Tapti, Mahanadi, and Godavari regions, and should be under higher alert because of the presence of seismogenic structures that may host moderate to large earthquakes. The fault plane solutions of the earthquakes in these regions corroborate the reactivation in compressional regime, combined with plate tectonic forcing. Stresses can be locally augmented in some areas by high pore-fluid pressure resulting from dehydration of serpentinites at lower crustal levels and weaker mantle or hydrological diffusivity through highly fractured rocks (Rao & Rao, 2006). For example, the Koyna earthquake (M_w 6.3) of 1967 is a classic case of seismicity triggered by a nearby reservoir (Gupta, 2002). Thus, the past earthquake occurrences indicate that moderate Killari-type earthquakes (M_w 5.5 to 6.5) within the continental interiors may also be expected in parts of Madhya Pradesh, National Capital Region (Delhi), Uttar Pradesh, Maharashtra, Karnataka, Kerala, Tamil Nadu, and coastal Andhra Pradesh where some of the geologically ancient structures prone to reactivation are known to exist. There may be other unknown structures elsewhere, and their potential would be known only from future earthquakes.

The Himalaya straddling the India-Eurasia plate boundary is directly subjected to a steady convergence stress field (with a yearly rate of 13 to 21 mm/yr; Stevens & Avouac, 2015). Consequently, the 2500 km-long, 100 km-wide plane separating the Indian and Asian plates, called the Main Himalayan Thrust, accumulates seismic moment at a rate of about 10^{20} Nma^{-1} (Newton meters per year), equivalent to that released by a $M_w = 7.3$ earthquake (Bilham, 2019). A corollary to this statement is that in 10 years' time the region would have accumulated enough energy to generate an earthquake of M_w 8. In a century, it could scale up to a M_w 8.6 earthquake and in about 350 years to one of M_w 9. These combinations of events would be required to release the accumulated 4 m slip, with the creeping component almost ruled out (Bilham, 2019). Thus, many parts of the Himalaya have accumulated enough strain energy to be released in moderate, large, or great earthquakes.

Paleoseismic investigations indicate a spatial and temporal non-uniformity in the production of great earthquakes, but these data sets could be incomplete (Rajendran et al., 2015; Wesnousky, 2020). It has been suggested that fault friction, geometry, and heterogeneities of the décollement would be critical in inhibiting or promoting seismic productivity (Zilio et al., 2019). Some segments of the Himalaya are reportedly undergoing a 'deceptively' quiet period although they have histories of large to great earthquakes. These segments have been identified as 'seismic gaps,' where future large/great earthquakes are likely to occur (Feldl & Bilham, 2006; Khattri, 1987). Among these gaps, the central Himalaya segment (extending from the eastern rupture margin of the 1934 earthquake towards the rupture boundary of the 1905 Kangra earthquake) stands out with its quiescence for at least 600 years. Consequentially this region is said to be ready for a massive earthquake(s) to release unspent accumulated slip (Stevens & Avouac, 2016, 2021; Bilham, 2019). Seismic gaps also exist in other areas like the Kashmir Himalaya or possibly the Bhutan segment in the eastern Himalaya. In the event of such earthquakes, the Gangetic plains face the largest threat. Past events imply that more than 80% of building stock in the cities of the Ganga alluvial plains may not be able to withstand ground shaking emanating from strong earthquakes. These projections should call for an action plan that may include updating the building code provisions for earthquake-resistant constructions and retrofitting the heritage structures.

A set of earthquakes that are associated with the compressional and flexural stresses of the subducting Indian plate also need to be mentioned here. An example is the 1988 Udaipur earthquake (M_w 6.7) sourced under the Nepal Terrai near the Bihar (India)-Nepal border at a hypocentral depth of 50 km (Ghimire & Minoru, 2007). Predominantly marked by strike-slip faulting, the 2011 M_w 6.9 Sikkim earthquake, sourced at a depth of about 50 km under the Kanchenjunga peak, about 130 km north of the main frontal thrust, is considered as a part of the ongoing the intra-slab deformation (Sunilkumar et al., 2019). The Indo-Gangetic alluvial plains distributed between the northern extremity of the Indian Peninsular shield and the Himalayan arc also host several subsurface north-northwest trending ridges bounded by faults (Valdiya, 1976; Dasgupta et al., 2000; Kayal, 2008). Blind step-out thrusts under the Gangetic Plains extending from the frontal Himalaya are also reported, and they pose an underestimated risk (Duvall et al., 2020). The National Capital Region that includes Delhi and surrounding areas is found to be hosting clusters of small magnitude earthquakes (e.g., M 3 to 4.5 during April-August 2020) and several nominally damaging local earthquakes have been reported from here (1720, 1831, 1956, and 1960). Historical data suggest that infrequent moderate earthquakes cannot be precluded in this area. In such areas proximal to the collisional zones, an increase in strain rate is expected because of the curving of the lithosphere as India underthrusts the Himalaya (Copley et al., 2011).

Another topic of interest is the interaction between seasonally induced non-tectonic hydrological cycles and tectonic deformation along the Himalayan plate boundary. Geodetic and seismic time series combined with GRACE gravity data have indicated connections between annual cycles of water loading by monsoonal rains and seismicity in the Himalaya (Bettinelli et al., 2008). The seismicity rate is reported to be twice as high in the winter as in the summer and it correlates with the stress rate variations. Panda et al. (2018) analyzed continuous Global Navigation Satellite System (GNSS) sites in the Nepal and Garhwal-Kumaun region of the Himalaya arc and found that four sites showed higher-amplitude seasonal variations in the horizontal component. These transients are attributed to changes in aseismic slip rate on the deep megathrust, controlled by seasonal hydrological loading. However, these observations derived from GNSS stations have been flagged stating that much of the residual seasonal horizontal and vertical signals are likely to be caused by the difficult-to-detect seasonal variations in the Earth temperature field (Chanard et al., 2020). Thermoelastic deformation of GNSS monuments is also cited as a source of systematic error, among other factors. Thus, the claim for seasonal slow slip on the Main Himalayan Thrust needs to be treated with caution as it hinges on the particularly high seasonal amplitude and H/V ratios, which are not unusual and not exclusively related to tectonics.

The spurts in seismicity around Delhi and the adjoining Himalayan region have also been attributed to large-scale anthropogenic seasonal groundwater extraction that influences aquifer contraction and basement rock expansion and thus modulates horizontal compression on the bounding faults (Kundu et al., 2015; Tiwari et al., 2021). The Indo-Gangetic Plain is marked by a steady mass loss due to excessive extraction of groundwater (Tiwari et al., 2009), which is also likely to augment the stability of faults. A recent work on the Tehri Reservoir (a 260.5 m-high earth and rockfill dam with a total storage capacity of 3,540 Mm^3) on the Bhagirathi River in the State of Uttarakhand is noteworthy in this context (Xie et al., 2021). Using a variety of data including multiple space geodetic observations, and poro-elastic models, the study shows that the seasonal loading cycles in the reservoir can modulate elastic deformation in the vicinity of the reservoir. These models argue that small changes in the hydrological regime can lead to variations in strain leading to earthquakes. This remains an idea that needs to be tested with future data.

16.1 The Way Forward

The discussion presented in this book reveals that India exhibits a wide range of seismic zones with markedly different seismicity patterns and is extremely prone to devastating earthquakes. The presentation given in this book provides a synthesis of our current state of knowledge of some of the major earthquakes within the interplate and intraplate settings in India and the adjoining areas. Among other inputs, an improved understanding of the earthquake source properties along with ground motion characteristics form the key database in seismic hazard modelling. The role of the society in the mitigation efforts is clear because damage from earthquakes is closely linked to engineering infrastructure and preparedness. Policy experts and administrators must keep this threat in focus and evolve a dynamic long-term strategy to deal with earthquakes. Such events are to be expected from the proximal plate boundaries, as well as plate interiors. The earthquake scientists must always grapple with the question about how to interact with the society at large so that they are encouraged to get involved in the process of mitigating earthquake damage. In turn a question that is often being directed at the seismologists from the other side is, how the database on the earthquake characterization can help in earthquake prediction.

In the conclusion of a classic paper dealing with physics of earthquakes, Kanamori and Brodsky (2004) say, "… that the quantitative prediction of earthquake initiation is an extremely complicated and perhaps impossible task. Even in the best-case scenario of a predictable fault nucleation length, the nucleation length of 1 m requires instruments to be too densely spaced to be practical. Perhaps one day we will be able to accomplish accurate earthquake prediction, but the current state of the science implies that that day is decades, if not centuries away. In the short term, it is more practical to save lives by using the detailed knowledge we have about the propagation of seismic waves and strength of seismic shaking to design buildings and infrastructure that will protect people during an earthquake."

Over the last decade, scientists have been able to deploy very dense seismic arrays with classical instruments, optical fibers, and large borehole sensors. Thus, the number and quality of observations have increased phenomenally. Knowledge about earthquakes has been improved, owing to an increase in the number and quality of available observations coming from continuous deployment of monitoring stations. In recent years, dense networks of multi-parametric stations have been deployed worldwide around faults and fault systems that are expected to generate data on large earthquakes (Festa et al., 2021). Near-Fault Observatories are being established in Europe where seismological, geodetic, and geochemical observations are used to reconstruct the physical and chemical processes occurring along the faults and to study the preparatory phase of large earthquakes. New approaches including machine learning methods for earthquake detection help in identifying microseismic events buried in the seismic noise, thus contributing to more detailed earthquake catalogs (Festa et al., 2021).

Despite many advances in seismology, short-term earthquake prediction remains a grand challenge. Recent satellite-observations have revealed transient-slip phenomenon over periods of days to years could shed light on the physics of earthquake cycles. In the publication, "Thriving on Our Changing Planet: A Decadal Strategy for Earth Observation from Space", issued by the National Academies of Sciences, Engineering, and Medicine (2018), it is explained that the progress on earthquake forecasting can be made by continuously observing areas prone to earthquakes using space based InSAR and high-resolution optical imagery. For very large earthquakes, temporal variations in gravity may reveal large-scale deformation not observable by other methods. It is encouraging that the long-term GPS measurements of surface deformation in some subduction zones like the Cascadia, where a great earthquake is expected, reveal that steady strain accumulation is punctuated by creep events at the down-dip limit of the locked fault. The M 9.0 Tohoku earthquake and many other recent large events, including the 2004 Sumatra-Andaman (M 9.1) earthquake, were preceded by seismic and aseismic precursors. The report suggests that understanding the details of these transients during the interseismic periods may provide insight into the major rupture. Thus, it is likely that combined with high-quality seismic data and knowledge of rupture histories of the fault zones, much of the advancement in earthquake forecasting is expected from space-based monitoring systems.

References

Arya, A. S. (2000). Recent developments toward earthquake risk reduction in India. *Current Science, Special Issue Seismology, 79*, 1270–1277.

Bilham, R. (2019). Himalayan earthquakes: a review of historical seismicity and early 21st century slip potential. In Treloar, P. J., Searle, M. P. (Eds.), *Himalayan tectonics: A modern synthesis* (Vol. 483, pp. 423–482). London: Geological Society. Special Publications.

Bettinelli, P., Avouac, J-P., & Flouzat, M. et al. (2008). Seasonal variations of seismicity and geodetic strain in the Himalaya induced by surface hydrology. *Earth and Planet Sci, 266*, 332–344.

Chanard, K., Métois, M., Rebischung, P., & Avouac, J-P. (2020). A warning against over-interpretation of seasonal signals measured by the Global Navigation Satellite System. *Nat Com, 11*, 1375. https://doi.org/10.1038/s41467-020-15100-7

Copley, A., Avouac, J.-P., Hollingsworth, J., & Leprince, S. (2011). The 2001 Mw 7.6 Bhuj earthquake, low fault friction, and the crustal support of plate driving forces in India. *Journal of Geophysical Research, 116*, B08405. https://doi.org/10.1029/2010JB008137.

Dasgupta, S., et al. (2000). *Seismotectonic atlas of India and its environs* (Vol. 59, 87 p.). Kolkata: Geological Survey of India. Special Publications.

Duvall, M., Waldron, J. W. F., Godin, L., & Najman, Y. (2020). Active strike-slip faults and an outer frontal thrust in the Himalayan foreland basin. *PNAS, 117*, 17615–17621.

Feldl, N., & Bilham, R. (2006). Great Himalayan earthquakes and the Tibetan Plateau. *Nature, 444*, 165–70. http://doi.org/10.1038/nature05199.

Festa, G., Piccozi, M., Scala, A., & Zollio, A. (2021). Earthquake Seismology. In: *Encyclopedia of Geology*. 2nd Edition. 575–586. Elsevier.

Ghimire, S., & Minoru, K. (2007). Source process of the Ms = 6.6, Udayapur earthquake of Nepal India border and its tectonic implications. *Journal of Asian Earth Sciences, 31*, 128–138.

Gupta, H. K. (2002). A review of recent studies of triggered earthquakes by artificial water reservoirs with special emphasis on earthquakes in Koyna, India. *Earth- Science Reviews, 58*, 279–310.

Kayal. J. R. (2008). *Microearthquake seismology and seismotectonics of South Asia* (503 p.). Springer, Berlin.

Kanamori, H., & Brodsky, E. (2004). The physics of earthquakes. *Rep. Prog. Phys., 67*, 1429–1496.

Kundu, B., Vissa, N. K., & Gahalaut, V. K. (2015). Influence of anthropogenic groundwater unloading in Indo-Gangetic plains on the 25 April 2015 Mw 7.8 Gorkha, Nepal earthquake. *Geophysical Research Letters, 42*(24), 10,607–10,613.

Khattri, K. N. (1987). Great earthquakes, seismicity gaps and potential for earthquakes along the Himalayan plate boundary. *Tectonophysics, 38*, 79–92.

Nath, S. K. & Thingbaijam, K. K. S. (2012). Probabilistic Seismic Hazard Assessment of India. *Seismological Research Letters, 83*, 135–149. http://doi.org/10.1785/gssrl.83.1.135.

Rajendran, C. P., John, B., & Rajendran, K. (2015). Medieval pulse of great earthquakes in the central Himalaya: Viewing past activities on the frontal thrust. *Journal of Geophysical Research: Solid Earth*, 1623–1641.

Rao, B. R., & Rao, V. K. (2006) Influence of fluids on deep crustal Jabalpur earthquake of 21, May 1997: Geophysical evidences. *Journal of seismology, 10*, 301–314. https://doi.org/10.1007/s10950-006-9018-y

Stevens, V., & Avouac, J. P. (2021). On the relationship between strain rate and seismicity in the India-Asia collision zone: implications for probabilistic seismic hazard. *Geophysical Journal International, 226*, 220–245.

Stevens, V. L., & Avouac, J.-P. (2016). Millenary Mw > 9.0 earthquakes required by geodetic strain in the Himalaya. *Geophysical Research Letters, 43*, 1118–1123.

Sunilkumar, T. C., Earnest, A., Silpa, K., & Andrews, R. (2019). Rupture of the Indian slab in the 2011 Mw 6.9 Sikkim Himalaya earthquake and its tectonic implications. *Journal of Geophysical Research: Solid Earth, 124*, 2623–2637.

Tiwari, D. K., Jha, B., Kundu, B., Gahalaut, V., & Vissa, N. (2021). Groundwater extraction-induced seismicity around Delhi region, India. *Scientific Reports, 11*, 10097.

Tiwari, V. M., Wahr, J., & Swenson, S. (2009). Dwindling groundwater resources in northern India, from satellite gravity observations. *Geophysical Research Letters, 36*, L18401.

Valdiya, K. S. (1976). Himalayan transverse faults and folds and their parallelism with subsurface structures of North Indian plains. *Tectonophysics, 32*, 353–355.

Wesnousky, S. G. (2020). Great pending Himalaya earthquakes. *Seismological Society of America*, XX, 1–9.

Xie, L., Xu, W., & Burgmann, R. et al. (2021). Tehri Reservoir Operation Modulates Seasonal Elastic Crustal Deformation in the Himalaya. *Jour Geophys Res, 126*. http://doi.org/10.1029/2020JB021122

Zilio, D. L., van Dinther, Y., Gerya, T., et al. (2019). Bimodal seismicity in the Himalaya controlled by fault friction and geometry. *Nature Communications, 10*, 48.